Frontiers in the History of Science

Series Editor

Vincenzo De Risi, Université Paris-Diderot – CNRS, PARIS CEDEX 13, Paris, France

Editorial Board Members

Karine Chemla, Lab SPHERE UMR 7219, bâtiment Condorcet, Univ Paris Diderot, Paris, France

Sven Dupré, History of Art, Science and Technology, Utrecht University, Utrecht, The Netherlands

Moritz Epple, Historisches Seminar, Goethe-Universität Frankfurt, Frankfurt, Germany

Orna Harari, Dept. of Classics, Ramat Aviv, Tel-Aviv University, Tel Aviv, Israel

Dana Jalobeanu, Faculty of Philosophy, University of Bucharest, Bucharest, Romania

Henrique Leitão, History and Phiosophy of Science, University of Lisbon, LISBOA, Portugal

David Marshall Miller, Philosophy, Auburn University, Auburn, United States Minor Outlying Islands

Aurelien Robert, CNRS, University of Paris, Paris, France

Eric Schliesser, Social and Behavioural Sciences, University of Amsterdam, Amsterdam, The Netherlands

Frontiers in the History of Science is designed for publications of up-to-date research results encompassing all areas of history of science, with a special focus on the history of mathematics and the exact sciences. The series is also committed to interdisciplinary approaches at the crossroads between history of science and history of logic, epistemology, philosophy of nature, figurative arts, and history of culture more in general. Graduates and post-graduates as well as scientists and historians will benefit from selected and thoroughly peer-reviewed research monographs. The series is curated by the Series Editor with the support of an international group of Associate Editors.

Richard T. W. Arthur • David Rabouin

Leibniz on the Foundations of the Differential Calculus

Richard T. W. Arthur
Department of Philosophy
McMaster University
Hamilton, ON
Canada

David Rabouin
Laboratoire Sphère, CNRS
Université Paris-Cité
Paris, France

ISSN 2662-2564 ISSN 2662-2572 (electronic)
Frontiers in the History of Science
ISBN 978-3-031-77258-0 ISBN 978-3-031-77259-7 (eBook)
https://doi.org/10.1007/978-3-031-77259-7

© The Editor(s) (if applicable) and The Author(s), under exclusive license to Springer Nature Switzerland AG 2025

This work is subject to copyright. All rights are solely and exclusively licensed by the Publisher, whether the whole or part of the material is concerned, specifically the rights of translation, reprinting, reuse of illustrations, recitation, broadcasting, reproduction on microfilms or in any other physical way, and transmission or information storage and retrieval, electronic adaptation, computer software, or by similar or dissimilar methodology now known or hereafter developed.
The use of general descriptive names, registered names, trademarks, service marks, etc. in this publication does not imply, even in the absence of a specific statement, that such names are exempt from the relevant protective laws and regulations and therefore free for general use.
The publisher, the authors and the editors are safe to assume that the advice and information in this book are believed to be true and accurate at the date of publication. Neither the publisher nor the authors or the editors give a warranty, expressed or implied, with respect to the material contained herein or for any errors or omissions that may have been made. The publisher remains neutral with regard to jurisdictional claims in published maps and institutional affiliations.

This book is published under the imprint Birkhäuser, www.birkhauser-science.com by the registered company Springer Nature Switzerland AG
The registered company address is: Gewerbestrasse 11, 6330 Cham, Switzerland

If disposing of this product, please recycle the paper.

Preface

This work owes its origin to the suggestion by Vincenzo De Risi in 2021 that the two of us should amplify our reading of Leibniz on infinitesimals (Arthur and Rabouin 2020) into a potential volume for his series with Birkhäuser. We thank him for his continued encouragement and feedback, and for welcoming our volume into his series. Thanks are also due to an anonymous Birkhäuser reviewer of a first draft for helpful comments and advice.

Some material in Chaps. 2, 4 and 6 is a reworking of parts of a paper in the *Archive for History of the Exact Sciences* (Arthur and Rabouin 2020; https://doi.org/10.1007/s00407-020-00249-w), and the latter two sections of Chap. 5 have material in common with a paper we published in *Historia Mathematica* (Arthur and Rabouin 2024; https://doi.org/10.1016/j.hm.2023.12.001). We thank the editors of these journals for kindly giving us permission to reuse and refashion this material.

We are grateful to members of the *Mathesis* group of scholars for their collaboration, in particular, Sandra Bella, Mattia Brancato, Valérie Debuiche, Anne Michel-Pajus, Ariles Remaki, and Claire Schwartz; to colleagues in the *Leibniz Archiv*, especially Siegmund Probst for his constant help in the dating of the manuscripts and the realization of some diagrams, but also Michael Kempe, Charlotte Wahl (who provided the letter to Gallois), Achim Trunk and Elisabeth Rinner; to Eberhard Knobloch for his constant support; to Simon Gentil for his help in preparing high-resolution versions of the figures; and also to Jeffrey Elawani and Filippo Costantini for their insightful discussions and talks bearing on matters we treat in Chap. 5.

During the writing of this book, David Rabouin has received funding from the *Agence Nationale de la Recherche* (ANR), project N° ANR-17-CE27-0018-01 AAP GENERIQUE 2017: *Mathesis. Editer les manuscrits mathématiques de Leibniz*, and from the European Research Council (ERC) under the European Union's Horizon 2020 research and innovation programme, grant agreement n°101020985 : *PHILIUMM: "The Philosophy of Leibniz in the Light of his Unpublished Mathematical Manuscripts"*.

Toronto, ON, Canada   Richard T. W. Arthur
Paris, France   David Rabouin

Note on Texts and Translations

All the translations of the texts we have chosen are our own.

Nos. 1, 3, 4, 7, 8, 9, 10, 11 are revised translations of texts appearing in Leibniz (2001), and nos. 30 and 33 are extracts from translations of texts to appear in Arthur and Ottaviani (forthcoming), given here by permission of the authors. Where there are already existing English translations by others (as in nos. 21, 22, 23, 34, 35, 36, 38, 45, 47 and 50), we have referenced these, but have provided fresh translations of our own.

Regarding the original language versions of Leibniz's texts, the great majority are available on-line either in the Akademie edition[1] or in Gerhardt's two series of Leibniz's mathematical and philosophical writings, and sometimes in on-line facsimiles of his original published journal articles. Consequently, we have not reproduced these here, although we have checked the transcriptions against the original sources and given references to available transcriptions. We have, however, included a transcription of the Latin source for text 12, for which no transcription is otherwise extant.

We systematically translate "*fingere*" as "feign" (and not "imagine", as is often proposed), in order to maintain the link between the verb (*fingere*), the noun (*fictio*) and the adjective (*fictitius*), which are used by Leibniz in this context.

We have adopted the convention of using "ôr" to translate the Latin *seu* or *sive*, the "or of equivalence", in order to discriminate it visually from "or" denoting an alternative.

Leibnizian texts are cited by the original language source first, followed by an existing English translation if one is readily available, after a solidus: thus (A VI, 3, 525/ LLC 61). If the same sources are repeated consecutively, they are abbreviated thus: (525/61).

---

[1] https://www.gwlb.de/leibniz/digitale-ressourcen/repositorium-des-leibniz-archivs (accessed April 30, 2024).

Contents

Part I   Interpretive Essay

1   Introduction 3

2   The Question of the Existence of Infinitesimals (1669–1676) 23
  2.1   Early Views on the Continuum (1669–1672) 24
  2.2   Infinite Series and the Infinitely Small in
        Mathematics (1672–1675) 29
  2.3   Leibniz's Rejection of the Existence of the Infinitely
        Small in 1676 40

3   Mathematical Fictions 51
  3.1   Leibniz's Fictions in Context: Fictive Entities in
        Sixteenth and Seventeenth Century Mathematics 51
  3.2   Mathematical Fictions in Leibniz 57
  3.3   Fictions and the Question of Impossibility 64

4   *The* De Quadratura Arithmetica *(DQA)* 79
  4.1   Quadratures as Riemannian Sums 79
  4.2   The *DQA* and the "Direct Method" 84
  4.3   The Posteriority of the *DQA* 92

5   Infinitesimals and Their Existence After 1676 99
  5.1   First Publications on Differential Calculus,
        and Foundational Reflections 99
  5.2   Mathematical Foundations: Quantity, Magnitude
        and Quasi-minima 111
  5.3   Further Thoughts on the Existence of Infinitesimals 119

6	**Leibniz's Mature Justifications of the Calculus**	133
	6.1 Leibniz's Syncategorematic Conception of Infinitesimals	133
	6.2 Leibniz's Mature Justifications of the Use of Infinitesimals	142
	6.3 The Justifications of the Differential Algorithm	149
7	**Conclusion**	157

Part II A Selection of Translations of Key Texts

8	**Texts for Chapter 2, The Question of the Existence of Infinitesimals (1669–1676)**	167
	8.1 (for 2.1) Early Views on the Continuum [1669–1672]	167
	Text 1. From Notes on Galileo's *Discorsi* [Fall 1672]	167
	Text 2. From *Accessio ad Arithmeticam Infinitorum* [ca. December 1672]	168
	Text 3. From *On Minimum and Maximum; on Bodies and Minds* [Nov. 1672-Jan 73]	169
	Text 4. *A Noteworthy Observation about the Infinite* [Feb. 1676?] ...	171
	8.2 (for 2.2) Infinite Series and the Infinitely Small in Mathematics [1672–1676]	171
	Text 5. From *De serierum summis et de quadraturis pars octava* [October 1674]	171
	Text 6. The Cancelled Scholium to Prop. 11 of the *DQA* [June 1676]	172
	8.3 (for 2.3) Leibniz's Rejection of the Existence of the Infinitely Small in 1676	174
	Text 7. From *On the Secrets of the Sublime* [February 1676]	174
	Text 8. On the Infinitely Small [26 March 1676]	176
	Text 9. From *On Motion and Matter* [1–10 April 1676]	176
	Text 10. From *Infinite Numbers* [ca. 10 April 1676]	178
	Text 11. From *Pacidius to Philalethes* [October 1676, OS]	182
	Text 12. *Geometry with the Metaphysics of the Continuum Omitted* [1679]	182
9	**Texts for Chapter 3, *Mathematical Fictions***	185
	9.1 (for 3.2) Mathematical Fictions in Leibniz	185
	Text 13. From *Little Tracts of a Mathematical Collection, Tract 3, Part 2: Quadrature of a logarithmic figure* [Late Spring 1673]	185

	9.2	(for 3.3) Fictions and the Question of Impossibility	186
		Text 14. From *Arithmetic Quadrature of the Circle and the Hyperbola* [3 May 1676]	186
		Text 15. From *New Elements of a Universal Mathematics* [Summer 1683?]	187
10	**Texts for Chapter 4, *De Quadratura Arithmetica* (*DQA*)**	**189**
	10.1	(for 4.2) The *DQA* and the "Direct Method"	189
		Text 16. Selections from the Scholia of *De Quadratura Arithmetica* [June-September 1676]	189
	10.2	(for 4.3) The Posterity of the *DQA*	194
		Text 17. Extracts from the *Compendium of the Arithmetical Quadrature* [1690]	194
		Text 18. From the Appendix to a Letter for Bodenhausen [28 October, 1690]	195
		Text 19. *An Arithmetical Quadrature common to conic sections that have a centre, and from which are deduced a canonical trigonometry for precision in numbers to any degree, and freed from the need for tables, with a special use for the nautical rhumb line, and appropriate for the planar projection of the sphere* [April 1691]	197
		Text 20. From a Letter to Otto Mencke for Johann Christoph Sturm [November-December 1695]	197
11	**Texts for Chapter 5, Infinitesimals and Existence After 1676**	**201**
	11.1	(for 5.1) Publication and the Lemmas on Incomparables	201
		Text 21. From *A New Method for Finding Maxima et Minima, As Well As Tangents, Unhindered by Either Fractions or Irrational Quantities, and a Singular Kind of Calculus for Finding Them* [1684]	201
		Text 22. From *Extract from a Letter from M. L[eibniz] on a General Principle, Useful for the Explanation of the Laws of Nature by the Consideration of Divine Wisdom; to Serve as a Reply to the Response of Rev. Father M[alebranche]*	205
		Text 23. The Lemmas on Incomparables, from *An Essay on the Causes of the Celestial Motions* (the *Tentamen*) [February 1689] . . .	206
		Text 24. *From a Letter of Mr. Leibniz to Mr. Huygens*, 1690	207
	11.2	(for 5.2) Mathematical Foundations: Quantity, Magnitude and Quasi-Minima	207
		Text 25. From *On Quantity* [1680s?]	207
		Text 26. From *On Magnitude and Measure* [1690s?]	209
		Text 27. From *Specimen Geometriae Luciferae* [1695?]	213

	11.3	(for 5.3) Further Thoughts on the Existence of Infinitesimals............	215
		Text 28. *Quicunque Attentius ea Considerabit...* [1677–1682?]....	215
		Text 29. From Annotations on Thomas White's *Euclides Physicus* [Summer 1689?]...........................	217
		Text 30. From Reflections on Libert Froidmont's *Labyrinthus Continui* [1693–94?]............................	218
		Text 31. *A Marvellous Axiom* [1695].......................	220
		Text 32. From the Leibniz–Johann Bernoulli Correspondence [1698–99]......................................	220
12	**Texts for Chapter 6, *Leibniz's Mature Justifications of the Calculus***		229
	12.1	(for 6.1) The Infinite as Syncategorematic....................	229
		Text 33. From *Towards a Science of the Infinite* [c. 1698].........	229
		Text 34. From *New Essays on the Understanding*, Book II, Chapter XVI [1704].................................	232
		Text 35. From the Leibniz-Des Bosses Correspondence...........	232
		Text 36. From *Essays on Theodicy* [1710], Preliminary Discourse......................................	234
	12.2	(for 6.2) Mature Justifications of the Use of Infinitesimals.........	234
		Text 37. A Memoir by Mr. Leibniz Concerning His Opinion on the Differential Calculus (From a Letter to Pinsson)...........	234
		Text 38. From the Correspondence with Varignon [1702].........	235
		Text 39. From *Quaestio De Jure Negligendi Quantitates Infiniti Parvas* [Autumn 1702-1703].......................	238
		Text 40. From a Letter to the Abbé Gallois [July, 1705]..........	242
		Text 41. *An Observation That Ratios or Proportions Do Not Hold for Quantities Less Than Zero, and on the True Meaning of the Infinitesimal Method*............................	244
		Text 42. From a Letter to Grandi [September 1713].............	245
		Text 43. *Regula de Transitu per Saltum Non Admittendo* [After 1710?]....................................	247
		Text 44. From a Letter to Christian Wolff....................	248
	12.3	(for 6.3) Justifications of the Differential Algorithm.............	248
		Text 45. From the Response to Nieuwentijt [July 1695]..........	248
		Text 46. From a Letter to Wallis [30 March 1699]..............	250
		Text 47. *Justification of the Infinitesimal Calculus by that of Ordinary Algebra* [Supplement to a Letter to Pinsson of April 1702]....................................	251

Text 48. From Leibniz's Responses to Jenisch's Criticisms
[October 1702]..................................... 253
Text 49. *Defense du Calcul des Differences* [1702?]............. 257
Text 50. *Cum Prodiisset* [1702?]........................ 259

Bibliography... 273

Author Index.. 281

Subject Index... 285

Abbreviations

A	G. W. Leibniz. *Sämtliche Schriften und Briefe*, (Leibniz 1923–); cited by series, volume and page, e.g. (A VI, 2, 229).
AT	Adam and Tannery (eds.), *Oeuvres de Descartes*, (Descartes 1964-76); cited by volume and page, e.g. (AT II, 490).
DSR	G. W. Leibniz, *De Summa Rerum:* transl. Parkinson (Leibniz 1992).
GM	Gerhardt, *Leibnizens Mathematische Schriften* (Leibniz 1849-63); cited by volume and page, e.g. (GM IV, 97).
Gotha FB	*Forschungsbibliothek Gotha der Universität Erfurt*: a library containing a (small) collection of Leibniz's manuscripts.
GP	Gerhardt, *Die Philosophische Schriften von Gottfried Wilhelm Leibniz* (Leibniz 1875-90); cited by volume and page, e.g. (GP VI, 90).
L	Loemker, *Leibniz: Philosophical Papers and Letters,* (Leibniz 1969).
LDB	G. W. Leibniz. *The Leibniz-Des Bosses Correspondence*, transl. Look and Rutherford, (Leibniz 2007).
LH	*Die Leibniz-Handschriften der königlichen öffentlichen Bibliothek zu Hannover*, (Leibniz 1895): a catalogue of Leibniz's handwritten manuscripts in Hanover, other than those of his correspondence.
LLC	G. W. Leibniz. *The Labyrinth of the Continuum*, transl. Arthur, (Leibniz 2001).
NE	*Nouveaux essais sur L'entendement humain*, translation in (Leibniz 1981), which has page numbers keyed to A VI, 6.

Part I
Interpretive Essay

Introduction

Gottfried Leibniz (1646–1716) is known for his innovations in a variety of different subjects, and the best known and most fruitful of these was undoubtedly his devising of the Differential and Integral Calculus. For a long time his invention of the calculus was called into question by the highly politicised dispute about whether he had plagiarised it from his rival Isaac Newton (1642–1729), whose Method of Fluxions, together with a battery of infinite series techniques, was effectively equivalent to Leibniz's calculus. The allegation of plagiarism, first launched by Nicolas Fatio de Duillier (1664–1753) in 1699[1] and then amplified by Newton's acolyte John Keill in 1708,[2] is now considered sufficiently refuted by Carl Gerhardt's publication, in the mid-nineteenth century, of the papers that led Leibniz to his formulation of the calculus prior to and independently of anything he had learned from Newton.[3]

---

[1] Fatio had made this insinuation in a publication, but Leibniz, on receiving assurances from Sloane, one of the secretaries of the Royal Society, that this did not have the Society's prior approval, had let the matter drop.

[2] Keill made the accusation of plagiarism explicit in (Keill 1708), which was then corroborated in the *Commercium Epistolicum* (the report drawn up by the Royal Society to adjudicate on the matter in 1712), whose author was none other than Newton himself!

[3] Gerhardt's publications included not only Leibniz's own account of his route to his discovery, the *Historia et Origo Calculi Differentialis* (ed. C. I. Gerhardt, Hanover 1846), completed (but not published) in the last year of his life, but also various mathematical manuscripts from the period 1673–1676 showing the stages by which Leibniz had arrived at his invention (*Die Entdeckung der Differentialrechnung durch Leibniz*, ed. C. I. Gerhardt, Hanover 1848); both works are included in English translation in J. M. Child's *Leibniz's Early Mathematical Works* (Child 1920).

Yet it has long been held that Leibniz was not very concerned about the foundations of the calculus.[4] This is an impression that could be given by some of his published statements on the matter. For instance, in the letter he sent for publication to François Pinsson (1612–1691), one of his correspondents in Paris, which appeared in the August 1701 issue of the *Mémoires de Trévoux*, he wrote: "I would even add that one has no need to take the infinite here in all rigour, but only as when one says in optics that the rays from the sun come from a point infinitely far away, and are thus judged parallel" (text [**37**]).[5] Leibniz wrote this letter when he was prevailed upon by his supporters in France to answer criticisms of the opponents of the calculus, and such a defence would hardly have seemed helpful to those expecting a vigorous defence of infinitesimals. It is similar with Leibniz's statement in his famous letter to Pierre Varignon (1654–1722), his main supporter in France, published in the *Journal des sçavans* of March 20, 1702 that "the rules of the finite are found to succeed in the infinite, as if there were atoms […] and conversely, the rules of the infinite succeed in the finite, as if there were metaphysical infinitely small things" (text [**38a**]). To those convinced of the reality of infinitesimals, this enigmatic formulation has appeared as something requiring rigorous justification. Thus, when Abraham Robinson devised his Nonstandard Analysis in the 1960s, he held that this work could be interpreted as "fully vindicating" Leibniz's theory of infinitesimals, since that theory implied "the introduction of ideal numbers, which might be infinitely small or infinitely large compared with the real numbers, but which were *to possess the same properties as the latter*" (Robinson 1966, p. 2, his emphasis).

As more of Leibniz's unpublished papers have come to light, however, it has become apparent that, contrary to such impressions of a lack of interest in foundations, Leibniz was deeply concerned with the status of the infinite throughout his career. With the publication of the philosophical papers he wrote during his stay in Paris from 1672 to 1676, especially the so-called *De Summa Rerum* of 1675–1676, it has been demonstrated that his typical allusion to his labours in the "labyrinth of the continuum" was not simply an idle trope, but indicative of a sustained effort to come to terms with the status and implications of the infinite and infinitely small, not just for mathematics, but also in connection with the

---

[4] See Lagrange (1797, p. 4). Likewise, Carl Boyer, in his *History of the Calculus*, opines that Leibniz's "justification of the infinitely small was not at first the result of serious effort" (Boyer 1949, p. 213); and Morris Kline, in his comprehensive history of mathematics, asserts that "Neither Newton nor Leibniz clearly understood nor rigorously defined his fundamental concepts. […] [Leibniz] saw the long-term implications of the new ideas and did not hesitate to declare that a new science was coming to light. Hence he was not too concerned about the lack of rigor of his calculus" (Kline 1972, vol. 1, p. 384).

[5] For instance, the acute Jesuit thinker Bartholomew Des Bosses found it hard to ascertain what Leibniz meant by this. See his letter of 2 March 1706 (LDB 27), and Leibniz's reply of 11 March (LDB 31–35).

development of his views in physics and metaphysics.[6] Thus, in a manuscript of February 1676 (text [7]) we find him investigating whether "there is something that is infinitely small and yet not indivisible", and wrestling with the consequences. But by April of the same year he has come to the conclusion that interpreting infinitesimals as existing would lead to "impossible" consequences, such as a continuous motion's involving a new endeavour (an assumed infinitely small element of motion) being impressed at every moment, and the composition of time out of instants. Such entities as polygons greater than any assignable and lines smaller than any assignable, he concludes, "are fictitious" (text [10]).

More light was shed on Leibniz's early views on the mathematics of the infinite with the publication in 1993 of Eberhard Knobloch's finely wrought edition (Leibniz 1993) of the treatise on quadrature (the calculating of the area under a curve) which Leibniz had written in Paris, the *De Quadratura arithmetica circuli, ellipseos et hyperbolae* (hereafter the *DQA*). In it Leibniz presented a sophisticated conception of quadrature which, Knobloch showed, anticipated Riemannian integration (Knobloch 2002). The basic idea was to show how demonstrations involving the assumption of infinitely small quantities (or, in the terminology of the time, "indivisibles") could be translated into ones involving only finite quantities by showing that, if the "difference" (or "error") between the approximating quantity and the area under the curve is taken as a variable quantity, it could be rendered as small as one wishes, and therefore null.

But these more recent developments have still left a number of problems of interpretation, which we will be concerned to resolve here. One is the objection that, if Leibniz had already arrived at his mature view of the infinite and infinitesimals as "fictions" before developing the calculus,[7] why do we find him still engaging with the question of the existence of infinitesimals in February 1676, when he was in possession of the calculus and composing his treatise on quadrature, the *DQA*? Might he not have changed his mind?[8] Or could he have been entertaining a different conception of fictions in 1676 than the one he was to present to his Parisian supporters at the turn of the century?[9] After all, the *DQA* is a

---

[6]See in particular the bilingual collections edited by G. H. R. Parkinson, *De Summa Rerum* (Leibniz 1992), and R. T. W. Arthur, *The Labyrinth of the Continuum* (Leibniz 2001), following the edition of the original texts from the Parisian Years (1672–1767) in the so-called Akademie edition in 1980 (series VI, volume 3; cited as A VI, 3).

[7]The first known appearance of the terminology of fiction in relation to the infinite in mathematics is in text [5] dated October 1674.

[8]Cf. Esquisabel and Raffo Quintana (2021) for this criticism.

[9]This is suggested by Douglas Jesseph: "I take Leibniz's claims about the fictionality of the infinitesimal to be his considered view on the subject, although I am not convinced that he held consistently to a 'fictionalist' position from his earliest writings on the calculus." (Jesseph 2008, p. 215); and "It is tolerably clear that Leibniz was a committed fictionalist about infinitesimal magnitudes by 1702. His view of infinitesimals as 'useful fictions' seems to have taken shape in the mid 1690s, although there are certainly traces of it as early as the 1670s, and a forthright statement of the fictionalist position seems to have come from Leibniz's pen only in the aftermath of the dispute in the Academie des Sciences over the foundations of the calculus." (Jesseph 2015, p. 195)

treatise about quadratures studied in a geometrical manner, and does not include any use of the differential algorithm; it depends only on the kind of argumentation described above. So, although it is plausible that the kind of strategy Leibniz developed in the *DQA* for translating talk of infinitesimals into finitist language is also what he had in mind when talking of a justification of the calculus "in the style of Archimedes", this has remained open to question.[10]

Then there is the awkward fact of Leibniz's persistent claims that the calculus is neutral with respect to the existence of infinitesimals, and that mathematicians could assert or deny their existence without this making any difference to their practice.[11] This has seemed to many interpreters to show that Leibniz accepted the reality of infinities and infinitesimals in his mathematics. It is thought that since in his mature philosophy he regarded mathematical entities as ideal, this accounts for his classifying them as fictions.[12] Questions of existence, on this reading, concern only whether such things as atoms, infinite wholes, or infinitely small velocities, exist *in rerum natura*.[13] Such interpretations derive support from modern views about the existential import of theories, according to which existence is to be evaluated solely in terms of whether they must be assumed in the domain of the theory in question: if they are *used* in mathematics, it is presumed, then they must have mathematical *existence*.[14]

We attempt to resolve such issues in our interpretive essay, as we will describe below. But before entering into specifics, we should first note some more general difficulties that occur in trying to get straight on Leibniz's position.

---

[10] For example, Paolo Mancosu, in his thorough analysis of reactions to Torricelli's infinite solid, writes: "[Leibniz's] claim of being able to recast any proof involving infinitesimals into a proof in the style of Archimedes—a proof using the method of exhaustion—was extremely suggestive, but it was never developed in a completely convincing way." (Mancosu 1996, p. 171).

[11] See, for example, Leibniz's notes on Bayle of 1702: "mathematicians do not need all these metaphysical discussions, nor need they embarrass themselves over the real existence of points, indivisibles, infinitesimals, and infinites in any rigorous sense." (GP IV, 569/ L 583–584).

[12] See, for example, John L. Bell: "Leibniz's attitude toward infinitesimals and differentials seems to have been that they furnished the elements from which to fashion a formal grammar, an algebra, of the continuous. Since he regarded continua as ideal entities, it was then perfectly consistent for him to maintain, as he did, that infinitesimal quantities themselves are no less ideal—simply useful fictions, introduced to shorten arguments and aid insight" (Bell 2006, p. 97). See also the quotations from Robinson given above.

[13] This is the main contention of Katz and co-authors in (Katz, Kuhlemann, Sherry, and Ugaglia 2021). They appeal for support to the paper of Esquisabel and Raffo Quintana (2021), who advocate a more nuanced position, claiming that fictions such as infinites and infinitesimals are only "accidentally impossible", and that the impossibility of their existence could be the result of their lack of geometric representability or their incompossibility, rather than their being contradictory.

[14] Cf. the memorable dictum attributed to Willard Quine, "to be is to be the value of a bound variable"; this paraphrases: "we are convicted of a particular ontological presupposition if, and only if, the alleged presuppositum has to be reckoned among the entities over which our variables range in order to render one of our affirmations true" (Quine 1953, p. 181).

1 Introduction

The first source of difficulty concerns the incomplete nature of the Leibnizian corpus. Although Leibniz published extensively, with over a hundred journal articles on mathematics, physics, and chemistry to his name, a rough estimate indicates that he published only around 5% of his mathematical writings during his lifetime. Since the end of the nineteenth century, tremendous progress has been made in rendering the unpublished material in his *Nachlass* accessible, and treasures have already been discovered in it. A large part of what is presented in this book depends on recent discoveries, most of them made in the last 50 years, and on further research we have ourselves conducted on the manuscripts kept in Hanover. As we will argue, this new material has profound implications for our understanding of Leibniz's thinking about infinitesimals. One goal of this book is thus to give access to this unpublished material. To this end we have provided here translations of a wide range of relevant texts, most of which have never before appeared in English.[15] Yet the reader should keep in mind that the situation is far from ideal, since more than half of Leibniz's mathematical manuscripts are still not properly edited.[16] Although we have spent considerable time exploring the *Nachlass*, we cannot exclude the possibility that new documents might be discovered that would create new questions for interpretation.

A second source of difficulty is that of the divergence between statements made by Leibniz on our topic at various times or in differing contexts. Thus many of the manuscripts he wrote during his stay in Paris have the status of "thinking on paper", during which he would change his mind on key issues. A case in point is the text of February 1676 described above, where he passes from a realistic interpretation of the infinitesimals of his calculus to its implications, including the implied existence of infinite number, only to contradict himself, and conclude that "an infinite number is impossible" (text [7]). This indicates the danger of taking quotations out of their full context. Other divergences in his expressed views may be due to the historical context in which he was expressing himself, especially in the case of his letters and their publication. For example, in gauging his replies to criticisms of the calculus requested by supporters of the calculus, we need to bear in mind that many of them held to a realistic interpretation of infinitesimals, contrary to his own considered view, thus putting him in a delicate situation.[17]

---

[15] Even where we have used our own previous translations, we have systematically revised them.

[16] Although the corpus has been already explored by renowned scholars such as Dietrich Mahnke, Enrico Pasini, Marc Parmentier and Heinz-Jürgen Hess, on which our work very much depends, it should be kept in mind that the publication of the manuscripts on differential calculus is now available in the *Akademie* edition only for the Parisian stay. The estimate for the incoming volumes on this topic is of 12 volumes of 800 pages each and the estimated end of this edition is around 2050.

[17] Thus in his letter to Pierre Dangicourt of September 11, 1716, Leibniz explains that his friends were not satisfied when he told them that infinitely small things were fictions, and politely asked him to stay mum on this issue, a plea with which he found it easy to comply (Leibniz 1768, vol. III, pp. 500–501). Among these friends was the Marquis de l'Hôpital, author of the first textbook explaining the calculus, who, Leibniz tells Dangicourt, accused him of "betraying the cause".

In response to such difficulties, we have sought to place the texts in this volume in their proper historical context. This includes reconstitutions of local contexts, such as the way in which Leibniz relates to this or that interlocutor, and also discussions of more global considerations, such as the way in which he was led to give up on the idea that infinitely small entities could be represented by proportions to physical endeavours, or the question of the relationship between his use of fictions in his work to that of his contemporaries and predecessors.

A third source of difficulty in interpreting Leibniz's views on the calculus stems from a widespread tendency to approach them according to how they measure up to modern standards of rigour. This is partly a result of the fact that authors, even already in the eighteenth century, have assumed that both Newton and Leibniz developed their mathematical methods primarily with an eye to application, without paying too much attention to providing them with "rigorous" foundations. Accordingly, many commentators had (and still have) the unfortunate tendency to take it as their first task to *judge* (either favourably or unfavourably) what they take to be at stake with Leibniz's differential calculus. Thus on one side we have those who see it as committed to the existence of dubious entities, which should be suppressed from "rigorous" mathematical discourse. Symmetrically, other mathematicians have taken it as their task to show that Leibniz's calculus could be set on a rigorous foundation by an interpretation of infinitesimals according to modern conceptions, as in Robinson's Nonstandard Analysis, as noted above. In contrast, in this work we will apply such terms as the "rigour" or "soundness" of foundations, or the "demonstration" of a result, only as actor's categories. When we present some "foundation" as "rigorous", it is only because Leibniz himself presented it as such (*fundamenta, demonstratio, rigorosa, solida* or *"vraies à la rigueur"*).[18]

Let us now turn to a summary of the specific arguments of the various chapters.

In Chap. 2 we examine how Leibniz understood the question of the existence of the infinite and the infinitely small during his Paris years, the period from the Fall of 1672 until the end of 1676. This is the period in which, in his own words, he "became a mathematician", and by the end of which he had put the finishing touches to the algorithm he had devised in October 1675, with the notation still used today and taught in modern calculus textbooks. This makes it curious, as we noted earlier, that Leibniz should still have been unsure about the status of infinitesimals in February 1676, surmising that the success of the hypothesis of the infinite and the infinitely small in geometry "increases the likelihood that they really exist" (text [7], A VI, 3, 477). But it is also problematic for those maintaining that the very use of infinitesimals and infinities in the calculus entails that they must exist as

---

[18] For instance, in the *Defense du calcul des differences* (text [49]), in order to justify his calculus Leibniz describes a situation in ordinary algebra where one has to take a limiting case of inequality as if it is an equality — "to put 0 instead of the vanishing magnitudes z and x" — in order for the reasoning to be made general. He comments: "the *foundation* of all of this can be explained by taking z and x [...] *in the very act of vanishing* and falling onto A", so that this becomes an example of applying his Law of Continuity.

elements in its domain of quantification.[19] This was not, however, how Leibniz understood the question of existence in mathematics, as we argue.

Thus, very soon after arriving in Paris, Leibniz jettisoned his earlier conception of infinitely small lines as indivisible points, but still retained his conception of them as intensive entities, defined in terms of endeavours (*conatus*), so that the ratio of the quantities of two infinitely small lines would be determined by a proportion to the endeavours (infinitely small elements) of the motions generating them. It is this conception that he abandons in the Spring of 1676, for a variety of reasons. Their existence as ultimate elements would, he argued, entail the actual division of matter "into an infinity of points", so that "any part of matter will be commensurable with any other", allowing an exact quadrature of the circle, and the actual division of time into instants — all of which consequences he came to regard as impossible. There is no consideration here that infinities and infinitesimals might exist mathematically, while failing to exist *in rerum natura*.[20] Leibniz is considering them solely insofar as they are possible.[21]

Yet, as we also explain in Chap. 2, Leibniz had already concluded early in his stay in Paris that the infinities and infinitieth terms in decreasing infinite series must be interpreted as *fictions*. But how could this be? If Leibniz had already come to regard the infinities and infinitely small occurring in geometry as fictions, then how could the issue of the existence of infinitesimals still be an issue that he thought he had to resolve years later in 1676? This difficulty shows that we need a deeper understanding of what it meant in a seventeenth century context for mathematical entities to be regarded as fictional, and that is what we seek to provide in Chap. 3. Here we show that the describing of certain mathematical entities as "fictional" or "imaginary" was not original with Leibniz, but an accepted feature of the mathematics of that time. Negative numbers, points at infinity, logarithmic elements, and square roots of negative numbers, to name some examples, were all treated by Leibniz's predecessors or contemporaries as imaginary or fictional entities, whose

---

[19] See for example the claim of Laugwitz, who asserted that Leibniz's statement of his Law of Continuity in his 1702 letter to Varignon ("*les règles de l'infini réussissent dans le fini*") "contains an *a priori* assumption: our mathematical universe of discourse contains both finite objects and infinite ones" (Laugwitz 1992, p. 145). In fact, Laugwitz's claim of ontological commitment is contradicted by the continuation of that sentence: "*comme s'il y avait des infiniment petits métaphysiques, quoiqu'on n'en n'ait point besoin*".

[20] In Leibniz's philosophy, mathematical existence is defined in terms of non-contradiction. Mathematical entities, like other ideal entities, are *possibilia*, and he defines possibility by non-contradiction (see A, VI, 4, 277; 862; 865; 867, etc.). Actual existence, or existence in the actual world, on the other hand, is characterized by compossibility, i.e. compatibility with all the things subsisting in the world which God chose to create because it was the best (*Possibilia sunt quae non implicant contradictionem; Actualia nihil aliud sunt quam possibilium (omnibus comparatis) optima* A III, 8, 63).

[21] Indeed, in a letter to Johann Bernoulli dated 29 July/8 August, 1698, Leibniz goes so far as to assert that "If I were to concede such things as the infinite and the infinitely small that we are discussing to be possible, I would also believe them to exist." (text [**32e**]; A III, 7, 855–858)

usefulness in solving problems and equations was well attested. What this meant was that it was not required that they be interpreted as themselves existing in order for correct results to be found by employing them. Indeed, such entities often led to contradiction when they were interpreted realistically. This raised questions about the legitimacy of using them, but, as Leibniz insisted in texts both early and late, such scruples could be satisfied provided the results obtained through their use could be interpreted realistically and also derived by other means without employing them. Although no item whose concept leads to a contradiction could possibly exist, it could still be treated by mathematicians *as if* it existed, under certain conditions. For instance, the interpretation of negatives as numbers gives rise to paradoxes concerning the ratio of 1 to -1; yet one may still manipulate expressions containing them to give sound results in problems in which all the quantities at stake at the end of the process are real (i.e. positive magnitudes). In the same way, the introduction of a logarithmic element could allow one to deal with all magnitudes as having a fictional common measure (although it would be contradictory to claim that all magnitudes are commensurable). "A similar fiction," Leibniz maintains in a text dating from the 1680s, "takes place in geometry, by conceiving the matter just as if all lines are made up out of infinitely many infinitely small straight linelets" (GM VII, 39; text [**25**]). As early as 1672, however, he denies the existence of infinite wholes, and suggests that "one ought not to say anything about the infinite as a whole, except where there is a demonstration" (A VI, 3, 168; text [**1**]). Thus we see that Leibniz did not offer a fictionalist interpretation of infinitesimals in order to parry objections from his critics, as is often alleged; rather he considered them as belonging to a pre-existing category of such "imaginary" mathematical entities, including negative ratios, negative numbers, and points at infinity, and for which the question of existence had to be addressed independently of their use in mathematics.

This interpretation of infinities and infinitesimals as fictions is already explicit in Leibniz's masterful treatise composed in 1675–1676, the *DQA*. That is the subject of our Chap. 4. There we describe how in this treatise Leibniz develops a method for justifying quadrature by adopting a certain principle occurring in the work of his contemporaries and predecessors. This principle, which we have dubbed the "Principle of Unassignable Difference", or *PUD*, is that *if a difference between two quantities can be assumed smaller than any assignable difference, it is null*. In Proposition 6 of the *DQA*, Leibniz uses it in demonstrating "how certain step-spaces, and likewise certain polygonal spaces, can be increased continuously, to the point where they differ from each other or from curves by a quantity smaller than any given". That is, for quadrature, Leibniz supposes that the step-space is made up of unequal rectangles, and shows that the difference between the area of all the rectangles and the area under the curve must be less than or equal to the area of a rectangle whose height is that of the tallest of the unequal rectangles. But since the height of this rectangle "can nevertheless be assumed smaller than any assigned quantity, for however small it is assumed to be, others can be assumed still smaller," it follows that its area can be made smaller than any assignable, so that, by the *PUD*, the difference between the two areas is therefore null.

In the face of objections that this method amounts to an inessential variant of the Method of Indivisibles or of the traditional Method of Exhaustion,[22] we explain in detail what Leibniz accomplished with his original approach in the *DQA*. Apart from the innovations of taking unequal rectangles and establishing inequalities, and also his theorem of transmutation, it showed how the Method of Indivisibles used by many of his contemporaries (and scorned by others for its lack of rigour) could be established according to traditional standards using only finite quantities. Yet, in distinction from the traditional Archimedean Method of Exhaustion, it required neither the geometric construction of inscribed and circumscribed figures as rectilinear approximations nor the two *reductio* proofs invoked in that method.

Although the method of Proposition 6 is direct, however, it does not involve infinitesimals, and relies only on the *PUD*. In Proposition 7 Leibniz establishes that a direct proof in terms of unassignable differences and the *PUD*, such as that given in Proposition 6, is equivalent to a *reductio* argument which proceeds by showing that a quantity is rendered smaller than itself. This is significant, in that it demonstrates the *PUD* in accordance with the standards of rigour demanded by those suspicious of proofs involving the neglecting of unassignable differences. But it is not until Proposition 8 that "infinitely small" quantities are finally introduced as fictions. What necessitates their introduction is that, in cases where one begins the quadrature at the origin of the curve, the ultimate element of the step-space at the origin does not have the same shape as the others, and so can only be included as a kind of fiction. Yet the relations established in Proposition 7 concerning arbitrarily small segments can be considered to hold, as Leibniz demonstrates, also in the case when the initial rectangle under the curve degenerates into a triangle. This use of infinitesimals, then, constitutes a direct proof, and it anticipates the Law of Continuity that Leibniz will appeal to later (according to which the degenerate case can be included in the general reasoning although it is of a different nature than the other terms). But what Leibniz does in Proposition 8 is to justify this inclusion of an infinitely small last element by translating the direct proof using the method of Proposition 7. Given the equivalence of that proposition with Prop. 6, Leibniz has thus established a justification for using infinitesimalist methods in quadrature. As he repeats in the famous scholium to prop. 23, the direct method, based on "fictions", allows one to bypass the introduction of inscribed and circumscribed figures, the use of reductio argument, and even of the demonstration that the error is smaller than any assignable, "although what we have said in *Props. 6, 7 & 8* establishes that it can easily be done by means of them."

Some historians of mathematics have sought to minimise the importance of Proposition 6 of the *DQA* for Leibniz's mature thought, and also its relevance to the foundations of the calculus. Noting that Leibniz's employment of infinitesimals and infinites as fictions in his

---

[22]Thus, Viktor Bläsjö has argued that "Leibniz didn't think of it [sc. Proposition 6] as a foundational innovation but as a rather pedantic and basically routine way of applying what is essentially the ancient Greek method of exhaustion" (Bläsjö 2017, p. 135).

DQA depends on a nascent version of his Law of Continuity, they have claimed that this is independent of the modified Method of Exhaustion by which Leibniz justified his quadratures in Proposition 6 of that treatise. Leibniz's establishment of an equivalence between the direct proof of Prop. 8 and that of Prop. 6 and 7, we argue, is sufficient to undermine those objections. But even though this equivalence makes it plausible that the methods of the *DQA* embodied the kind of strategy which Leibniz had in mind later in defending the calculus by talking of a justification "in the style of Archimedes", this connection with the calculus remained to be substantiated by other sources. In §4.3 we provide such sources by retrieving texts from the 1690s showing that this is indeed the case.[23] Chief among these is the *Compendium* of the *DQA* (text [**17**]), which Leibniz had prepared for publication in the 1690s. It is clear that in this piece Leibniz had no trouble translating the results of the *DQA* into the differential calculus. The same thing is even clearer in the Appendix he prepared for Bodenhausen in October of 1690 (text [**18**]), where he translates the proof of Prop. 8 into the differential calculus. Moreover, it is to the arguments of the *DQA* that Leibniz reverts in that case, as well as in the reply he prepared in 1695 for Johann Christoph Sturm, in response to the difficulty that these two astute thinkers had expressed in understanding his calculus and he presents explicitly these argument as following the "Archimedean method".

That brings us (Chap. 5) to the differential calculus itself, and Leibniz's publication of it in two ground-breaking articles in the *Acta eruditorum* of October 1684, a substantial extract from the first of which, the *Nova methodus*, is our text [**21**].[24] One of the puzzling features of that article was that Leibniz did not explicitly introduce his differentials as infinitely small quantities, but rather stipulated that the differences dv of variables such as v are defined by their proportions to an "arbitrarily assumed straight line" dx. As we explain, he did not do this because he was afraid of becoming embroiled in foundational disputes if he defined them as infinitely small[25]; on the contrary, it was a deliberate strategy. The dx stand for lines of arbitrary length, and so may be taken as finite and as small as

---

[23] Some of these sources were already published by Gerhardt in his *Mathematische Schriften*, but remained unnoticed due to the fact that Gerhardt was not precise with the dating, for instance, misconstruing the *Compendium* as probably dating from the late 1670s.

[24] We have published that part of the 1684 article which is more germane to the issue of foundations. For an English translation of the whole paper, including Leibniz's examples of application of the method, see Struik (1986, pp. 272–80). The second article, (*De geometria recondita et analysi indivisibilium atque infinitorum Acta eruditorum* 1686), written by Leibniz as a friendly response to John Craig's having adopted his notation for differentiation, shows the inverse relationship between 'sum' (i.e. integral) and difference, introduces the integral sign $\int$, and shows how easily transcendental curves may be treated in the calculus. See Struik (1986, pp. 281–82) for an English translation of an extract.

[25] This is a common view. Thus, John Bell writes "Although the use of infinitesimals was instrumental in Leibniz's approach to the calculus, in 1684 he introduced the concept of differential without mentioning infinitely small quantities, almost certainly in order to avoid foundational difficulties" (Bell 2006, pp. 95–6). Cf. also Mancosu (1996, p. 156).

needed either for the problem at hand, or as infinitely small; in order for the error in the calculation to be less than any pre-assigned error. All that is important is that the dx are differences of what can be taken fictionally as "neighbouring" values, not whether they are finite or infinite. Leibniz takes the same stance in his *Tentamen* ("An Essay on the Causes of the Celestial Motions"), published in 1689,[26] in a paragraph whose contents, he asserts, constitute his "Lemmas on Incomparables". Here Leibniz prefers the term "incomparable" to denote the infinitely small difference between two "common" (i.e. finite) quantities. He explains that this can be accepted even by those rejecting the existence of infinitely small magnitudes, where a difference is incomparable if it is either so small as "to produce an error of no importance", as in representing the diameter of the Earth as infinitely small with respect to the heaven, or is so small that the error is smaller than any that can be assigned, and thus zero, as explained above. These differences can themselves have further differences which will be incomparably small with respect to them, as for instance the versed sine of an infinitely small angle $d\phi$ will be infinitely small in this sense by comparison with $d\phi$ itself. Incomparables, Leibniz maintains, are not fixed quantities in the continuum; rather they are variables, like the magnitudes of which they are the differences.

Between these two publications, Leibniz had also published an article explaining his famous Law of Continuity, as an extract from a letter to Malebranche (text [**22**]). In this article, Leibniz generalises his considerations about differences that can be taken as vanishing into a general law, one that can be considered as applying not only to variable magnitudes in mathematics, but quite generally to any continuous transitions: "When the cases (or what is given) continually approach and finally disappear one into the other, it is necessary that the consequences or results (or what is sought) do so also". These continuous transitions between cases may occur in mathematics, in nature, or even at the meta-level, ruling out discontinuities between adjoining cases of relative motions governed by Descartes' and Malebranche's rules of collision. Thus a parabola can be considered as an ellipse as one focus is moved to an infinitely distant point, and "rest can be considered as an infinitely small speed".[27] But as in the *DQA*, Leibniz was careful to assure readers that this does not require commitment to the existence of infinitesimals or infinities, and that in imagining a parabola as an ellipse one of whose focuses is at a point at an infinite distance, all that is necessary is to imagine "a figure which differs from a certain ellipse by less than any given difference".

---

[26] *Tentamen de Motuum Coelestium causis, autore G.G.L*, Acta eruditorum, February 1689, 82–96; translated in (Leibniz 2023, pp. 95–111); the Lemmas on Incomparables are in paragraph (5), reproduced here as text [**23**].

[27] There exists an interesting attempt by Leibniz to reformulate his Law of Continuity in the previously unpublished manuscript of uncertain date, the *Axioma mirabile*, which we include as text [**31**]. It was included by Enrico Pasini in his dissertation (Pasini 1985-86), a goldmine for anybody wishing to access unpublished materials on Leibniz's foundational reflections.

Consistently with this status of magnitudes as indeterminate and variable, differentiation is for Leibniz an operation on variable magnitudes producing differences or differentials that are themselves indeterminate. Differentials are, as he explains to Huygens (text [**24**]), "affections" of these magnitudes, analogous to their powers, such as x^3 as a power of x. Incomparability denotes the relationship between variable quantities and their infinitely small, variable differences, where two homogeneous quantities are comparable "if one can be made greater than the other when multiplied by a finite number", in keeping with the Archimedean axiom (as Leibniz explains in his reply to Nieuwentijt, text [**45**]); otherwise they are incomparable.

The definition of incomparability in terms of violation of the Archimedean axiom has suggested to some that Leibniz's infinitesimals are non-Archimedean quantities, and that they are therefore vindicated by the twentieth century creation of Nonstandard Analysis, as noted above. In Sect. 5.2 we subject that claim to examination by investigating Leibniz's own views on quantity, to which he gave careful consideration. Some texts outlining these views were published by Gerhardt in the nineteenth century, but they are not widely known, being in Latin; accordingly, we have given substantial relevant extracts here in English translation.[28] According to the accounts Leibniz gives in these texts, a quantity must be comparable to another quantity taken as a unit, that is, they must have a finite ratio; similarly, a number is something having a ratio to 1. Although, as Leibniz argues, even irrationals can qualify as numbers according to these definitions (we explain how in Sect. 5.2), the infinitely small and infinitely large cannot be quantities or numbers in this sense. Thus, we conclude, there are no non-Archimedean magnitudes according to Leibniz's theory of magnitude or quantity.

This conclusion, however, does not itself rule out infinitesimals being interpreted as geometric entities of some other kind. In fact, the geometric tradition deriving from the Ancient Greeks acknowledged certain entities that do not satisfy the Archimedean axiom, such as the angle of contact or "horn angle" between a circle and a straight line, which was regarded as smaller than any rectilinear angle.[29] As we show, in his writings on mathematics from the 1680s onward, Leibniz allowed for quantities in this broad sense (*lato sensu*, in Latin). But although one could give sense to the idea of a point being smaller than a line

---

[28] These texts are respectively two tracts, *De quantitate* and *De magnitudine et mensura*, and the important essay *Specimen Geometriæ luciferæ* (texts [**25**], [**26**] and [**27**]). Gerhardt published the first two as though they were parts of the same manuscript, but recent research has shown that the first dated from the 1680s, and the latter from the 1690s. The *Specimen* has been provisionally dated as 1695, but that date may be revised as it is further investigated and there are arguments to date it from the end of the 1680s too.

[29] That an "angle of contact" is smaller than any rectilinear angle was proved by Euclid (*Elements* III, 16). Accordingly, some ancient authors such as Proclus considered that "angles" were not magnitudes in and of themselves, but should be considered under the category of "form" or "quality", rather than "quantity". This is the position still expressed by Wallis in his exchange with Leibniz, as we shall see in Sect. 5.3.

using this looser meaning of quantity, it is not a quantity in the strict sense (*stricto sensu*) that could allow a line to be conceived as an infinite number of congruent "points" as parts. The consideration of such entities, we maintain, throws interesting light on Leibniz's earlier theories, prior to mid-1676, in which he had appealed to angles and horn angles to justify his notions of one point being greater, and even infinitely greater than another. In those years, as noted above, he had held that infinitely small lines could be interpreted in terms of their proportionality to endeavours or infinitely small motions conceived as intensive quantities. As he came to recognize in 1676, however, the infinitieth of a common quantity (such as an infinitely small part of a finite line segment) cannot be interpreted as actually existing without giving rise to a contradiction. For an infinitieth part of a finite whole will be in the same ratio as a finite part of an infinite but bounded whole, so that a finite line will be the mean proportional between an infinitieth part of a line and a bounded infinite line.[30] But Leibniz took the latter to be a contradictory notion. Thus, if an infinite bounded line is contradictory, then so too will be an infinitesimal part of a finite line: both are mere "chimeras". Leibniz had the idea for this argument already in 1676 (text [**11**]), but he does not seem to have been able to put it into a form that satisfied him until the early 1690s as we show by an examination of a succession of unpublished manuscripts (texts [**28**], [**29**] and [**30**]).

Leibniz offers the same argument in response to his friend Johann Bernoulli (1667–1748) (text [**32a**]), who had presented his own views on the infinite and infinitely small in an exchange with Pierre Varignon. Bernoulli had argued—precisely as Leibniz himself had in February of 1676, before deciding against the reality of infinitesimals—that infinitude is a relative notion. Just as we, regarding ourselves as finite or "ordinary", would be regarded as infinitely small by putative beings infinitely greater than ourselves, so beings that were to us infinitely small would regard themselves as finite, and us as infinitely large. In response, Leibniz asserts that an infinite division could be interpreted as issuing, not in actually infinitely small parts, but only in ones "perpetually smaller, and yet ordinary" (text [**32e**]). In support of this, he recalls the mean proportional argument of his *DQA* (texts [**6**] and [**11**]), according to which "it is possible to doubt whether there could be infinitely long straight lines that were nevertheless also in fact bounded", and his conclusion that "it suffices for the calculus that they be taken as fictions, like imaginary roots in algebra" (text [**32e**]). But Bernoulli was unmoved by the argument, insisting that in a decreasing infinite series of fractions, there must be an infinitieth fraction or actual infinitesimal. Leibniz reaffirms his position that this does not necessarily follow: while there are indeed infinitely many terms in an infinite series, this means only that however many terms are taken, there are still more; but that there is no infinitely small magnitude, and no infinitely large one.

---

[30]This "mean proportional" argument is the subject of an interesting analysis by Brad Bassler in his (2008).

Thus, we argue, on the question of the existence of infinitesimals as genuine magnitudes, Leibniz's stance from 1676 onwards is unambiguous: they are not possible existents, even though they can be treated *as if* they are to solve mathematical problems. That is, like imaginary roots and negative ratios, one can *use* his *dx* and *dy*, and, more generally, supposed quantities that are infinitely small or infinitely large with respect to other given quantities, as "fictions", independently of the answer given to the question of their existence.[31] Consequently, although Leibniz *uses* infinitesimals, and even on occasion refers to them as "infinitesimal", "unassignable", or "incomparable *magnitudes*", one cannot infer merely from this that they are for him genuine magnitudes.[32] Leibniz's practice was precisely conceived as being *neutral* between interpretations advocating the existence of the infinite and infinitely small and those rejecting it.[33]

Of course, it's all very well saying that infinitesimals can be used as fictions independently of their existence. But how can we be sure this will not lead to error? What justifications did Leibniz offer in his maturity for their use? This is the topic of our Chap. 6. Here, in keeping with our resolve to avoid anachronism, we do not interpret justification according to today's standards, but use that term solely as an actor's category. Accordingly, we begin by first setting the issue of the justification of the calculus in its historical context.

Leibniz's differential calculus did not attract a lot of attention until it was enthusiastically adopted and developed by Jacob Bernoulli (1654–1718) and his brother Johann, who had followed Leibniz's lead in applying it successfully to a number of challenge problems in the 1690s. Subsequently, it was taken up by mathematicians in Malebranche's circle in

---

[31] Tzuchien Tho had earlier urged the same distinction between use and existence: "We should be careful, as Leibniz is, in clearly separating the debates concerning the metaphysics of the infinite and the mathematical use of fictional infinite and infinitesimal terms." (Tho 2012, p. 74). See also (Bos 1974, p. 54) on the distinction between the metaphysical question of existence and the justification of the use of infinitesimals in differential calculus, and Samuel Levey's discussion of Leibniz on fictions (Levey 2008).

[32] As a case in point, in his article "*De la chaînette*" (*Journal des sçavans*, 31 March 1692), Leibniz writes: "*L'analyse nouvelle des infinis ne regarde ni les figures, ni les nombres, mais les grandeurs en général, comme fait la Spécieuse ordinaire. Elle montre un algorithme nouveau, c'est-à-dire une nouvelle façon d'ajouter, de soustraire, de multiplier, de diviser, d'extraire, propre aux quantités incomparables, c'est-à-dire celles qui sont infiniment grandes ou infiniment petites en comparaison des autres.* [The new analysis of infinites concerns neither figures not numbers, but magnitudes in general, as does ordinary algebra. It shows a new algorithm, that is to say a new manner of adding, subtracting, multiplying, dividing, and extracting roots, proper to incomparable quantities, that is to say, those that are infinitely great or infinitely small in comparison with others.]"

[33] This is a subtle point. From Leibniz's claim that the practice of the calculus is neutral with regard to existence, it is tempting to infer that infinities and infinitesimals are possible objects that may or may not exist; we are saying, on the contrary, that they are for him not possible objects, yet they can be used in mathematical practice *as if* they exist. As we argue in our 2020 paper (Rabouin & Arthur 2020), this can be compared with the use of V as the fictitious "set of all sets" in standard ZFC. This is a contradictory notion, if interpreted literally, but which can be used without harm with proper caveat.

Paris receiving guidance from Johann Bernoulli, including Pierre Varignon, and Guillaume Marquis de l'Hôpital (1661–1704), both of whom corresponded about it extensively with Leibniz, and also Bernard Le Bouyer de Fontenelle (1657–1757).[34] Apart from the specious criticisms of his implacable opponent, the Abbé de Catelan (active 1675–1710), which were dealt with on Leibniz's behalf by L'Hôpital, the calculus was also criticised by Detlef Clüver (1645–1708) in 1687 and, at much greater length, by the Dutch theologian and mathematician Bernard Nieuwentijt (1654–1718).[35] In articles published in 1694 and 1695, Nieuwentijt objected to Leibniz's neglecting of incomparably small quantities, proposing instead that infinitesimals be defined in such a way that their squares were identically zero, so there would be no higher-order differentials. These criticisms were parried in part by Jakob Hermann in 1700, but subsequently Leibniz was also moved to make a published reply (text [**45**]).

In 1696 L'Hôpital published his *Analyse des Infiniment Petits* (L'Hôpital 1696),[36] the first textbook on the calculus, and the fruit of his collaboration with Johann Bernoulli, whom he had hired as his private tutor. This publication stirred their opponents in the *Académie Royale des Sciences* in Paris, provoking a dispute in the *Académie* that lasted until 1706, with the noted algebraist Michel Rolle (1652–1719) the main champion of traditional methods, while Varignon took up the cause of methods using infinitesimals. It began in May-June 1701 when Father Gouye published an anonymous review of an article by Johann Bernoulli in the *Mémoires de Trevoux*, charging that the new analysis lacked "in its demonstrations the evidence that one expects from them" (Gouye 1701, p. 430). This prompted Varignon to ask Leibniz for a response (although it was forbidden to mention the schism in the *Académie*), and Leibniz obliged with his letter to Pinsson (text [**37**]), published in the *Mémoires de Trevoux*. As noted above, this did little to assuage the worries of his supporters, and in response to Varignon's pleas for further clarification in the face of the criticisms of Rolle, Leibniz wrote a series of letters from 1702 (text [**38**]). This includes his famous letter to Varignon, (text [**38a**]), published in the *Journal des sçavans* of March 20, 1702, in which he clarifies what he had said in his letter to Pinsson.

As Varignon reports (text [**38b**]), Gouye was surprised by Leibniz's clarification of what he had written to Pinsson, understanding Leibniz now to be affirming that the only "infinite

---

[34] Others in the Malebranchian circle were Montmort, Carré, and Charles Reyneau (1656–1728). For details of the savants in this circle, see Malebranche (1958–1968, vol. 20) and Bella (2022). Among the opponents of the calculus were Father Thomas Gouye (1650–1725), Michel Rolle (1652–1719), Philippe de la Hire (1640–1718), and the Abbé Gallois (1632–1707), one of the editors of the *Journal des sçavans*. See Paolo Mancosu's historical overview of the critical reactions to Leibniz's calculus in chapter 6 of (Mancosu 1996), which largely follows (Mancosu 1989), as well as Bella (2022).

[35] For details of the objections of Clüver and Nieuwentijt and responses to them, see Mancosu (1996, pp. 156–164).

[36] Leibniz had planned to write his own book on the new approach to analysis, the *De Scientia Infiniti*, and to enlist other collaborators for it, but gave L'Hôpital his permission to publish first, after which his own work did not eventually materialise.

in all rigour" is the actual infinite (the "real and existent infinite"), which is of no use in the differential calculus, for which one needs only the mentally inexhaustible one. In fact, however, Leibniz accepts that the actual infinite occurs in mathematics, for instance in the actually infinite multiplicity of terms in an infinite series, but denies that such a multiplicity can be interpreted as forming a collective whole. Instead, as we saw him explain to Bernoulli above, he conceived an actually infinite division of a magnitude as a division into ever smaller parts, with no number of such parts and no ultimate part, except as fictions; and with no wholes formed from an infinity of parts, also except as fictions.

Now, it is in his first letter to Varignon of 1702 that we find him for the first time describing his conception of the infinite as "syncategorematic". Since this notion of the syncategorematic infinite has engendered controversy ever since Hidé Ishiguro drew attention to it,[37] we subject it to examination at the beginning of Chap. 6. The term is a scholastic one, having its origins in a grammatical distinction. It was applied to the infinite by thinkers like Ockham, with whom Leibniz was familiar, to denote the idea that something—for instance, the parts of a continuum—could be infinite without there being a particular thing that is infinite (which would be a "categorematic" use of the term). Leibniz appears to have been prompted to use the term by his reading of the mathematician Juan Caramuel y Lobkowitz (1606–1682) in 1689.[38]

The first thing to note, however, is that Leibniz's use of this term does not constitute a new position. The example which he gives in introducing it in his letter to Varignon is the same example that was the subject of his protracted debate with Johann Bernoulli in 1698, namely that of an infinite series with decreasing terms, in this case the series $\frac{1}{1} + \frac{1}{2} + \frac{1}{4} + \frac{1}{8} + \frac{1}{16} + \frac{1}{32}$ etc., which he equates with 2. He stresses, as he had done with Bernoulli, that each of the terms is a finite fraction, and "only ordinary numbers are used in it, and no infinitely small fraction, or one whose denominator is an infinite number, ever occurs in it" (text [**38a**]). In fact, he had explained this conception at length in a Latin manuscript that has hitherto been neglected, now titled "Towards a Science of the Infinite".[39] We reproduce a section of this piece here (text [**33**]). In it Leibniz explains (*inter alia*) how an infinite series in the sense of an indefinite finite series gives only an approximation, and

---

[37] This syncategorematic conception of the infinite in Leibniz was first noted by Hidé Ishiguro in the second edition of her *Leibniz's Philosophy of Logic and Language* (1990). There she affirms that for Leibniz "there are infinitely many substances and infinitely many numbers in his sense of infinitely many, i.e., actually more than any finite number of them, and not merely potentially more" (Ishiguro) 1990, p. 80. This needs a slight modification: it is not that there are "more than any finite number" but that 'for any finite number of them, there are more'.

[38] A VI, 4, 1342.

[39] This (unfinished) essay was published by Gerhardt as the second of two texts, the first of which appears to be a historical preface for Leibniz's intended work, *De scientia infiniti* (Gerhardt 1876, 594). It has now been re-transcribed and translated in, *Leibniz on the Metaphysics of the Infinite*, Arthur and Ottaviani (eds.) forthcoming. In their commentary on the piece, the editors tentatively date it as c. 1696–1698, although it could have been written slightly earlier.

distinguishes it from one that is truly infinite. He gives as an example of the latter the same series as he uses in his letter to Varignon, explaining that when that series is said to equal 2, this means that "if each of these fractions is assumed and none besides, then neither more nor less is assumed than what is in 2". So, although there is no infinite series as a collective whole properly speaking, one can regard it as one, although in fact it is a distributive whole in which each term of the series is included and nothing else. This is the same distributive sense of whole that Leibniz later describes in an unsent addendum to his letter to Des Bosses of 1 September 1706:

> There is also an actual infinite in the sense of a distributive whole, not a collective one. Thus, something can be stated of all numbers, though not collectively. In this way it can be said that for every even number there is a corresponding odd number, and vice versa; but it is not therefore accurately said that there is an equal multitude of even and odd numbers. (text [35c])

A second point of interest in this connection concerns the potential infinite. It is often said that this is the only infinite appearing in Leibniz's mathematics, and it is identified with the syncategorematic.[40] This is suggested to commentators by his description of the syncategorematic in the same addendum for Des Bosses as "a passive power having parts" (text [35c]). On the contrary, we argue, this does not refer to the potential infinite, and Leibniz nowhere appeals to the Aristotelian designation of this type of infinite, except when quoting others. The meaning of the term is well described in the *New Essays*: "to say that there is an infinity of things" in this syncategorematic sense is to say "that there are always more of them than can be assigned. But there is no such thing as an infinite number, or line or other infinite quantity, if if we take them to be true wholes, as is easy to demonstrate" (text [34]).

As we have seen, Leibniz had justified using infinities and infinitesimals in geometry in the *DQA* by appealing to the Principle of Unassignable Difference (*PUD*). In Propositions 6, 7, and 8, he demonstrated that one can justify the vanishing of an "infinitely small" difference by replacing it by an assignable difference that can be shown to be smaller than itself and thus, by *reductio ad absurdum*, null. In all his defences of the use of these fictions in his calculus, Leibniz consistently takes the same line. Thus, as he writes to Pinsson,

---

[40]Cf. Fabio Bosinelli: "*Für Leibniz sind sein synkategorematisches Unendliches und das potentielle Unendliche ein und dasselbe.* [For Leibniz the syncategorematic infinite and the potential infinite are one and the same.]" (Bosinelli 1991, p. 168); and Herbert Breger: "*Leibniz will die unendlichen Reihen also potentialistisch verstanden wissen. Daraus folgt inbesondere, dass unendliche Dezimalbrüche potentialistisch aufzufassen sind.* [Leibniz therefore wants infinite series to be understood as potentially infinite. From this it follows in particular that infinite decimal fractions are to be understood as potentially infinite.]"; and "*Ein Aktual-Unendlich gibt es in der Leibnizschen Mathematik nicht.* [There is no actual infinite in Leibniz's mathematics.]" (Breger 2016, pp. 122, 124). Even Mancosu suggests that Leibniz needed the potential infinite in order to handle the case of Torricelli's infinite solid (1996, pp. 144–45); Breger, too, interprets Leibniz's fictional *infinitum terminatum* as a potential infinite (2016, p. 125).

"instead of the infinite or infinitely small we take quantities as great or as small as is needed for the error to be less than the given error, so that we differ from the style of Archimedes only in the expressions" (text [**37**]); and to Varignon, "… since whenever an opponent tries to contradict this statement, it follows from our calculus that the error will be less than any error he could assign, it being in our power to take this incomparably small as small enough for this purpose, inasmuch as one can always take a magnitude as small as one wishes" (text [**38a**]). This allusion to incomparables underlines the fact that this is also the foundation for his Lemmas on Incomparables. Almost identical formulations are given in his *Essais de Théodicée* of 1710, (text [**36**]), in his explanation to Des Bosses (text [**35b**]), and in his letter to Gallois (text [**40**]).[41]

In making this case we have referenced published articles and letters that have long been known, but it is interesting that Leibniz takes the same line in another reply to criticisms from 1702–1703 (hitherto practically unknown), the *Quaestio de jure* (text [**39**]).[42] In it Leibniz reiterates that the infinite and infinitely small appearing in the calculus are "imaginary", and may be used by mathematicians independently of controversies about their existence. He also stresses that one can never use them to abbreviate reasoning "without an explanation of what is substituted, by which the matter reduces to a rigorous demonstration in the style of Euclid or Archimedes", after which he presents a succinct formulation of such demonstration precisely as above, describing it as "the principle of our differential calculus":

> Therefore, when it can be shown that in the assertions of the geometers the error is smaller than any assignable, that is, null, we say by way of abbreviation that the error or discrepancy is infinitely small, and we employ it in our calculus as a quantity, and neglect it in comparison with that of which it is the discrepancy. For the reasoning is immediately made rigorous by changing the indefinitely small into one determinately small, but less than an assigned quantity, which a demonstration shows can be held to be zero, so that the matter reduces in a way to that peculiar kind of demonstration, examples of which are to be found just about everywhere, where from the supposed inequality of two quantities we directly conclude their equality.

In the Response to Nieuwentijt (text [**45**]), and also in correspondence with Wallis (text [**46**]), Leibniz stresses how the same justification can be given for second order differentials that are incomparably smaller than first order ones, just as he had in his Lemmas on Incomparables, and in general to any supposed quantities that are incomparably smaller than those of a higher degree of infinity, for "the error which could result from this is

---

[41] For instance, to the Abbé Gallois: "…it is evident that the error will be less than any given magnitude, and consequently that there is no error whatsoever, which renders the reasoning entirely rigorous, just as Archimedes' reasoning was, without going through the detours that he took" (text [**40**]).

[42] This is another text we owe to Enrico Pasini (1985-86, App. pp. 40–47).

smaller than a given, if someone wished to translate the calculation into the Archimedean style" (text [**46**]).

"The Archimedean style", however, was not likely to win much approval from the "modernists" among the Cartesians. As we point out, Descartes himself was not enamoured of either the method of indivisibles (which he knew), or proofs using *reductio* arguments. With this in mind, Leibniz had devised another strategy to convince the Cartesian algebraists, namely to show that what was needed in order to justify his algorithm was no different from what was needed for "ordinary algebra". This is the main point of his *Justification du calcul des infinitésimales par celui de l'algèbre ordinaire* (text [**47**]), sent to Pinsson and Varignon in 1701, but not subsequently published. The basic idea is that underlying the law of continuity: it is often the case that, in order to have a general rule, the rule must be extended to cases where a certain magnitude vanishes. It is in this way that the rules of collision, in order to be universal, have to apply not only when one of the colliding bodies has a finite velocity, but also in the case where the velocity is diminished to zero. Similarly, in the algebraic case Leibniz adduces, the ratio between the two lines c and e in his diagram must still be finite as c and e both vanish, in order for the case to be considered under the same general rule. In the same way, in geometry a circle can be regarded as a fictional ultimate polygon, a "termination" of a sequence of regular polygons with an increasing number of sides, and equality can be regarded as a terminating case of inequality, even though it is not rigorously true that rest is a kind of motion, a circle a regular polygon, or equality a kind of inequality. Moreover, Leibniz adds, even though such terminations "are not comprised in all rigour in the variations which they bound, nonetheless they possess the same properties as if they were so comprised, in accordance with the language of infinities and infinitesimals, which takes the circle, for example, as a regular polygon with an infinite number of sides" (text [**47**]). Finally, anyone who is not satisfied by this appeal to including fictional terminating cases, "can be made to see in the style of Archimedes that the error is not assignable, and cannot be given by any construction".

In fact, though, this strategy based on the law of continuity is not a new strategy, as it is also the one underlying the use of the fictions of infinitesimals and infinite bounded lines in the *DQA*. As we explain, there Leibniz claims that the paradoxical nature of Torricelli's infinitely long solid with a finite volume is defused when the infinity in question is treated syncategorematically (to use his later term). Thus the length of the solid is not a categorematically infinite magnitude, but one whose magnitude can be made greater than any given magnitude: each segment of the length corresponds to one of the elements of the solid of revolution and one of the elements of the cylinder with which it is compared, but although there is no last or terminating element, one can be included as a fiction in order to make the reasoning general.

All of this receives further confirmation in the unpublished paper, *Defense du Calcul des Differences*, (text [**49**]) of c. 1702.[43] We argue that this paper, clearly a rewriting of the

---

[43] This text was already transcribed by Enrico Pasini in his dissertation (Pasini 1985-86).

Justification, appears to have been composed as a preface to the well known *Cum prodiisset* (text [**50**]), the extensive justification of the calculus composed by Leibniz that has been the subject of several studies since it first appeared in Gerhardt's edition.[44] For the *Defense* very aptly summarises the contents of *Cum prodiisset*. Against the accusation that the calculus is necessarily committed to the existence of infinitely small magnitudes, it responds that "everything that is concluded by this calculus can be proved by a *reductio ad absurdum* in the style of Archimedes, and by using the Lemmas on Incomparables proposed in the Leipzig *Acta*". This alludes to the reasoning we summarised above, originating in Propositions 6, 7, and 8 of the *DQA*, and epitomised in the Lemmas. Secondly, it responds to the accusation that elisions are made arbitrarily, with the rejoinder that "the elisions are made according to certain rules, and not as we see fit". Then, for the benefit of those who are unconvinced by "Archimedean-style demonstrations, which are extremely long and not sufficiently appropriate to enlighten the mind", it presents the argument from the Law of Continuity, showing that precisely the same kinds of fictions are appealed to in ordinary algebra in order to make the reasoning more general by including fictional terminating cases in all continuous transitions. Finally, for those suspecting that this commits one to the existence of infinitesimal quantities, it presents a formulation of this latter strategy without recourse to fictions and by recourse to ordinary quantities only.

In sum, we can see that it is very far from the case that Leibniz was not much concerned about the foundations of his differential calculus. On the one hand, he had always been acutely aware of all of the implications for physics and metaphysics of the existence of the infinite and the infinitely small and the composition of the continuum. But, on the other hand, having decided against the possibility of their *existence*, he had sharply separated that question from the questions of their *use* in mathematics, and the *justification* of that use. This separation was *possible* precisely because the justification he devised for their use was entirely finitist, and in keeping with the Archimedean tradition, and yet *necessary* because the concepts of infinite wholes and infinitieth parts or fractions implied contradiction, given the dependence of the notion of quantity on the part-whole axiom. And far from being insouciant in the face of the criticisms of the calculus by his peers, he had well thought-out strategies for defending the use of infinities and infinitesimals, and knew precisely how to demonstrate the rules for differentiation on that basis. The absence of a published document demonstrating all this in detail was most unfortunate, given the subsequent history of disputes about foundations. But that is largely the result of contingent circumstances, such as papers he prepared for publication not being published by their recipients, or his not wishing to undermine his supporters. The existence of several such documents among his papers is a testament to his serious concern for foundations, and also to the subtlety and profundity of his thinking on that question.

That gives the general outline of our arguments in this book. Now let us turn to their detailed elaboration.

---

[44] Gerhardt (1846), pp. 39–50.

The Question of the Existence of Infinitesimals (1669–1676)

Introduction

Leibniz regarded the problem of the composition of the continuum as one of the most profound problems in philosophy, one that occupied him throughout his career. It was for him a *metaphysical* problem, a classic problem bequeathed by the Scholastics, which the moderns ignored at their peril. Is space composed of geometrical points? Is matter fundamentally atomic? Are there infinitely small elements of motion? Does time consist in a succession of instants?

Leibniz's thinking on these problems underwent substantial evolution, especially in the formative period of his philosophy between 1669 and 1676, his last year in Paris. We will not rehearse here all the details of these early changes of position, which have been treated at length elsewhere.[1] But there are certain aspects of these early views that are of great importance for understanding his mature positions on the foundations of the calculus, and it is these aspects on which we will concentrate in this chapter.

In the first section we will examine Leibniz's early views, which are interesting in that he advocated a conception of the infinitely small as actual but unextended indivisibles occurring everywhere in the continuum. Defining them by a proportionality with infinitely small motions, these actual infinitesimals appear to have been conceived on the model of intensive magnitudes, not conforming to what is now called the Archimedean Axiom. As he became more proficient in mathematics after arriving in Paris, Leibniz worked with infinities and infinitesimals as fictions without worrying about the question of their existence, as we discuss in the second section. But shortly after inventing his algorithm for the Differential Calculus in late 1675, Leibniz once again concerned himself with the

---

[1] See Arthur (2009) and Arthur (2018).

question of their existence. To begin with he appears to have been tempted by a conception of the infinite as relative to what is taken as finite, precisely analogous to the position his friend Johann Bernoulli would later advocate in their correspondence from 1698–1699. But then he was led to decide the question of their existence in the negative. Precisely what motivated him to this conclusion is the topic of our discussion in the fourth section.

2.1 Early Views on the Continuum (1669–1672)

In his earliest attempts on the problem of the composition of the continuum, Leibniz held that the positing of indivisibles everywhere in it gave a way of solving various problems of metaphysics: indivisibles in body, interpreted as principles of action, had the right properties to be classed as substances; as such they could serve to ground the doctrine of transubstantiation, to provide "the true distinction between body and mind", and even to explain how cannibalism was not in conflict with the doctrine of bodily resurrection.[2] Various documents—letters to his former teacher Jakob Thomasius, and sundry unpublished manuscripts—testify to his preoccupation with the problem in the years prior to his arrival in Paris in the Spring of 1672. These attempts culminated in a treatise, *Hypothesis Physica Nova*, composed in two parts in the winter of 1670–1671. The part dealing with the composition of the continuum, the *Theoria Motus Abstracti* (*TMA*), he sent to the Académie Royale des Sciences and the Royal Society of London in 1672.

One feature of the theory contained there that would remain important to Leibniz is the central role of the concept of *endeavour* (*conatus* in the Latin), which was inspired by his studies of Thomas Hobbes' *De corpore* in the late 1660s and early 70s.[3] Hobbes had defined endeavour as "a motion through a space and time smaller than what is given, that is, … through a point", with the endeavours at any time of two uniform motions being in the same ratio as the motions. Hobbes conceived a point as a body whose length, breadth and depth is "not considered" in the problem at hand, and thus smaller than what is given, though still finite, and defined endeavours and times analogously. Leibniz, on the other hand, took points, instants and endeavours to be indivisibles, contrary to Hobbes's own conception.[4] He wrote: "*A point is not that which has no part* [*contra* Euclid], *nor that whose part is not considered* [*contra* Hobbes], *but that which has no extension.*" (A VI,

---

[2] For an introductory account of these issues, see Arthur (2014), esp. ch. 3, "Natural Philosophy and the Science of Life"; more extended treatments may be found in Beeley (1996) and Arthur (2018).

[3] Thomas Hobbes (1588–1676?) is now much better known as a political philosopher. But prior to his disastrous controversy with the Oxford professor of mathematics John Wallis (1616–1703), he was taken very seriously by his contemporaries as a natural philosopher and mathematician.

[4] Hobbes defines conatus or endeavour as "*a motion through a space and time smaller than what is given, that is, is determined ôr is assigned an expression or number,* that is, *through a point*" (*De corpore*, III, ch. 15; Hobbes 1655, p. 177). But he insists that "a point is not held to be an indivisible, but as undivided", and likewise an instant (pp. 177–178).

2, 264–5/LLC 339–40). And contrary to Hobbes's claim that a point has no ratio to a line, Leibniz boldly declared that it has a ratio, albeit an infinite one: "(10) An endeavour is to a motion as a point is to a space, i.e. as one to infinity, for it is the beginning and end of a motion" (265/340). By this means he was able to claim that a point in a moving body occupies a space greater than when the body is at rest, so that the endpoints of bodies colliding with one another will mutually penetrate, thus making them continuous with one another. Such mutually penetrating indivisibles will then serve to make two colliding bodies continuous, rather than merely contiguous.[5]

In justification of this recourse to "indivisible" elements lacking extension, Leibniz appealed to Scholastic sources. One involved analogy with angles. As commentators on Euclid's *Elements* observed, one rectilinear angle could be contained in another even though they lack extension. Leibniz held angles to be portions of a circle which can be taken smaller than any assignable (because the angle remains the same whatever the length of the radius of the circle is), so they served as a model for his extensionless points.[6] Moreover, he saw further justification of his idea of extensionless points containing parts in the Scholastic doctrine of *signs*, according to which something instantaneous may still have parts (the signs) that are "indistant", in that they may have an order of precedence "by nature" even if not one in time.[7]

Not long afterwards, however, Leibniz rejected his previous characterisation of points as indivisibles. In a manuscript written in the winter of 1672–1673, when he was in Paris, he was moved to claim that he had "excluded indivisibles from nature", and even "from the realm of intelligible things" (*De minimo et maximo*; text [**3**]). By that time, at the suggestion of the eminent Dutch mathematician Christiaan Huygens (1629–1695), he had undertaken a serious study of the writings of Galileo, Pascal, Wallis, and Grégoire de Saint-Vincent, among others. In his reading notes on Galileo's *Discorsi* of late 1672, he recorded the Italian savant's remark that "geometrical definitions are nothing but abbreviations of speech [*compendia loquendi*]", and a similar remark by Pascal (A VI, 3, 167),[8] remarks that he certainly took to heart. Galileo made his remark in the course of

---

[5] Cf. Hobbes: "Two spaces or times are said to be *continuous* with each other when they have some part in common." (*De corpore*, II, ch. 7, §10, p. 87); "*Two bodies are continuous with each other when they have a part in common; and several are continuous when any two that are next to each other are continuous.*" (III, 97/LLC 359).

[6] Thus Leibniz wrote: "An *angle* is the quantity of a point of intersection, i.e. a portion of a circle smaller than can be assigned", so that "the whole doctrine of angles is about quantities of unextended things" (LLC 342).

[7] Thus Leibniz writes that the parts of time "do not cease in an instant, but are indistant. In this they are like the angles at a point, which the Scholastics (whether following Euclid's example, I do not know) called *signs*, as there appear in them things that are simultaneous in time but not simultaneous by nature, since one is the cause of the other" (A VI, 2, N.41/LLC 342).

[8] See Appendix 2b of Leibniz (2001) for a translation and discussion of the relevant passages from Galileo's *Discorsi*, Dialogue 1 (LLC 352–357).

providing a construction which purported to show that a point is equal to a circle. Leibniz reproduced his reasoning, adding: "These things demonstrate well enough that points are nothing, and that only bodies smaller than any given must be used" (A VI, 3, 167). As should be evident, to regard points as mere "nothings" represents a profound shift from Leibniz's earlier view, where mathematical indivisibles were enlisted in defence of the indestructibility of the substances contained in them. In our text [3], Leibniz rejects indivisibles as fixed infinitely small parts of a line, arguing that in that case the diagonal of a rectangle would have the same number of them as the side. Yet he continues to uphold the existence of actual infinitesimals in the continuum, stressing that if these "points" were instead defined by their proportionality to endeavours, they could be "smaller than any given sensible thing", and yet not minima or "indivisibles", because they would be further divisible. He still appeals to the case of angles to justify his claim that one such point could be smaller than another, and even holds that "one point can be infinitely smaller than another", justifying this by reference to Euclid's proof that the angle of contact—the curvilinear angle between an arc of a circle and its tangent—is smaller than any rectilinear angle at the same point.[9]

In text [1] Leibniz extends a similar treatment to infinite wholes, which he also declares to be "nothings": "infinity itself is nothing, ôr not one and not a whole" (168). That remark is occasioned by Leibniz's notes on the discussion in Part I of the *Discorsi*, where Galileo demonstrates what has come to be called "Galileo's Paradox". Leibniz here describes the demonstration as "noteworthy", and it will subsequently feature prominently in his own discussions of the infinite. The argument proceeds from the fact that there is a square of every (natural) number, so that there are as many squares as numbers; but this yields a contradiction, since there are non-square numbers, so that the number of numbers, square and non-square together, is greater than the number of squares. From this contradiction Galileo infers that "in the infinite, the whole is not greater than the part", so that the part-whole axiom does not apply to the infinite. The same conclusion is advocated by Grégoire de Saint-Vincent, but it is one which, Leibniz says, "I cannot accept". Rather, he says, it follows that

> infinity itself is nothing, ôr that it is not One and not a whole. Or we will say: distinguishing among infinites, that the most infinite, ôr all the numbers, is something implying a contradiction, for if it is a whole it can be understood as made up of all the numbers continuing into infinity, which will be far greater than all the numbers ôr the greatest number. Or we will say that one ought not to say anything about the infinite as a whole, except where there is a demonstration. (text [1]; A VI, 3, 168).

Notice the last sentence, which allows one to say something about a notion implying contradiction, the "infinite as a whole", "when there is a demonstration". As Leibniz states

---

[9]"For an angle of contact is a point, and a rectilinear angle is a point, and yet any rectilinear angle, however small, is greater than any angle of contact whatsoever." (text [3]; A VI, 3, 99)

2.1 Early Views on the Continuum (1669–1672)

in another text at the time, "there is indeed a certain reason or truth about impossible and false things" (A VII, 3, 69—end of 1672)—a conviction that will play a crucial role in his handling of infinites as "fictions" as we will see in the next chapter. Leibniz expanded on this idea of the infinite as "nothing" in the *Accessio ad Arithmeticam Infinitorum* (text [2]), a paper he prepared for publication in the *Journal des sçavans*, sending it to the editor, Jean Gallois, at the end of 1672.[10] This paper contains results he had obtained from a new method for summing infinite series, which he had devised in solving the problem Huygens had set him of finding the sum of the series of reciprocal triangular numbers. This depended on noting patterns among consecutive series of fractions, whose terms were the reciprocals, respectively, of unities ($1/1 + 1/1 + 1/1 + 1/1 + 1/1$ etc.), of the natural numbers ($1/1 + 1/2 + 1/3 + 1/4 + 1/5$ etc.), of the triangular numbers ($1/1 + 1/3 + 1/6 + 1/10 + 1/15$ etc.), of the pyramidal numbers ($1/1 + 1/4 + 1/10 + 1/20 + 1/35$ etc.), of the triangulo-triangular numbers ($1/1 + 1/5 + 1/15 + 1/35 + 1/70$ etc.), and so on, as he showed in a Table of Fractions. Leibniz called the "exponent" of each series the number immediately following unity in each progression: 1 for the unities, 2 for the natural numbers, 3 for the triangular ones, and so on. He noted that just as each number in the series of reciprocal triangular numbers is twice the difference between consecutive terms of the reciprocal natural numbers, so each number in the series of reciprocal pyramidal numbers is three halves of the difference between consecutive terms of the reciprocal triangular numbers, and so on as we ascend through the sequence of series. In each case the multiplier is the quotient of the exponent of the preceding series over the one preceding it. Leibniz then formulated the following rule, without proof: the sum of each series is a fraction whose numerator is the exponent of the immediately preceding series, and whose denominator is the exponent of the series before that. Thus, he claimed, taken in order these series sum to 0/0, 1/0, 2/1, 3/2 and 4/3.

After presenting this result, Leibniz takes the opportunity to offer his views on Galileo's paradox. According to his results, the sum of all unities, of the "number of all numbers" (since any number can be resolved into unities) is equal to 0/0, which he takes to be equal to 0. From the contradiction that the number of natural numbers is both equal to and greater than the number of squares, he claims,

> it follows that in this infinite number the axiom that the whole is greater than the part fails (just as Father Grégoire de Saint-Vincent contends that it fails for the angle of contact); but it is impossible for that axiom to fail, ôr what is the same thing, that axiom never fails and only fails for nought or *Nothing*. Therefore an infinite number is impossible, not one thing, not a whole, but *Nothing*. (text [2]; A II, 1, 349)

Leibniz supports his argument that it is "impossible" for the axiom to fail by providing a proof of it from the definitions of 'part', 'whole, 'greater than' and 'smaller than'.[11] He

---

[10] For an insightful discussion of the *Accessio*, see Esquisabel and Raffo Quintana (2017).

[11] "*Nam si (defin. 1)* partes sint *a, b*, totum *(defin. 2)* erit *a + b*. Item si minus *(defin. 3)* sit idem *a*, majus *(defin. 4)* erit *c = a + b*" (A II, 1, 355). For an insightful discussion of the argument, see

notes Galileo's argument that unity has more in common with the infinite than any other natural number: just as there are as many squares and cubes in infinity as there are natural numbers, so "unity contains in itself as many squares as there are cubes and natural numbers [*tutti i numeri*]", leading Galileo to assert that "unity is the only infinite number" (EN 82–83). To this Leibniz responds that 0 not only has this property in common with the infinite, "but also all others, for the square and the cube of 0 is 0, and the double or triple of 0 is 0, and $0 + 0 = 0$, the whole equal to the part" (text [**2**]).

The gist of this paradoxical claim is that neither 0 nor infinity have magnitude. This is the key to interpreting Leibniz's contemporary claim in the Scholium to the *De minimo et maximo* where he writes:

> *Scholium*: We therefore hold that two things are excluded from the realm of intelligibles: minimum, and maximum: the indivisible, and what is entirely *one* and *everything*; what lacks parts, and what cannot be part of another. (A VI, 3, 98)

He will later retract the claim that the maximum is not intelligible, although not that a maximum *magnitude* is unintelligible, which is how this should be understood. Nevertheless, Leibniz is quite clear that in denying indivisibles he is not denying the Cavalierian Method of Indivisibles (which he himself had applied as early as 1670–1671 in an attempt at the quadrature of a parabola).[12] In the company of even Cavalieri's earliest followers, he had interpreted these indivisibles as being infinitely small actual parts of the same dimension as the whole.[13] Indivisibles in this sense, however, would have to be understood as having quantity "smaller than any given", in contradistinction from those he rejected in *De minimo et maximo*, which, equated with minima (or things lacking any parts), would be nothing. As he wrote in another manuscript dating from the end of 1672, *De progressionibus et de Arithmetica Infinitorum*,

> Therefore the infinite is nothing, having neither a whole nor parts, and one infinite is neither greater, nor smaller, nor equal to another, because the infinite has no magnitude. But the arithmetic of the infinite and the geometry of indivisibles no more fail than do surd roots and imaginary dimensions and numbers less than zero. (A VII, 3, 69)

Esquisabel and Raffo Quintana (2017). After explaining Leibniz's philosophical proof of the part-whole axiom, they conclude: "Thus, Leibniz has an analytical demonstration of the whole-part axiom (that is, from definitions). Thereby, the universal scope of the axiom, that is, its universal validity for every quantity, either finite or infinite, is guaranteed. Therefore, it is incorrect to say that the axiom 'fails' in the infinite…" (1329). A proof of this axiom, we might add, was already in Leibniz's possession before he arrived in Paris, cf. A VI, 2, 482–483.

[12] See Siegmund Probst (2008) for a pellucid discussion of this early attempt of Leibniz using the Method of Indivisibles.

[13] As Probst has observed, it is noteworthy that even in this early manuscript Leibniz distinguishes such an approach using an infinity of indivisibles, "aimed at obtaining a geometrically exact determination", from one using only finite elements, which would only yield an approximation (Probst 2008, p. 100). More on this below.

2.2 Infinite Series and the Infinitely Small in Mathematics (1672–1675)

But what did Leibniz mean by saying in his notes on Galileo that "one ought not to say anything about the infinite, as a whole, *except where there is a demonstration*" (text [1])? We can get some idea of what he had in mind by examining his work in mathematics, more specifically on infinite series in connection with his analysis of quadratures during his sojourn in Paris, to which we now turn.

At the beginning of Leibniz's serious study of mathematics in Paris in the Fall of 1672, Huygens had set him the problem of finding the sum of the reciprocal triangular numbers, $A = {}^1/_1 + {}^1/_3 + {}^1/_6 + {}^1/_{10} + {}^1/_{15}$ etc.[14] In the *Scheda exigua* of 1675, Leibniz presented the method he had devised to solve the problem. He had noticed that if this series is compared with the series of reciprocal natural numbers, $B = {}^1/_1 + {}^1/_2 + {}^1/_3 + {}^1/_4 + {}^1/_5$ etc., we see that each term of the A-series is twice the difference between successive terms of the B-series. Thus ${}^1/_1 = 2({}^1/_1 - {}^1/_2)$, ${}^1/_3 = 2({}^1/_2 - {}^1/_3)$, ${}^1/_6 = 2({}^1/_3 - {}^1/_4)$, ${}^1/_{10} = 2({}^1/_4 - {}^1/_5)$, and so on. So if we form the series $C = {}^1/_2 A = {}^1/_2 + {}^1/_6 + {}^1/_{12} + {}^1/_{20} + {}^1/_{30}$ etc., then each of its terms will be the difference between corresponding terms in the B-series, $({}^1/_1 - {}^1/_2)$, $({}^1/_2 - {}^1/_3)$, $({}^1/_3 - {}^1/_4)$, $({}^1/_4 - {}^1/_5)$, $({}^1/_5 - {}^1/_6)$, and so on. Thus the first five terms in the C-series sum to $({}^1/_1 - {}^1/_6)$, the difference between the first term and the sixth and last term in the B-series. Leibniz realized that this result will hold for any two series one of which is the difference series for the other: If, from a given series A, one forms a difference series B whose terms are the differences of the successive terms of A, the sum of the terms in the B series is simply the difference between the last and first terms of the original series: "the sum of the differences is the difference between the first term and the last".[15] This principle, which as far as we know Leibniz did not name, we may call his "Difference Principle".[16]

Leibniz showed no hesitation in applying it to continually decreasing infinite series of fractions, taking the last term to be zero: "Note that in all decreasing differences the last term is to be taken as 0. For it is the last term (*terminus ultimus*) even though the series decreases to infinity".[17] Thus the sum of the infinite series A is easily found. For the sum of the infinite series C will be the difference between the first and last terms of the infinite B-

---

[14] Leibniz solved this problem in September 1672. See *De summa numerorum triangulorum reciprocorum* (A VII, 3, 3), as well as the note he makes 2 years later in *Summa fractionum a figuratis, per aequationes* about when he solved it (A VII, 3, 365), and the editors' introduction to volume VII 3, p. xxv.

[15] In *De progressionibus et de arithmetica infinitorum* (A VII, 3, 95), dated Fall-December 1672.

[16] See Arthur (2006) for discussion.

[17] "Nota in omnibus differentiis decrescentibus terminus ultimus censendus est 0. Is enim est terminus ultimus etsi decrescat series in infinitum." (A VII, 3, 95)

series, $(^1/_1 - ^1/_\infty) = 1$, and since $C = ^1/_2 A$, the sum of the infinite series of reciprocal triangular numbers $A = 2$.

As we shall see, it is Leibniz's extrapolation of the relationship involved in the Difference Principle from discrete differences to infinitesimal differences that he will exploit to arrive at what we call the Fundamental Theorem of the Calculus. As he expresses it in the *Historia et Origo Calculi Differentialis* of 1714, "differences and sums are the inverses of each other, that is, the sum of the differences of a series is a term of the series, and the difference of the sums of a series is a term of the series; the former of which I express as $\int dx = x$, the latter as $d \int x = x$".[18]

At the end of 1672, however, Leibniz was not yet able to apply such methods to the summing of infinitely small elements to produce continuous geometrical quantities. He made some progress towards this, though, in the following spring. After studying his copy of Christiaan Huygens' just published *Horologium oscillatorium* (1673), as well as Hendrick van Heuraet's letter on the rectification of curves[19] and Honoré Fabry's *Synopsis geometrica* (1669), Leibniz turned to a close study of Blaise Pascal's *Lettres de Dettonville* (1659). As noted by Pasini (1993, p. 53) and Probst (2008, p. 102), his reading of Pascal resulted in his introduction of the concept of a *unity* in the method of indivisibles. Leibniz writes:

> Note: just as when in the equations of geometry lines are compared with surfaces, or surfaces with solids, or lines with solids, it is necessary that a unity be given (from which equations between dimensions of different degrees are freely admitted in numbers), so in the geometry of indivisibles, when it is said that the sum of lines is equal to some surface, or the sum of surfaces to a solid, it is necessary that a unity be given, that is, that some line be given to which they are understood to be applied, that is to say, into one of whose equal infinitely many parts they are multiplied, which part exhibits a *unity*, so that infinitely many surfaces are made from this, even though they are smaller than any given surfaces.[20]

The idea is that where the dimensional homogeneity of equations is preserved in algebraic geometry by introducing a constant—so that a parabola, for instance, would be represented

---

[18] "Fundamentum calculi: *Differentiae et summae sibi reciprocae sunt, hoc est summa differentiarum seriei est seriei terminus, et differentia summarum seriei est ipse seriei terminus, quorum illud ita enuntio: ∫dx aequ. x; hoc ita: d ∫x aequ. x.*" (Leibniz 1846, p. 36)

[19] Heuraet's letter to van Schooten (*Epistola de transmutation curvarum linearum in rectas*) was reprinted in an edition of Descartes's *Geometria* that also contained letters by Jan Hudde: Descartes (1659, pp. 517–520).

[20] "*Nota quemadmodum in aequationibus Geometriae quando comparantur lineae cum superficiebus, vel superficies cum solidis, vel lineae cum solidis, necesse est dari unitatem (unde in numeris aequationes inter dimensiones diversorum graduum libere admittuntur), ita in Geometria indivisibilium, cum dicitur summam linearum aequare cuidam superficiei, vel summam superficierum cuidam solido, necesse est dari unitatem, dari scilicet lineam quandam cui applicatae intelligantur, seu in cuius partium infinitarum aequalium, unam, quae unitatem exhibet, ducantur, ut infinitae unde fiant superficies, etsi qualibet data minores.*" (A VII, 4, 135)

as $ay = x^2$—the constant a can without loss of generality be set equal to 1. Similarly, Leibniz argues, when in the geometry of indivisibles, lines (or surfaces) are added to make a surface (or volume), they can be understood to be multiplied by an indivisible which can be taken as exhibiting a unity. Thus "applying" a variable ordinate line y, say, to an indivisible part of the abscissa x—which Leibniz will later denote dx—the latter infinitely small part of x is taken as a unit. The area under the curve is represented as a sum of infinitely many infinitely small rectangles of unit width (corresponding in modern terms to a definite integral of y as a function of x between, say, $x = 0$ and $x = a$, where a may then be taken as 1 without loss of generality).[21] Below, we will give an example of this idea of the "application" of a variable ordinate line to a "unity", in Leibniz's successful quadrature of the hyperbola.

At this juncture (the Spring of 1673), however, Leibniz has not yet discovered the characteristic triangle from his studies of Pascal, and consequently he does not have the means to achieve the arithmetical quadrature of the circle. He writes:

> From which it is clear how necessary it is to undertake that more profound contemplation of indivisibles and the infinite without which it is impossible to meet the difficulties that occur in the doctrine of indivisibles and the infinite. Note: *indivisibles* are to be defined as infinitely small, ôr things whose ratio to a sensible quantity [...] is infinite.[22]

It is only after he has discovered the characteristic triangle in the Fall of 1673, and the Transmutation Theorem he formulates by reflecting on it, that Leibniz succeeds in obtaining his arithmetical quadrature of the circle. He obtains the result geometrically by establishing a ratio between the elementary triangles under the curve to be squared and rectangular elements under a second curve, as we will explain in Chap. 4 below.

The essence of what he achieved can be expressed analytically using the notation of the Differential Calculus that he developed in October 1675. Suppose we wish to find the quadrature Q of a given curve y; that is, the area under the curve whose ordinate is y between, say, $x = 0$ and $x = b$. If we represent this area as an infinite sum of infinitesimal rectangles whose width is dx and whose height is $y(x)$ the total area will be $Q = \int y(x)\,dx$. Now if one is able to find an expression for the general term of another series z for which the difference between its successive terms is $dz = y dx$, one may apply the Difference Principle: the sum of the differences equals the difference between the last term and the first—what we would now call the definite integral evaluated between first and last terms, $\int_0^b y\,dx = z(b) - z(0)$. But as Leibniz had learnt from Pascal, the slope of the tangent to a

---

[21] This "translation" of the indivisibles into unities is already what Wallis is doing in his *Arithmetica infinitorum*, from whom Leibniz might have taken inspiration here.

[22] "*Unde apparet quam necessaria sit ista profundior contemplatio indivisibilium atque infiniti, sine qua occurrentibus in infiniti atque indivisibilium doctrina difficultatibus occuri non potest. Nota*: indivisibilia *definienda sunt infinite parva, seu quorum ratio ad quantiatem sensibilem [...] infinita est.*" (A VII, 4, N.16)

curve $z(x)$ at an arbitrary point x can be expressed as the quotient of the two sides of an infinitesimal right triangle, the "characteristic triangle" dz/dx. Now, by the definition of z, the slope of the tangent to the curve z at any point on the abscissa x is equal to the value of the original curve y at the same point on the abscissa. This gives us the fundamental connection between the problems of finding the tangent to a curve, and determining quadratures, the areas under curves. Leibniz's great innovation was his transmutation theorem, which can be expressed in his later notation as $\int_0^b y\, dx = [xy]_0^b - \int_0^b x\, dy$. This enabled him to compute the quadrature of the circle as $Q = \frac{1}{2} \int_0^b y\, dx = b - \int_0^b \frac{t^2}{1+t^2}\, dt$, and using a power series expansion of $\frac{1}{1+t^2}$ as $1 - t^2 + t^4 - t^6 + \ldots$, to obtain by term-wise integration (with $b = 1$) the result $Q = 1 - 1/3 + 1/5 - 1/7 + 1/9 - 1/11$ etc., his arithmetical quadrature of the circle.

It is important to recognize, though, that these studies culminating in the development of his calculus did nothing to dissuade Leibniz from his opinion about the infinite not being a true whole. In fact, the expressions he found for the quadrature of a hyperbola in terms of infinite series gave a graphic representation of an infinite whole being equal to its part, and at the same time seemed to him to confirm the fictional nature of such a whole. This can be seen in a paper Leibniz wrote in October 1674, our text [5], which, as promised earlier, also illustrates his early "proto-calculus" technique of performing a quadrature by "applying" a variable ordinate line to a "unity".

To perform the calculation, Leibniz gives a symmetrical diagram (see the Fig. 8.2 in text [5]) of Apollonius' hyperbola GBE with centre A, vertex B, and "radius" $AC = BC = a$, which, without loss of generality, he will set equal to 1. DE represents the variable abscissa, x; positive y goes downwards from A to F. M represents the imaginary point where the curve "meets" each line AF... "at infinity".[23]

So we have $AD = AC - CD = 1 - y$, $DE = 1/AD = 1/(1-y)$. That is, the equation of the curve is $x = 1/(1-y)$.

Leibniz now sets about discovering the area $ACBEM$ under the curve between the horizontal line CB and the line at the top of the figure, ACF..., by "applying" the variable line DE to the line $AC = 1$. Taking $a = 1$, he writes, "$DE = 1 + y + y^2 + y^3 + y^4$ etc. and the sum of all the DE, applied to AC will give $1 + 1/2 + 1/3 + 1/4 + 1/5$ etc. = the space produced to infinity, $ACBEM$".

What Leibniz has done is to use a power series expansion[24] of $DE = 1/AD = 1/(1-y)$:

---

[23] Leibniz will make the fictional character of this point more explicit in a draft for the *DQA* written in May–June 1676. See Proposition XI of this earlier draft, A VII 6, 210. It is in the Scholium to this proposition that Leibniz develops his important distinction between unbounded and bounded infinities, as we will discuss further below.

[24] Leibniz had learned the power series expansion from his studies of Wallis (*Mathesis universalis*, 1657, ch. XXXIII, prop. 68, p. 303 (WO I, p. 175 f.)) and Mercator (*Logarithmotechnia*, 1668 in prop. XIII, p. 25). At A VII, 3, 127 he wrongly attributes it to Wallis's *Arithmetica Infinitorum* (see the editors' informative note there).

2.2 Infinite Series and the Infinitely Small in Mathematics (1672–1675)

$$DE = 1/(1-y) = 1 + y + y^2 + y^3 + y^4 + y^5 + \ldots$$

Then "applying" the variable line DE to the line $AC = 1$, he has effectively performed a term-wise integration to obtain the "sum" $\int (1 + y + y^2 + y^3 + \ldots)\, dy$, between 0 and 1. This gives

$$ACBEM = \left[y + y^2/2 + y^3/3 + y^4/4 + \cdots\right]_0^1 = 1 + {}^1/_2 + {}^1/_3 + {}^1/_4 \ldots$$

By a similar argument, to calculate the area between CB and FG Leibniz uses the power series expansion of the variable line $HL = 1/AH = 1/(AC + CH) = 1/(1 + y)$ and applies it to CF, obtaining for the finite area $CFGLB$ the expression

$$CFGLB = 1 - {}^1/_2 + {}^1/_3 - {}^1/_4 + {}^1/_5 - {}^1/_6 \text{ etc.}$$

(In modern terms, the same result is obtained by integrating $x = 1/(1 - y)$ with respect to y between -1 and 0.) Now Leibniz subtracts this finite area $CFGLB$ from the infinite area $ACBEM$, to get

$$\begin{aligned}
ACBEM - CFGLB &= 1 - (1) + {}^1/_2 - (-{}^1/_2) + {}^1/_3 - (+{}^1/_3) + {}^1/_4 - (-{}^1/_4) + {}^1/_5 - (+{}^1/_5) \text{ etc.}\\
&= {}^2/_2 + {}^2/_4 + {}^2/_6 + {}^2/_8 + {}^2/_{10} + {}^2/_{12} \text{etc.}\\
&= 1 + {}^1/_2 + {}^1/_3 + {}^1/_4 + {}^1/_5 + {}^1/_6 \text{etc.}\\
&= ACBEM
\end{aligned}$$

That is, subtracting the area $CFGLB$—an area that is perfectly determinate and visible in the figure—from the area under the curve, leaves that area the same! Leibniz comments:

> Which is rather amazing, and shows that the sum of the series 1, ${}^1/_2$, ${}^1/_3$, etc. is infinite, and consequently that the area of the space $ACGBM$ remains the same even when the finite space $CBGF$ is taken away from it, that is to say, nothing noticeable is taken away.
>
> By this argument it is concluded that the infinite is not a whole, but only a fiction, since otherwise the part would be equal to the whole. (text [5])

One finds here again the argument about the infinite not being a whole, because it would contradict the part-whole axiom (with what seems to be the first appearance of the term "fiction" in this context).[25] But what is remarkable is that this whole is not here the

---

[25] Before that period, Leibniz was more apt to use the vocabulary of the "imaginary", which he will never abandon (see next chapter). In 1673, he already mentions the infinitesimal rectangle taken between two ordinates as something which cannot be thought, but only "feigned" (*fingi*), see note 36 in Chap. 3.

paradoxical "collection" of all natural numbers, but a geometrical entity (the area comprised between the hyperbola and one of its asymptotes). This will remain in later years one of Leibniz's favourite examples of the danger of accepting a true infinite in mathematics. Commenting on this result in the draft of the *DQA*, he will explain in more detail that the paralogism here comes from the fact that one confuses the truly infinite space under the hyperbola and its asymptote, which is without bound, and the bounded area obtained when one considers a parallel situated at an infinitely small distance to the asymptote (A VII, 6, 210–211). This will be the matrix for the famous (but cancelled) scholium to Prop. XI of *DQA* in which Leibniz explains that all of this relies on the "fiction" of a line at the same time infinite and bounded—an example to which we will return later in this chapter.

Of course, if the infinite is not a whole, one might wonder about the legitimacy of calculating with it in mathematics, as Leibniz himself was doing in his manipulations of infinite series. Connected with this is the problem of assuming a last term (*terminus ultimus*) in an infinite series. As the above arithmetical quadrature of the hyperbola shows, to take an infinite area like *ACBEM* as a whole is to take an infinite series as a whole, in this case taking the series $1 + 1/2 + 1/3 + 1/4 + 1/5 + 1/6$ etc., *as if* it had a last, infinitieth term—even though in this case the series diverges and has an infinite sum. As we have seen, in extending the "Difference Principle" that he had articulated for finite series to apply also to infinite series with continuously decreasing terms, Leibniz had assumed a last term for such infinite series, equal to 0. In so doing, he was following the practice of John Wallis, as we will argue below. Unlike Wallis, however, Leibniz was aware that calculating with infinite series of descending fractions as if they have a last term of zero, was something that needed justifying.

Thus in one of the fragments in which he calculates the sum of the reciprocal triangular numbers, the *Theorema Arithmeticae Infinitorum* of Fall 1674, Leibniz had argued as follows. Let us call the sum of the reciprocal triangular numbers *A*, that of the series of reciprocal natural numbers *B*, and that of the series each of whose terms is half those of *A*, *C*. Now Leibniz forms a fourth series *D* by subtracting *B* from *C*, then simplifying to give the series *E* whose sum is "$1/2 + 1/3 + 1/4 + 1/5 + 1/6$ etc. *in infinitum*", which is $B - 1$. Therefore $B - C = B - 1$, so $C = 1$ "because when subtracted from the same series [*B*] they are left the same. Therefore twice the series [*C*], that is to say, [*A*], will be equal to two. Which is the demonstration we had undertaken." (A VII, 3, 362).

But Leibniz immediately acknowledges that such subtracting of one series from another "ought to be demonstrated to come out in the infinite" (*ibid.*), and sets about providing such a demonstration. To this end he calculates an expression for the general or y^{th} term of each series concerned, where "*y* signifies any number whatever" (362). Thus the general term of the *B* series is $\frac{1}{y}$ and that of the reciprocal triangular numbers, $\frac{2}{y^2+y}$. So the general term of the *C* series is $\frac{1}{y^2+y}$. Subtracting the term $\frac{1}{y^2+y}$ from the term $\frac{1}{y}$ gives $\frac{y^2+y-y}{y^3+y^2}$ or $\frac{1}{y+1}$. "Thus if half of the triangular fraction is subtracted from the corresponding natural fraction, the remainder is the natural fraction that corresponds to the next greater triangle" (363). Leibniz does not

2.2 Infinite Series and the Infinitely Small in Mathematics (1672–1675)

complete this demonstration in these terms, preferring to proceed to a geometrical depiction in terms of the triangles. But his Difference Principle entails that if y is taken as the *terminatio* or fictional last term, then the sum of the series C, which is ½ the sum of the series A, is $1 - \frac{1}{y+1}$. Consequently, the sum of the reciprocal triangular numbers, A, approaches 2 arbitrarily closely as y is taken arbitrarily large.

Such considerations prompt Leibniz to give a rule for calculating the sum of a converging infinite series, which he states in April 1676 (text [10]):

> Whenever it is said that a certain infinite series of numbers has a sum, I am of the opinion that all that is being said is that any finite series with the same rule has a sum, and that the error always diminishes as the series increases, so that it becomes as small as we would like.

Moreover, as he explains in another paper of the same period concerning the same example of "the infinite space between Apollonius' Hyperbola and its asymptote" and its corresponding infinite series $A = {}^1/_1 + {}^1/_2 + {}^1/_3 + {}^1/_4 + {}^1/_5 + {}^1/_6$ etc., whose sum is the reciprocal of 0, we "must understand this 0, or nought—or rather, in this place, a quantity infinitely or unassignably small—to be greater or smaller according as we have assumed the last denominator of this infinite series of fractions, which is itself also infinite, smaller or greater" (A VI, 3, 282/LLC 115). Underlying this justification for summing a converging infinite series is the general principle we mentioned in the introduction, which we call the Principle of Unassignable Difference, or *PUD*, that if the difference between two quantities can be made smaller than any pre-assigned quantity, then they are equal.[26] Leibniz implicitly appeals to this principle in text [8] of 26 March, 1676, suggesting it would demonstrate that "in quadratures, a difference [or differential] is, however, not infinitely small but nothing at all", since when a curve is represented as a polygon with ever smaller sides, the difference between successive vertices may be made so small that "even when the difference is assumed infinitely small, the error will be smaller" (text [8], A VI, 3, 434).

Accordingly, the use of an "infinite number" of terms in a series does not entail the existence of a "last" or "maximal" term strictly speaking, but is a way of speaking of a variable term (the "infinitieth" one) which is a finite number taken as large as one wants. In Leibniz's own terms: "the infinite number cannot coincide with an infinitieth of z, since this varies, and differs according to the value of z assumed".[27] Accordingly, "an analytic method can already be obtained by converting expressions of this kind that use infinite

---

[26] Cf. Leibniz's comment in his 1675–1676 treatise *De Quadratura* that he prefers a justification "which simply shows that the difference between two quantities is nothing, so that they are then equal (whereas it is otherwise usually proved by a double *reductio* that one is neither greater nor smaller than the other)" (text [16]; Leibniz 1993, p. 35).

[27] "numerus autem infinitus non potest coincidere cum numero infinitesimarum ipsius z quia is variat, alia atque alia assumta z, nisi et infinitesimas proportione putes" (A VII, 3, 769; May 1676, text [14]).

number into other expressions in which there is no infinite number".[28] So a fictional last or infinitieth term can be included in calculations involving the Difference Principle, and the series can be treated as complete, and thus as delivering an exact value, which is justified by the argument that the error can be made smaller than any assignable.[29]

Let us elaborate further on this last point about completeness. Leibniz was, of course, inspired by Wallis's arithmetical approach, and appreciated his arithmetical expression for the ratio of the square of a circle's diameter to its area,[30] $4/\pi$, namely

$$\frac{1.3.3.5.5.7.7.9.9.11.11 \; etc.}{2.4.4.6.6.8.8.10.10.12.12 \; etc.}$$

But he insisted that this only gives an approximation: it is an "indefinite finite series", as opposed to a properly infinite series with an actual infinity of terms. In comments on Wallis, *De seriei Wallisiana* of late summer 1676, he noted that if the numerator and denominator of the above expression are continued to infinity, then by cancellation of terms one can obtain the absurd result that the square root of this expression is infinitely small. He continues:

> But I infer from this that the Wallisian expression can indeed be regarded as an approximation, whose continuability to infinity is clear, but not as an exact expression through an infinite series considered all at once. For by expressing it through a supposed continuation to infinity there arises an absurdity....[31]

"And therefore," he concluded, "the Wallisian series should properly be called an indefinite finite series, ôr one which should be conceived as ending somewhere, although it is not expressed where" (A VII, 3, 825).

That is, Leibniz agrees with Wallis that an infinite series has infinitely many terms even though there is no such thing as an infinite number; and also that the series can be considered as giving an exact value *on the supposition* that there is a last term.[32] But he

---

[28] "Et hinc iam haberi potest modus analyticus expressiones eius modi, infinito numero utentes convertendi in alias in quibus nullus sit numerus infinitus. Nam quae numerum habent infinitum hoc modo, earum ne partes quidem tractabiles sunt" (*ibid.*, text [**14**]).

[29] Cf. Leibniz's exchange with Jenisch in 1702 [**48**], where he says exactly the same thing, and explicitly calls the last term a fictional one.

[30] See Wallis (1656), *Arithmetica infinitorum*, Prop. 191, pp. 178–182.

[31] "Sed hinc colligo expressionem Wallisianam posse quidem haberi pro appropinquatione, cuius apparet continuabilitas in infinitum, sed no pro exacta expressione per infinitam seriem consideratam semel in universum, nam exprimendo per suppositam continuationem in infinitum, oritur absurdum, ..." (A VII, 3, 824).

[32] Here see Raffo Quintana's illuminating analysis of this relationship between the views of Wallis and Leibniz (Raffo Quintana 2018).

finds that Wallis's own series are only finite, albeit infinitely extensible.[33] As he will later explain explicitly in a piece which its editors have titled "Towards a Science of the Infinite" (text [**33**]), an indefinite finite series is incomplete in that it does not contain all its infinitely many terms, since there are always more. But there is no question of taking "all" the terms if "all" is taken to connote an infinite collection, since that involves the contradictory concept of infinite number. A properly infinite series, on the other hand, does contain all its terms, but the 'all' must be taken distributively: that is, it must contain each term given by the law of the series, and none which are not.[34] In order for an infinite series to express its sum exactly and not approximately, it must be "complete" in the sense of including all its terms in the latter sense. This, Leibniz claims, is precisely the case for his own arithmetical quadrature of the circle. Thus, as Raffo Quintana justly concludes, "The significance of an exact arithmetical quadrature is based precisely on taking series as wholes, since otherwise we would have an approximate value, but not an exact one." (2018, p. 71).

To give some background, Leibniz was aware of James Gregory's proof that when two sequences of circumscribed and inscribed polygons to a circle converge in what Gregory calls a *terminatio*, the latter cannot be expressed using the basic operations of addition, subtraction, multiplication, division and root-extraction.[35] And already in the Fall of 1673 Leibniz had coined the term "transcendent" to describe figures for which a precise mathematical quadrature is given, but which is not expressible by a number.[36] Such

---

[33] Cf. "Wallis gave an elegant means of approximating in rational numbers; but adapted only to the whole circle, and not indeed to its parts. [*Wallisius elegantem dedi tmodum appropinquandi in numeri srationalibus; sed toti tantum Circulo non vero et partibus aptatum.*]" (A VII, 6, 439)

[34] See our discussion in chapter 6, §1 below. For the distinction between the two types of wholes, see also Leibniz's notes on Nizolius (A VI, 2, 430), where Leibniz criticizes Nizolius for missing the distinction. In this connection we should note, however, that the term "complete" is not used by Leibniz himself in his notes on Wallis (although he speaks of series "considered all at once", *semel in universum*), but is introduced by Raffo Quintana in order to express the idea of a whole expressing an exact value. In other places, Leibniz uses the expression *tota series*.

[35] On this matter one should consult Davide Crippa's authoritative account of Leibniz's early work on quadrature, *The impossibility of squaring the circle in the Seventeenth Century*, (Crippa 2018a), especially in relation to the controversy between James Gregory and Christiaan Huygens. Crippa writes: "Leibniz was indeed particularly adamant in considering his arithmetical quadrature to constitute an exact quadrature. Thus we read in the *De Quadratura arithmetica* that the infinite series for $\pi/4$ was constructed according to a law that 'impresses the mind', or a law such that the 'mind pervades it [namely the series] all at one blow.' It seems that Leibniz's core idea of exactness is hidden behind these suggestive metaphors." (Crippa 2018a, p. 134). See also the same author's "One string attached…" (Crippa 2018b) for a detailed study of the issue of exactness in geometry in the period 1673–1676.

[36] "At some point we must inquire into the natures of continuously varying figures, which, namely, are not of a definite dimension, but transcendent, such as 1. 2. 6. 24. 120. Tangents and other things are sought for figures of this kind. For I see that the squared figure of the circle and of the hyperbola will be of such a nature, that is, transcendent; and no one has considered figures of this kind in geometry…. [*Inquirendum aliquando in naturas figurarum continue variantium, quae scilicet non*

considerations led him to declare, in the last Proposition of the *DQA* (Prop. LI), that "it is impossible to discover a better general quadrature of the circle, the ellipse, and the hyperbola, ôr relation between the arc of a curve and its side, or a number and logarithm, which is more geometrical than ours is".[37]

In contrast to Wallis's approximation, as well as Brouncker's one by continued fractions,[38] Leibniz therefore asserts that his own arithmetical quadrature gives an exact ratio of the area of the circle to its diameter (which we would denote as $\pi/4$)[39]:

> For, when I say that the circle is equal to this infinite series $1 - \frac{1}{3} + \frac{1}{5} - \frac{1}{7} + \frac{1}{9} - \frac{1}{11}$ etc., supposing that the circumscribed square is 1, I certainly say something more than if I were to exhibit a way of approximating it. Every infinite series indeed contains an approximation as exact as desired; but further, a true and exact equality.[40]

In sum, we see that Leibniz's approach to infinite series does not commit him to there being an actually *infinite number* of terms, nor to the existence of a last term in such a series. Infinite series are to be considered as infinite wholes in the distributive sense of wholes, and this is what grounds the notion of exactness and "completion" of a series ("*tota series*"). Nevertheless, one can treat these wholes *as if* they were collections and had a last term (on the model of genuine mathematical wholes, i.e. finite collections of things), in order to

---

sunt certae dimensionis sed transcendentes, ut: 1. 2. 6. 24. 120. huiusmodi figurae quaerantur tangentes, aliaque. Video enim talis naturae aut similis id est transcendentem fore figuram circuli et hyperbolae quadratricem; et nemo huiusmodi figuras consideravit in geometria...." (A VII, 3, N. 23, 266–267; cf. Probst 2012, p. 4).

[37] "*Impossibile est meliorem invenire Quadraturam Circuli Ellipseos aut Hyperbolae generalem, sive relationem inter arcum et latera, numerumve et Logarithmum; quae magis geometrica sit, quam haec nostra est.*" (A VII, 6, 674).

[38] See Stedall (2000) for an illuminating account of the attempts of Wallis and Brouncker to square the circle using continued fractions.

[39] In their later correspondence, Wallis protests (in a letter of 6/16 April 1697) against Leibniz's characterization of his value for $4/\pi$ as an approximation: "But also all my series of tables (in *Arithmetic of the Infinite*) are *infinite series*, and most of them are of the kind you call (by the new name) *transcendental series*" (A III 7A, 373). In his reply of 28 May/7 June 1697, Leibniz concedes that Wallis's expression (as also Brouncker's continued fraction expression) is exact, and that he "did not perceive their origin sufficiently", but adds that it is "nonetheless distinguished by me from an exact value through infinite series properly speaking, which are formed through a mere collection of terms [*Et valor ille exactus quails tuus ... vel Brounkerianus... cujus non satis perspexi originem; a me tamen distinguitur a valore exacto per series infinitas proprie dictas, quae per meram terminorum collectionem conflantur*]" (A III, 7A, 427–8). (Thanks here to Jeffrey Elawani for bringing this exchange to our attention.)

[40] "*Nam cum Circulum ajo aequari huic seriei infinitae $1 - \frac{1}{3} + \frac{1}{5} - \frac{1}{7} + \frac{1}{9} - \frac{1}{11}$ etc. posito quadratum circumscriptum esse 1, utique plus aliquid dico, quam si modum appropinquandi exhiberem. Omnis enim series Infinita continent appropinquationem quantum licet exactam; sed praetor ea. veram quondam exactam queaequalitatem.*" (A VII, 6, 439; cf. Raffo Quintana 2018, p. 71). This is from *De operis argumento et auxiliis*, dated as July–27 August 1676.

2.2 Infinite Series and the Infinitely Small in Mathematics (1672–1675)

calculate their value. This is the general strategy involved, as we will show in the next chapter, in the handling of "fictions" in mathematics. The proposal that such a series sums to an exact value is justified by the argument that by whatever quantity Δ it may be supposed to fall short of this value (the "error"), this error may be shown to be less than Δ by taking sufficiently many terms in a finite series with the same law, so that the error is therefore 0, and the value is exact. The same argument justifies taking an infinitieth term in an infinite series with continuously decreasing terms to be 0: for again the error in taking it to be anything other than zero can be shown to lead to contradiction. Likewise the method of indivisibles, under Leibniz's interpretation, is not committed to indivisibles in the sense that infinitely many points of zero magnitude will add to a line, or infinitely many indivisible lines will add to a surface. Rather the infinitely small and large are assumed as a hypothesis, as fictions: one takes the infinitely small or large as variable quantities that one can make as small or as large as one pleases, but not as fixed ingredients of the continuum.

Leibniz discusses this matter further in the cancelled Scholium to Proposition XI of the *De quadratura arithmetica circuli ellipseos et hyperbolae*, written between June and September 1676 (text [**6**]). The proposition concerns the hyperbola. He notes that Cavalieri's Geometry of Indivisibles is fallacious if the indivisibles are interpreted as literally indivisibles or points, and that they must instead be understood as infinitely small but divisible lines. By analogy, he says, the unbounded must also be distinguished from the infinite. "For the magnitude of an unbounded line is in no way subject to geometrical considerations, any more than the magnitude of a point." But, he says,

> it is otherwise with a bounded yet infinite line, which can be understood as constituted by some multitude of finite lines even if this multitude exceeds every number. And just as an infinite bounded line is composed out of finite ones, so a finite line is composed out of infinitely small yet divisible ones. Hence it cannot be said that a bounded line is the mean proportional between a point ôr minimum line and an unbounded ôr maximum line. But it can be said that a finite line is the mean proportional, not in a certain way, but truly and exactly, between some infinitely small line and some infinite one; and it is true that a rectangle made up of infinitely many infinitely small ones can be equal to a certain finite square; and this in fact occurs in the conic hyperbola. (text [**6**]; A VII, 6, 549)

For, he says, it can be understood to have an "infinitely small abscissa $\mu(\mu)$" and "an ordinate $(\mu)\lambda$ that is infinitely long, that is, greater than any designatable straight line", so that the area under the curve and its abscissa, "an infinite rectangle $\mu(\mu)\lambda$ comprised of infinitely many infinitely small ones will, by the nature of the squared hyperbola, be equal to some constant finite square". That is, denoting the infinitely small abscissa $\mu(\mu)$ by dx, and denoting by b_∞ the infinitely long ordinate that is bounded at both ends, the fact that the conic hyperbola is squarable on this assumption means that the infinite rectangle $\mu(\mu)\lambda$ comprised of infinitely many infinitely small ones, the product of dx and b_∞, will equal the area of a square of finite side x, namely x^2. So $dx: x = x: b_\infty$, and the finite length x is the mean proportional between the infinitely small dx and the infinitely long b_∞. Calling a quantity infinite "whether bounded or unbounded, provided that we understand it to be greater than any quantity that we can assign or that is designatable by numbers", Leibniz then remarks:

> Now, whether the nature of things tolerates quantities of this kind is for the metaphysician to discuss; it is enough for the geometrician to demonstrate what follows from supposing them. (text [**6**])

This ties back nicely to the example from Galileo with which we began this section. One can use such *compendia loquendi* to derive true and exact conclusions in mathematics without having to decide first whether they exist *in rerum natura*.

From this we can see that Leibniz did not regard the fictional status of infinitely small and infinite magnitudes in mathematics as deciding the question of their existence, either positively or negatively. This raises two sets of questions: (1) how, then, should we understand Leibniz's use of fictions, and more generally, what is their status in mathematics? and (2) what caused him to reconsider the question of the existence of infinitesimals as he was putting the finishing touches to his calculus, and how did he decide this question?

The answer to both sets of questions is complex, the first so much so that we will postpone its treatment to a separate chapter, following this one. Before that, let us return to the question of existence.

2.3 Leibniz's Rejection of the Existence of the Infinitely Small in 1676

As we saw in Sect. 2.1 above, Leibniz abandoned his earlier view of the infinitely small as properly indivisible already on arriving in Paris in 1672, preferring to work with a conception of points as "bodies smaller than any given". But in his mathematics, as we have just seen, although he seemed content to treat the infinitely small or large as variable quantities that could be made as small or as large as needed, this did not yet decide the question for him of whether infinitely small or large things could exist. This, we will show in the next chapter, is consistent with his views on the use of "fictions" in mathematics, which does not preclude an interpretation of them as if they exist, even in cases where expressions in the surface language that rely on them result in contradiction.

Leibniz's first return to a consideration of the problems of the continuum appears to have been in December 1675 in *De materia, de motu, de minimis, de continuo* (A VI 3, 466–70/LLC 30–41), although his reasoning about it there seems inconclusive.[41] But his

---

[41] Having demonstrated to his own satisfaction that "it is absurd for there to be minima in the continuum", he declares: "Hence, if there are instants in time, there will be nothing but instants, and time will be nothing but the sum of instants" (A VI, 3, 470/LLC 39), which hardly seems to follow. At this point he seems still to think that anything that "is in" something else should be

2.3 Leibniz's Rejection of the Existence of the Infinitely Small in 1676

study of Spinoza's thought in the company of his younger German colleague Ehrenfried Walther von Tschirnhaus (1651–1708) seems to reawaken his interest in the metaphysics of the infinite. This takes place in early 1676, on the heels of his invention of the differential calculus.

As we have seen, Leibniz had previously justified the possible existence of the infinitely small in terms of the Hobbesian notion of endeavour. This is turn underlay his conception of the actual infinite division of matter. The idea is that the differing motion at any instant of each of the parts of matter to infinity would issue in what he called "perfect points". Each such point is distinguished from the others by its own endeavour, where that endeavour is understood as an actually infinitely small motion. Thus, in "On the Secrets of the Sublime" of February 1676 (text [7]), Leibniz characterizes matter as a *perfect fluid*, that is, as consisting of a discrete multiplicity of points, each one a body "smaller than any that can be assigned". Leibniz speculates there that in this way "matter alone is explicable by a multiplicity without continuity". There would be an "interspersed vacuum" but this would be "a metaphysical one, which is not contrary to a physical plenum". "A metaphysical vacuum", he explains, "is an empty space, however small, only true and real" (text [7]). This is a nod to his earliest account of the continuum, where the assignable points or instants are separated by unassignable gaps, understood as actually infinitely small intervals. Here it is space that would be continuous, whilst matter, when reduced to a state of liquidity by the absence of any "cement", would "be a discrete entity, not a continuous one; it is only contiguous, and is united by motion or a mind of some sort".[42]

A major motivation for taking infinitesimals to exist is, of course, the success of their use in geometry—not only in quadrature, but in the calculus that he had just developed the previous October. Leibniz notes exactly this in text [7]: "Since we see the hypothesis of infinite and infinitely small things is splendidly consistent and successful in geometry, this also increases the likelihood that they really exist". What he argues there is very interesting, especially in the light of Johann Bernoulli's taking much the same position in the 1690s. If there exists something infinitely small yet not indivisible, he suggests, we will be infinite in comparison with creatures of that world. Then, by the same token, "it is clear in turn that we could be feigned [*fingi*] as being infinitely small in comparison with another world that is of infinite magnitude, and yet bounded". But that would make the infinitude or

---

considered its part, a notion that he will reject in his mature writings on the *inesse* relation from the mid-1680s onwards.

[42] The same reasoning is repeated in "On the Plenitude of the World" (early March (?) 1676), on the assumption that matter is composed of spheres as small as desired: "For there is no place so small that we cannot imagine a smaller sphere to exist in it. Let us suppose this is so, then there will be no assignable empty place. And yet the world will be a plenum, from which an unassignable quantity is understood to be something" (A VI, 3, 525/LLC 61). Assuming only that "there is no place so small that there does not exist some body equal to or smaller than it", Leibniz infers that "body and space are not co-extended, because however many spheres there are, they will not fill the whole. All the vacuities collected into one would not have a greater ratio to some assignable space than an angle of contact has to a rectilinear angle" (525/61).

finitude of our world entirely *relative*, even though it is bounded.[43] This, he notes, makes it clear that the infinite should be distinguished from the unbounded infinite, which should more properly be called the "immensum". Judging from what he says in the following months, his argument is that the immensity of God is not relative, but neither is it bounded.[44] So we must distinguish the infinite (which could be relative to us) from the unbounded infinite (which is absolute, like all God's attributes).

In the same text Leibniz notes an important consequence of this position: if "any part of matter, however small, contains an infinity of creatures, i.e. is a world, it follows also that matter is actually divided into an infinity of points", and from this it will follow that "any part of matter will be commensurable with any other" (text [7]). But, he cautiously adds, "it must be seen whether this truly follows", as must the related consequence he has derived "elsewhere"[45] that "a circle, if it exists, has a ratio to the diameter as one number to another" (A VI, 3, 474).

To summarise, at this point Leibniz has noted three main consequences of interpreting the infinitesimals of his calculus as actually existing: (1) that this would entail a kind of relativity of the infinite, where a finite line would be infinitesimal with respect to an infinite one, but would itself be infinite in comparison to a second order infinitesimal, for instance; (2) that matter would be subdivided into an infinity of points, rendering every part of matter commensurable with any other, since each part would consist in a determinate infinite number of points; (3) that consequently any existing circle would have a determinate ratio to its diameter—the circle would be squarable.[46]

---

[43] Leibniz makes a similar point with respect to time in our text [7], comparing an infinite time in existence with eternity, the former being "not absolutely infinite, but only in relations to our [times]".

[44] Thus in a piece from 15 April 1676 Leibniz distinguishes the spaces of various possible worlds from divine immensity, which would be the same in all of them: "*Caeterum satis hinc patet spatium differe a Deo, quia plura spatial esse possent, Deus Unus, eadem tamen in omnibus Dei immensitas*" (A VI, 3, 512); and in another written shortly afterwards, "it is the immensum that persists during continuous change of space; therefore it does not have and cannot have bounds, and is one and indivisible" (A VI, 3, 519).

[45] As Crippa explains, Leibniz had hoped "that an infinite series of rational numbers like the one obtained for $\pi/4$ (or, concerning the hyperbola, $\ln(2)$) might be equal to a known rational or radical number, so that the the [sic] definite quadrature of the central conic sections could be solved by a construction", but he gradually came to realize "that the indefinite and definite geometrical quadratures of the circle were both impossible" (Crippa 2018a, p. 134). He refers to Leibniz's attempts to find a solution in terms of a rational or radical number in (A VII, 3, N. 24), (A VII, 6, 90), and (A VII, 6, 111) (Crippa 2018a, p. 135).

[46] Compared with a modern point of view, a striking feature of these reflections is the way that Leibniz moves back and forth between infinitesimals of matter and motion and infinitesimals in geometry. As noted in the previous chapter, this seems to be because the existence he is concerned with is *possible existence*: if infinitesimals and infinities could possibly exist in nature, then they are acceptable in mathematics; whereas if positing them leads to contradiction, then they are not possible existents.

2.3 Leibniz's Rejection of the Existence of the Infinitely Small in 1676

Immediately after noting these implications, however, Leibniz raises a doubt about the idea that the actually infinite division of matter would in fact issue in points, taken as perfect or true minima:

> It must be seen, on the other hand, whether there does not follow a subdivision in a liquid that is now greater, now smaller, in accordance with the various motions of a solid in it; and so it must be rigorously examined whether there follows a perfect division of a liquid into metaphysical points, or only into mathematical points. For mathematical points could be called Cavalierian indivisibles, even if they are not metaphysical points, or minima. But if a liquid can be shown to be divided to a greater or lesser degree, it will follow that a liquid is not resolved into indivisibles. (text [7], A VI, 3, 474)

Leibniz returns to a consideration of the question of the existence of infinitesimals in two papers from the first 10 days of April 1676, papers which may be regarded as a first attempt at the rigorous examination of the "whole labyrinth of the composition of the continuum" that he had urged himself to undertake in "On the Secrets of the Sublime" (text [7]; 475). These papers in fact mark something of a watershed in Leibniz's thinking about the composition of the continuum. For the first, *De motu et materia* (text [9]), begins with the explosive claim that he has "demonstrated elsewhere, very recently, that endeavours are true motions, not infinitely small ones" (A VI, 3, 492).[47] This reversion to essentially Hobbes's finitist understanding of endeavours necessitates a wholesale rethinking of his position. The first consequence Leibniz sees is that "it will follow that there is no really curvilinear motion in things that endeavour along tangents" (492). That is, what appears to the senses as a continuous motion in a curve will in fact be a polygon. Moreover, if a new endeavour is impressed on a moving body at any moment, as "could happen if there were such a thing as a perfect fluid", then time would be divided into instants, "which is not possible". Leibniz concludes:

> So there will be no uniformly accelerated motion anywhere, and so the parabola will not be describable in this way. And so it is quite credible that circles, parabolas, and all other things of this kind, are all fictitious entities. (text [9]).

Explaining this, he comments:

> Assuming there are no curved lines, what is said about them will be properties of polygons. And a particular circle will not be an entity, but [should be] taken in general, that is, what is demonstrated about it will have to be understood of any polygon inscribed in it, or of one having a greater number of sides than we have employed. (text [9]).

---

[47] We do not know where Leibniz might have demonstrated this, or what form the demonstration might have taken. Perhaps a manuscript containing such a demonstration will eventually come to light.

The other consequence Leibniz had seen (in "On the Secrets of the Sublime") of the hypothesis of a "perfect division of a liquid into metaphysical points" was the commensurability of any part of matter with any other, with the consequence that "a circle, if it exists, has a ratio to the diameter as one number to another" (text [**7**]; 474). That is, if matter can be resolved into a perfect fluid, composed of points as minima, then every curve will be commensurable with a straight line since they will have an infinitely small common measure. This is precisely the problem Leibniz sets out to examine in the paper *Numeri infiniti* (text [**10**]; A VI, 3, 496–504).

In *Numeri infiniti*, Leibniz begins by assuming that two infinite lines, if they are commensurable, must stand in the ratio of a finite number to a finite number (A VI, 3, 496). Then he considers the area under a curve (a quadrature) by assuming it is divided into infinitely many squares, and compares the area of this mixed curvilinear figure with the area of a rectilinear figure with the same number of ordinates divided into squares of the same size, and where the ordinates of the mixed curvilinear figure are in rational proportion to those of the rectilinear figure, "which will be possible if the equation of the curve allows it" (497). Then "the number of the infinitely many squares of one will be commensurable with the number of infinitely many squares of the other, and if one is exhausted by the repetition of squares, so will the other be", so that the two "whole figures will also have a common measure, namely, the assumed square". But since he has proved elsewhere (by his transmutation theorem) that such a curvilinear figure could be made congruent with the circle, it follows that "the circle is to the square as finite number to finite number, which is absurd".

At this point Leibniz realises the mistake on which this absurd conclusion rests, namely, his assumption that two infinite lines would be commensurable only if they stood in the ratio of a finite number to a finite number. For "two infinite numbers which are not as two finite numbers can be commensurable, namely, if their greatest common measure is a finite number—as if both are prime." That is, the two infinite numbers could have a common measure, which is of a different nature (because it is finite), in the same way that two prime numbers always have a common measure in unity (which was not considered by Euclid as a "number"). By the same token, Leibniz obtains a way to demonstrate that the quadrature of the circle cannot be obtained by algebraic means:

> Hence now at last there seems to be a way open for a marvellous demonstration that it is impossible for there to be a quadrature of the circle of the kind we are seeking: namely one which would express a relation by an algebraic [*equabile*] equation. And in order for this to be done, it must be shown that the diameter and the side do not have even an infinitely small common measure, even in the kind of line which is as an irrational root, whether quadratic or cubic—as, for example, the side of a double cube, or of some higher power. Hence here we have a splendid use for demonstrations about linear incommensurables, for they can also be transferred to the infinitely small, which arithmetical ones cannot. Supposing this, it follows that the magnitude of a circle cannot be expressed by an equation of any degree. By the same argument it is proved that not even any portion of a circle can be squared by this means; and it is the same with the logarithm and the hyperbola. (text [**10**]; A VI, 3, 497–98)

2.3 Leibniz's Rejection of the Existence of the Infinitely Small in 1676

That "the magnitude of the circle cannot be expressed by an equation of any degree" is equivalent to the later assertion that π is transcendental. Leibniz does not claim to have proved it, noting in the margin that the reasoning "should not be relied on as long as it has not been proved that the diagonal cannot—at least, by subtracting an infinitely small quantity—be rendered commensurable to the side, assuming an infinitely small measure." Nevertheless, given his theorem in the *DQA* that there is no quadrature of the circle more geometrical or exact than his arithmetical one (with the same going for the quadrature of the hyperbola), and that the figures representing these areas are therefore transcendent (i.e. not expressible as roots of an algebraic equation), it is clear that Leibniz believes that this also disproves the possibility of lines or surfaces truly having infinitely small common measures. According to what we have explained above, this would immediately dispel one of the strongest arguments in favour of the infinitely small (that is, the fact that they exist in nature as endpoints of the infinite division of matter).

This seems confirmed by the continuation of this passage, where Leibniz addresses the implications of there being no actual infinitesimals for his polygonal representation of curves, for instance, the circle. Thus, he continues, even though the circle can be conceived as an infinite polygon, "there is no image of a perfect circle in the mind"; rather,

> if certain polygons are able to increase according to some law, and something is true of them the more they increase, our mind feigns some ultimate polygon; and whatever it sees becoming more and more so in the individual polygons, it declares to be perfectly so in this ultimate one. And even though this ultimate polygon does not exist in the nature of things, one can still give an expression for it, for the sake of abbreviating propositions. (text [**10**]; A VI, 3, 498)

This conception of curves as polygons with an indefinite number of sides was for Leibniz a consequence of his re-evaluation of endeavours as finite (or at least, indefinitely small) motions along the tangent at successive moments. This obliges Leibniz to modify his theory of the continuity of motion. No longer can he account for the cohesion of bodies in terms of the overlapping of their extreme parts. Instead he is forced to conclude that the state of the impinging body tending towards a certain place is incompatible with that of the second body which is endeavouring to leave it, "and incompatibility does not admit of more or less" (A VI, 3, 493). The successive states of motion of bodies must therefore be strictly incompatible.

Moreover, as explained above, Leibniz had previously based his conception of actual infinitesimals not only on their proportionality with endeavours, but also on the analogy with angles. The failure of the analogy with endeavours thus obliges him to revisit his former argument. This he proceeds to do in the continuation of our text, presenting his former argument for taking an angle as actually infinitely small as follows: "for the length of the sides is irrelevant to it, and it remains even if you cut them back forever. Therefore there is quantity in a point, for it is the quantity of the angle" (text [**10**]; A VI, 3, 498). To this argument he now responds that without the addition of lines "there is no angle in a point by itself". And if you now hold that "the lines are infinitely small, yet are lines, the difficulty will remain, for I will cut back from them in the same way."

> Moreover, an angle is not the quantity of a point. For we have supposed a point to be that whose part is nothing, an extremum; for we have already shown that there is nothing else unassignable. Therefore the quantity of the angle will be nothing but the quantity of the proportion of the sine, which is the same however far you might produce it, so that the angle itself, it seems, is a fictitious entity. ... [Taken as] the space comprised between two intersecting lines that are smaller than any assignable, ... such an entity is fictitious, since lines of this kind are fictitious. (text [10]; 498–99)

Meanwhile, Leibniz writes, there remains this difficulty, that "diagonal to square is a certain ratio, since the diagonal is a line, a real quantity, and so too is the side." Since this ratio is that of two incommensurable lines, it means that the usual algorithm for the determination of their common measure, the Euclidean *anthyphairesis*, will go without end, to infinity. This brings Leibniz back to the difficulty with infinite number; to which he answers: "but to say all numbers is to say nothing; and for this reason that ratio also means nothing, unless it is something as close as desired." He concludes:

> Still, the ratio of these two lines is not thereby eliminated, even if no measure is assigned. Unless (there being no measure) you also say of the ratio what you said of the angle, that in itself it is nothing but the very agreement of the divisions; an agreement that always remains, as did the sine above. Indeed it seems that the ratio always subsists, since it is through this ratio that two figures are similar. (text [10]; A VI, 3, 503)

Coming back to the need for an infinity of terms in an infinite series, Leibniz comments that the number of terms in such a series "cannot go absolutely to infinity, since then there would be a greatest number." But, noting that this might entail that there is no multitude of things in the world (not an acceptable consequence for Leibniz), he argues that instead "it must be said that an infinity of things is not one whole, i.e. that there is no aggregate of them". Therefore, he concludes, "if you say that in an unbounded [series] there exists no last finite number that can be written in, although there can exist an infinite one: I reply, not even this can exist, if there is no last number." (text [10]; A VI, 3, 504).

Another argument that is of great relevance to the interpretation of infinitesimals as existents is developed by Leibniz in papers written in Holland after his encounter with Spinoza on 18–21 November (New Style) 1676. We saw that in "On the Secrets of the Sublime" he had argued that the spaces of the "worlds within worlds to infinity" would be bounded, but that each of them would be finite or infinite *only relative to the world from which it was considered*. But in *Catena Mirabilium Demonstrationum De Summa Rerum*, dated December 12, 1676, Leibniz argues that "there is but one kind of world", and "no bodies but those having a definite distance from us", because "if there were any, it could not be said whether they exist now or not". For, as he says in another paper of the same date, "existence, as it is conceived by us, involves a determinate time". So to introduce another

2.3 Leibniz's Rejection of the Existence of the Infinitely Small in 1676

genus of existing things, and as it were another world that is also infinite, would be to abuse the term 'existence'" (A VI, 3, 581). This opposition to interpreting worlds within worlds as an infinity of infinite worlds embedded one within the other will assume great importance in his later exchanges with Johann Bernoulli about the reality of infinitesimals. For what Leibniz has rejected here is the very same interpretation of worlds within worlds that Bernoulli presents in 1698 as supporting the reality of infinitesimals, as we will see in Chap. 5 below.

In the papers of April 1676 Leibniz constantly exhorts himself to investigate some of these issues more closely. This he finally finds time to do in late October 1676 at the end of his second visit to England. In the dialogue *Pacidius Philalethi* (text [**11**]), he gives a detailed analysis of the questions he had raised earlier about the composition of the continua of motion, matter, space, and time. Worth noting here, given its earlier centrality to his thought on the continuum, is the complete absence of any mention of the notion of endeavour. Taking on board his earlier insistence that successive states of motion (or of existence) must be mutually incompatible, Leibniz denies that there can be a state of change, since that would entail something being in two incompatible states at the same time, for instance, the state in the last moment of being alive and that of the first moment of death. So, the interlocutors conclude, "it cannot be said that 'something is moving now', unless this 'now' is interpreted as the sum of two neighbouring moments, ôr the point of contact of two times having different states" (A VI, 3, 545).

This reasoning suggests construing a continuous motion as consisting of successive states of being at rest at distinct places that are merely contiguous to one another at each 'now'. But in that case the problem resurfaces for the continuous states of rest in these intervals, as Pacidius (speaking for Leibniz) objects: "temporary rests cannot be interposed in any motion, otherwise we will necessarily arrive at leaps". To this Charinus (perhaps representing Tschirnhaus[48]) suggests a reversion to Leibniz's earlier theory that the rests could be interpreted as infinitely small or unassignable, so that a geometrically continuous motion would be preserved:

> Perhaps leaps through infinitely small spaces are not absurd, nor little rests for infinitely small times inserted between these leaps. For assuming the spaces of the momentaneous leaps to be proportional to the times of the rests, taken together they will correspond together in the way in which we explained the leaps and rests through ordinary times and lines above. (text [**11**]; A VI, 3, 564)

Pacidius rejects this suggestion (and Leibniz subsequently encloses the exchange between them in parentheses):

---

[48] See Arthur's introduction to (Leibniz 2021) for an explanation of this identification. Tschirnhaus was collaborating with Leibniz in Paris, especially in their study of Pascal's manuscripts, from October 1765 until the summer of 1676.

I would indeed admit these infinitely small spaces and times in geometry for the sake of discovery, even if they are imaginary. But I am deliberating whether they can be admitted in nature. For there seem to arise from this infinite straight lines bounded at both ends, as I will show elsewhere; which is absurd. Moreover, since some infinitely small things can also be assumed smaller than others to infinity, again no reason can be given why some should be assumed over others; but nothing happens without a reason. (text [**11**])

The second of these objections is one we will find Leibniz frequently making to the idea of actually infinitely small elements resulting from an infinite division. As he writes in the postscript to his fourth letter in his famous controversy with Samuel Clarke, "The least corpuscle is actually divided *ad infinitum* and contains a world of new created things, which the universe would lack if this corpuscle were an atom, that is, a body all of a piece and not subdivided. ... What reason can be assigned for limiting nature in the process of subdivision?"[49]

The first objection, however, is more specific to infinitesimals in geometry. It alludes to the argument we saw him giving at the end of the cancelled Scholium of Proposition XI of the *DQA*, based on the mean proportional. This is that to admit the existence of infinitesimals making up a finite line seems to entail the existence of finite lines making up an infinite line bounded at both ends—which he here declares to be "absurd". We will return to the argument he gives for its absurdity in Chap. 5 below. For now we may simply note its conclusion: *the assumption that infinitesimals exist seems to lead to absurdity.*

But how can infinitesimals be used if they do not exist? This is addressed by Leibniz in a cancelled passage from the first draft of the *Pacidius Philalethi* concerning leaps through "metaphysical" spaces:

> And these kinds of spaces are taken in geometry to be points or null spaces, so that motion, although metaphysically interrupted by rests, will be geometrically continuous—just as a regular polygon of infinitely many sides cannot be taken for a circle, even though it is taken for a circle in geometry, on account of the error being smaller than we can express by any number. It is by no means to be feared that the reasonings of geometry and mechanics are subverted by metaphysical considerations.[50]

---

[49] (Leibniz 1956, pp. 43–44). As the editor Alexander notes, although this was published by Clarke as a postscript to the fourth letter, it was actually a postscript to Leibniz's letter to Caroline of 12 May, 1716. Leibniz gives an extended version of this argument in the *Pacidius*, suggesting that we could perhaps imagine that the leaps only occur in "the realm of much smaller things, where we wouldn't be able to notice them". But, he objects, if we imagine these much smaller bodies as sentient, they would reason in the same way, and "would be justified in saying that this leap should have occurred in smaller things than in them", and so on down, so "they could never be established [*consistere*] in the nature of things"; "yet there is no reason why these miraculous leaps should be ascribed to this rather than that grade of corpuscles" (A VI,3, 560–61/LLC 199).

[50] "*Qualia spatia in Geometria pro punctis sive nullis habentur, unde motus Metaphysice quidem quietibus interruptus Geometrice continuus erit. Quemadmodum polygonum regulare infinitorum laterum pro circulo metaphysice haberi non potest, tametsi in Geometria pro circulo habeatur, ob errorem minorem quam ut numero ullo a nobis exprimi possit. Minime tuendum est, ne*

2.3 Leibniz's Rejection of the Existence of the Infinitely Small in 1676

Here the infinitely small rests are likened to the infinitely many infinitely small sides of the circle in the polygonal representation which, as we have seen, Leibniz regards as fictional. In each case one can proceed with the mathematical representation relying on the *PUD* (i.e. on the understanding that the error can be made smaller than any that can be assigned). This licenses proceeding in mathematics *as if* there are infinitely small entities, without being committed to their existence. As we shall discuss further below, this is the essence of the *Lemmas on Incomparables* that Leibniz was to publish in 1689 in the *Acta eruditorum*, according to which "if someone does not want to employ *infinitely small* quantities, one can take quantities as small as one judges sufficient for them to be incomparable, so that they produce an error of no importance, indeed an error smaller than any given" (text [**23**]; GM, VI 168).

In the above bracketed passage from the *Pacidius* Leibniz assures us that whatever position is taken on the metaphysical question of the existence of infinitesimals, this will not adversely affect the construal of a circle or other curve in geometry as a polygon of infinitely many sides—an assumption that he takes as foundational in the calculus. Although that passage is bracketed, that it expresses Leibniz's view is confirmed by a recently edited fragment, the *Geometria omissis metaphysicis de continuo* (text [**12**]), provisionally dated as 1679 by watermark evidence. There Leibniz takes the same line, stressing that one can proceed in geometry and mechanics without first having to decide the question of whether infinitely small lines exist. Just as one can treat a circle as an infinitangular polygon, so too a continuous motion can be treated as one consisting in infinitely many infinitely small leaps between neighbouring places, without that committing you to the reality of such infinitely small *quietulae*: "it is sufficient that as many leaps as possible be taken, all of them congruent, that is to say, equal and similar, to each other", and "so that the leaps are equal to one another in equal times." "By the motion of a straight line of this kind, a circle can also be described, and everything else in Geometry is achieved with equal certainty, for it will always be demonstrated rigorously that the error is as small as we wish at the beginning of the construction of each problem" (text [**12**]).

In conclusion, two things are clear. By 1676 Leibniz does not believe that infinitely small or infinitely large numbers or magnitudes exist; yet they can be used as fictions in mathematics in such a way as to reveal truths. In order to explain this state of affairs, we need to provide a thorough explanation of how Leibniz's mathematical fictions should be understood. That is the subject of our next chapter.

---

contemplationibus Metaphysicis Geometricae ac Mechanicae ratiocinationes subvertuntur" (A VI, 3, 569). Our translation here corrects the one previously given by Arthur in (LLC 409–410).

Mathematical Fictions

3.1 Leibniz's Fictions in Context: Fictive Entities in Sixteenth and Seventeenth Century Mathematics

A large part of the difficulty relating to the interpretation of infinitely large and infinitely small quantities as "fictions" comes from the fact that the latter term is often considered as an invention of Leibniz's with a technical meaning that the commentator would have to reconstruct.[1] Yet the idea that some mathematical expressions are "fictions" was very widespread in the mathematical literature from sixteenth and seventeenth centuries and Leibniz's usage seems in perfect continuity with this tradition. Moreover, Leibniz refers on several occasions to this tradition. Since this context is not very well documented in other historical studies (to say the least), we will first spend some time contextualizing this use of "fiction" in mathematics and show how it matches with Leibniz's conceptual background.

The recourse to "fictions" typically appeared when one had to qualify an expression which seemed to denote something "absurd" or "impossible" (like a quantity "less than nothing" or a "fraction of the unit"), but was nonetheless useful in the handling of other entities considered as "real" or "true". This is the way in which negative numbers were famously designated by Michael Stifel in his *Arithmetica integra* as early as 1544, in relation to a more general strategy which he saw at work in the introduction of "surd" (irrational) or "broken" (rational) numbers (Stifel 1544, *Liber* I, cap. 11, p. 8 and *Liber* III,

---

[1] See the preliminary inquiry ('sondage') undertaken by Costabel (1990) which is in fact limited to late texts dealing only with infinitesimals. Katz and Sherry (2013) acknowledge the case of negative numbers, but limit themselves to the study of the *Observatio quod rationes* (1712), leaving aside all the other texts in which Leibniz deals with negative numbers; Esquisabel and Raffo Quintana (2021) distinguish three meanings of "fiction", but in a corpus of texts in which many examples of mathematical fictions in Leibniz (such as negative numbers or logarithms) are absent.

cap. 5, pp. 248–249). Indeed, these new types of "numbers", which were already in widespread use in arithmetic at the time, directly contradicted the definition of what a "number" was in the Euclidean tradition, i.e. a multiplicity of (indivisible) units. They were not "true" numbers, but yet could be manipulated as if they were. In the same period, Cardano also called negative solutions to algebraic equations *fictae* or *falsae* and introduced them on a par with "true" solutions in the very first chapter of his *Ars Magna* (1545). As he explained later in the book (chap. XXXVII), when a problem admits only of negative solutions, it is always possible to reinterpret it so that it gives a "true" solution, typically by reinterpreting the data as asking for a debt (i.e. algebraically: to rephrase a problem about x as a problem about $-x$).[2] It is important to note on this occasion that one and the same expression, say "-3", can thus indicate sometimes an "impossible" entity/solution (when one is asked to find a "real" solution and stumbles upon this "impossibility"), sometimes a "possible" one (when one can reinterpret the data of the problem, for example in terms of debt, a debt of "-3" being a positive amount of money) – a situation we'll find again in Leibniz and is rarely taken into account by commentators. Here, the crucial distinction is between an intended (or "literal") meaning, which seems to refer to an entity (a "negative quantity" or a quantity "less than nothing"), and a figurative meaning in which the expression denotes something different (in this case: a positive quantity).[3]

As is well known, the resolution of third degree equations by mean of a general formula led Cardano to encounter even more absurd expressions, so to speak, since they involve square roots of such fictive solutions. He called them *sophisticae*. In this case too, he proposed a reinterpretation (in terms of "negative" surfaces), but said that this was "as subtle as it is useless" (*adeo est subtile, ut sit inutile*).[4] Although Cardano himself did not consider *sophisticae* as a subclass of *fictae*, later authors would have no difficulty doing so.[5] Note in passing that the introduction of "fiction" thus entails a form of relativity not only in the sense that some interpretations can lead to possible solutions and some others not, but also in the sense than some expressions can appear as "more absurd" than others (which in turn, appear as more "real"). A beautiful example of this situation is given in the exchange between Leibniz and Wallis in 1698/1699. As the latter claimed, if negative quantities are already "imaginary" (another name, as we will see, to designate fictitious

---

[2]This algebraic formulation of the transformation was made explicit by Stevin in his *Arithmétique* (1585, p. 332). Later reinterpretations by Albert Girard or John Wallis would include the fact that we are looking for a "negative" distance, such as when a man goes back in the other direction. Descartes explains in his *Géométrie* how all the "false" solutions can be transformed into "true" ones in this setting (Descartes 1954, p. 168).

[3]This is the argument developed by Cardano in chapter XXXVII with the example of a man's possession as compared to his wife's dowry. After having posited the initial property to be "in minus", he then solves the equation and concludes that the sought quantity is hence a debt.

[4]*Ars Magna* (1545), cap. XXXVII, *De regula falsum ponendi*.

[5]Van Schooten for example, although writing after Descartes who coined the term "imaginary" for complex solution to equations, calls them "fictitious" (Van Schooten 1651, p. 353).

quantities), then square roots of negatives would be some kind of "second order" (*secondo gradu*) imaginaries.[6]

The relativity of the couple fiction/reality (or fictitious/real) can also be seen very well when one takes into account more general uses of this terminology. Thus, in the same period as Stifel or Cardano, other authors designate by "fiction" the quantity introduced in the method of "false position" (or *Regula falsi*).[7] Indeed, the *Regula falsi* commands one to "feign" a certain number as solution to a system of linear equations and then correct the input in order to correct the error made.[8] A similar situation occurs in the kind of number-theoretic analysis originating in Diophantus, who is rediscovered during the same period.[9] Indeed, in Diophantine problems, one has also sometimes to "feign" that a certain sought number can be expressed as a sum or a product of other numbers or of their powers (but, from there, we may still sometimes arrive at the conclusion that the solution is "impossible"). In view of this more general context, it is not surprising to see authors such as Gosselin, Alsted, Schott or Gottignies going as far as considering that the entire art of algebra consists in "feigning" numbers as solutions to given equations.[10] In this setting, if we allow the appearance of negative solutions to equations, we even end up with fictitious

---

[6]"Quod $-bb$ quantitatem dixi Imaginariam; id intellige, quo sensu omnis Negativa Quantitas est (stricto sensu) imaginaria; quippe quòd non possit quicquam existere, quod sit *Minus quand Nihil*. Sed b√-1 est duplici nomine Imaginarium, quasi in secundo gradu remotum." (A III, 8, 42; this letter was published in 1699 in Wallis' *Opera mathematica* III).

[7]The method of false position is an arithmetic technique in which one solves a linear equation by introducing a "fictive" value, which we guess is close to the intended result. Performing the operation on this "false" value allows then to estimate the "error" made in the manipulations and correct the initial value in order to get the "true" one. Authors in this tradition talking of "fictive" numbers include Gemma Frisius, as early as 1540, Jacques Peletier, Jean Trenchant, Pierre Forcadel, amongst others (see Rabouin (2024) for detail). Note that it has echoes in the very title of the chapter of the *Ars Magna* in which Cardano tackles the way to handle negative and imaginary solutions to equations: *De Regula falsum ponendi*.

[8]Interestingly enough, the idea that one introduces infinitely small quantities and negative ones by a kind of "false position" is still the one put to the fore by D'Alembert in order to dismiss what he takes as false debates surrounding these apparently paradoxical quantities (*L'Encyclopédie* (1765), vol. 11, p. 72). Lagrange also considers the methods using the infinitely small as relying on a method by "fausse supposition" (Lagrange 1797, p. 3).

[9]In his translation from 1575, Xylander uses on several occasions the verb "to feign" (see, for example, Diophantus 1575, pp. 61; 75; 98; 132). Many authors from this Diophantine tradition in seventeenth century, such as Bachet, Fermat or Billy, will talk about "fictitious numbers" in this context.

[10]The last example, Gottignies *Logistica universalis* (1687, p. 12), is a very interesting case, since in order to vindicate the use of fiction, the author makes an explicit parallel with the fact that one can refer to human beings either in a real or in a fictive mode, for example in paintings. Both share some properties, although they do not share all of their properties. For example, explains Gottignies, fictitious human beings do not have the need to eat or to breathe.

quantities (such as x or y) possibly denoting... fictitious quantities (such as "–3" or "$\sqrt{-3}$"). We'll find traces of this relativity and this variety of uses in Leibniz again.

Considering this framework, it should be clear that the introduction of fictions took place in the larger context of the rediscovery of the "Analysis of the Ancients", a fact which was explicitly emphasized by both Wallis and Leibniz.[11] At the time, this model was attached to two proper names: Diophantus for arithmetic and Pappus for geometry. Indeed, in Pappus' description of the ancient art of "analysis", one acts "as if" (*pros ei, tanquam, quasi, ac si*) the sought solution was already found. Yet, as explained by Pappus, the process could lead to a proposition known to be false or to a construction known to be impossible. In that case, what was shown was precisely that the statement to be proved was false or that the problem was not constructible.[12] According to Proclus, this last situation was a natural way to obtain a *reductio ad absurdum* by inverting the process of analysis.[13] Another strategy, when stumbling upon an impossibility, was to attempt at "correcting" the hypothesis by specifying the conditions under which the problem *would have been constructible*—what Greek mathematicians classified as a kind of *diorismos*.[14] The various forms of *Regula falsi* were sometimes understood in the continuation of this tradition (where one introduces a "false" solution and then corrects the input in order to get the "true" one), as was Diophantus' approach to number theoretical problems. "Impossibility" thus offered various faces. Sheer contradiction (what Leibniz will call "absolute contradiction") was just one case. Other impossibilities arose as a consequence of the omission of some conditions in the formulation of the problem and were, for this reason, only "apparent" impossibilities. Another case occurred when the impossibility stems only at the level of the expression itself, i.e. when the sought quantity was expressed through a "fiction" although it was denoting in fact a "real" entity (think of the example of the negative quantity interpreted as a debt).

Finally, a last context in which the appearance of "fictions" in mathematics could be of interest to us was that of the development of logarithms, another kind of "fictitious numbers".[15] It was also related in this context to the introduction of infinitely small quantities. Indeed, in his quadrature of the hyperbola, Mercator worked with an infinitieth

---

[11] See Rabouin (2024) for more detail.

[12] "In the case of the problematic kind [*scil.* of analysis], we assume the proposition as something we know, then proceeding through its consequences, as if true, to something established, if the established thing is possible and obtainable, which is what mathematicians call 'given', the required thing will also be possible, and again the proof will be the reverse of the analysis; but should we meet with something established to be impossible, then the problem too will be impossible" (Pappus 1986, pp. 82–83).

[13] Proclus, *In Eucl.* 255–256

[14] This is what Pappus recalls just after the passage mentioned in note 106: "Diorism is the preliminary distinction of when, how, and in how many ways the problem will be possible".

[15] *Logarithmi sunt ficti numeri*, as Caramuel puts it in his *Mathesis biceps* (1670, p. 102). Leibniz uses the same expression in the mémoire (written in the third person) on "A New Binary Arithmetic": "les

3.1 Leibniz's Fictions in Context: Fictive Entities in Sixteenth and ...

(*infinitessima*) part of the unit segment, which he called *ratiunculae*. As we will see, this source is all the more important to keep in mind in that it is explicitly mentioned by Leibniz as the origin of the term "infinitesimals" and that he also worked with fictitious infinitieth parts in the framework of the determination of quadratures. In the case of logarithms, the designation as "fictitious number" was strongly related to the fact that these numerical expressions had a different denotation than the intended one. Another name for this type of fiction used by Napier or Mercator was *numeri artificiales* (as opposed to "natural" numbers, as Leibniz puts it).

As can been gathered from this brief overview, mathematical "fictions" were in rather widespread use at the time of Leibniz. Their meaning fits quite well with that of the verb *fingere*, which was used for what can be imagined, fantasized, or, more pragmatically, posited for the introduction of hypotheses or suppositions, typically in physics.[16] It corresponds to an expression for which we do not know (or feign not to know) whether or not it corresponds to something possible. Another broad context in which "fictions" were in widespread use and which was well known to Leibniz was the study of Law (*fictio juris*). That the two meanings were connected is attested very early in his work. In the discussion of the *casus perplexus* from 1666, a typical case where the *fictio juris* was useful,[17] Leibniz makes a parallel with demonstrations of impossibility allowed by symbolical algebra. He even mentions on this occasion that one rendering of the Greek ἄπορον was not *perplexus*, but *caecus*—a quite remarkable term under his pen, which we will encounter in connexion with fictions 10 years later.[18]

This general context was also indicated by a mathematician who Leibniz read very early: Juan Caramuel y Lobkowitz. Indeed, Caramuel dedicated a whole section of the chapter on Algebra in his *Mathesis biceps* to deal with the following question: "whether Dialecticians, Jurists, Astronomers, Geometers and Arithmeticians, etc., posit the false and deduce the true from it?".[19] This question was prompted by the fact that Caramuel shared the view that algebra was an extension by other means of the *Regula falsae positionis*.[20] He argued first that truth could not be deduced from falsity, but then objected to the widespread

---

Logarithmes, qui sont des espéces de nombres feints & supposés, sont un circuit que l'on prend pour arriver aux *Nombres Naturels*" (Leibniz 1705, p. 62).

[16] A familiar example was given by the use of fictitious circles as hypotheses for describing the trajectory of planets in astronomy cf. Ursus 1597. As is well known, Osiander insisted on this use of fictions (he uses at several occasions verbs like *fingere* or *confingere*) in the Preface he inserted (anonymously) to Copernicus's edition of his *De revolutionibus orbium coelestium* (1543). Leibniz often puts hypothesis and fiction (or a use of the verb *fingere*) on a par in his works in natural sciences.

[17] On the fact that *casus perplexis* belongs to *fictio juris* see A VI 1, 143.

[18] A VI, 1, 235-236. *Disputatio de casibus perplexis in jure*, November 1666.

[19] *An Dialectici, Jurista, Astronomi, Geometra, Arithmetici, etc. falsum ponant, et verum ex illo deducant* (Caramuel 1670, p. 99).

[20] He introduces his chapter on algebra by a *Prooemium de regula falsae positionis, quae est basis algebrae* (Caramuel 1670, p. 99).

use of "fictions" in various domains of knowledge. What is of interest to us is that it testifies for a general category in which were included the various situations we came across: hypotheses of the astronomers, the fiction of Law, logarithms, the rule of false position in arithmetic and *reductio ad absurdum* in geometry. Caramuel then studies each of these "fictions" case by case in order to show that they do not contradict his view. As a conclusion, he inserts a note in which he tackles a last case of fiction: that of infinitely small quantities, as posited in the *methodus indivisibilium* (in reference to a discussion by Maignan on the composition of the continuum).[21]

Before turning to the various uses of the term "fiction" (or "fictitious") in Leibniz' mathematics, let us make a last comment about the terminology. Although all the cases mentioned so far make recourse to a term related to "fiction", one should keep in mind that the perspective is even broader. In ordinary language, as in mathematical language, fiction is related to the domain of the "imaginary", a term which can have a technical meaning in mathematics after Descartes, but not necessarily. Thus Descartes himself, discussing the quantity "*e*" introduced by Fermat in his "*méthode d'adégalisation*" and which has to be cancelled at the end of the calculation, calls it an *imaginary* quantity.[22] We already pointed to the way Stevin saw negative solutions as being "real" if the problem was reformulated in terms of $-x$: the term he used was "*songées*" ("dreamt"). Cossic algebraists such as Faulhaber and Roth designated such solutions as *gedichte*.[23] Jacques Hume, editor of Vieta, called negative solutions "imaginary", and this expression can be found even after Descartes in authors such as Gottignies, whom Leibniz read with great attention. Leibniz himself, in the *Observatio quod rationes*, calls "imaginary" the ratio of a negative to a positive (which he calls fictitious elsewhere). On this occasion, he mentions other entities such as: logarithms of negative numbers, the angle of contact, infinite and infinitesimal numbers, infinite and infinitely small lines and, of course, imaginary roots themselves (text [**41**]; GM V, 387–389). In the letter to Wolff from 1713, he says that the fact of taking a point as an infinitely small line and other expressions of that sort utilized in his famous "Law of Continuity", contains "something fictional and imaginary" (*aliquid fictionis et imaginarii*) (text [**44**]; GM V, 385). Moreover, one of the most famous appearances of infinitesimals as fictions was made by reference to a discussion in which Leibniz first called

---

[21] *Ibid.*, p. 109.

[22] "Car une quantité réelle étant multipliée par une autre quantité imaginaire, qui est nulle, produit toujours rien. Et ceci est l'élision des homogènes de M. de Fermat, laquelle ne se fait nullement gratis en ce sens-là.". Interestingly enough, Fermat himself designates the "adégalité" (adequality) obtained by introducing the imaginary quantity a "fictive comparison" (*comparationes fictas*), see *Ad eamdem methodum* (Fermat 1679, p. 66).

[23] Already Stifel, in his *Deutsche Arithmetica* from 1545 (p. 61), which may be the source here for this terminology, called "erdichten" the kind of numbers which others called irrationals.

them "imaginary".[24] It is henceforth important to include in our survey expressions belonging to this extended terminology.

3.2 Mathematical Fictions in Leibniz

Let us now go back to the use of "fictions" (and related terms such as *fingere, imaginaria,* etc.) in Leibniz. As we have already seen, the idea that infinite quantities are fictions appears very early under his pen. The paradox related to the fact that an infinite whole has to be equal to its part, which Leibniz stumbles upon as early as 1672 in his work on series, led him to consider that "infinite wholes", which contradict Euclid's axiom, must be treated as fictions.[25] This statement will be repeated over and over until the end of his career: the infinite can never be a whole or a unity in mathematics (*nec totum, nec unum*), because such a notion, taken literally, entails a contradiction. For this reason, it cannot be treated as a genuine quantity: there is no such thing as an infinite number or an infinite magnitude.[26] But it can be handled as a fiction or "by analogy". Now what is of particular interest is that the discovery of the paradoxical character of infinite wholes was *immediately and explicitly* related to the larger framework which we recalled in the preceding paragraph. This appears in the text from the end of 1672 mentioned in Chap. 2 in which Leibniz puts on equal footing infinite wholes, surd roots, "imaginary" dimensions in geometry and negative numbers.[27] He then adds the example of the ratio between a line and a point, as was feigned in the treatment of logarithm à la Mercator and concludes: "there is indeed a certain reason or truth about impossible and false things" (A VII, 3, 69; end of 1672). Moreover, the text is presented as a defence of the kind of strategy deployed in the Arithmetic of the Infinites (à la Wallis) and the Geometry of the indivisibles (à la Cavalieri).[28]

But what does it mean to treat infinite wholes as imaginary entities or fictions? We saw that this is explained in various texts from this early period in which Leibniz indicates that

---

[24] Writing to Varignon in 1702, Leibniz claims: "I wrote some years ago to Mons. Bernoulli of Groningen that the infinite and infinitely small could be taken as fictions" (GM IV 93–94); but when one turns to the exchange with Bernoulli, what Leibniz said at first was that they were imaginary (*Fortasse infinita quae concipimus et infinite parva **imaginaria** sunt sed apta ad determinanda realia ut radices quoque imaginariae facere,* GM III, 499). See also the quote above p. 48 (from *Pacidius Philalethi*).

[25] A VII, 3, 468 (1674). Note that the term "fiction" is introduced in this framework to designate something which is explicitly contradictory.

[26] See Sect. 5.2 for an explanation of the link between the exclusion by Leibniz of infinite numbers, quantities and magnitudes and the rejection of infinite wholes.

[27] See quote given on p. 27 above.

[28] In this text, only dimensions greater than three are called "imaginary". But other texts from the same period use the same term to refer to negative numbers or fractions. See, for example, A VII, 3, 40. Another text from 1673 uses the same term to qualify Mercator's *ratiuncula,* see A VII, 4, 298.

one can work with infinite series by handling them *as if* they were wholes. This amounts to considering all of their terms as forming a fictitious totality and, for this reason, as including a last term (*ultimus terminus*)—although these various notions are contradictory when taken literally. This is made possible by interpreting the ultimate term of a decreasing infinite series, "or rather, in this place, a quantity infinitely or unassignably small, to be greater or smaller according as we have assumed the last denominator of this infinite series of fractions, which is itself also infinite, smaller or greater".[29] As should be clear from this passage, the "ultimate term", when not taken literally, denotes a *variable* quantity. This conception will still be at the centre of the exchange with Bernoulli in 1698, since the latter thought that in any converging series there ought to be an "infinitieth term", after which all the other terms would be of infinitely small size. Leibniz objected that the talking in terms of "infinitieth term" is just a "way of speaking", because such a conception of an infinite number, taken literally, entails a contradiction. Thus the "infinitieth term" should be understood as a finite variable one.[30] This meaning of fiction persists in exactly the same form in the exchange with Philipp Joseph Jenisch in 1702 (text [**48**]):

> In my judgement, the demonstration does not proceed sufficiently rigorously when we conceive a certain term x as the last one in a geometrically decreasing series; for in fact there is no last term when it is continued to infinity, therefore through *the fiction of a last ôr infinitesimal term* (to reduce things to a rigorous demonstration), one has to conceive an error smaller than a given one.[31]

Another way of treating infinite whole as fictions, besides the handling of a fictitious *ultimus terminus*, was to ascribe them a fictitious quantity or "number" by analogy to what was done in ordinary algebra. This is how Leibniz proceeds when demonstrating the general rule for obtaining the sums of the series of the inverse of figurate numbers (the famous "harmonic triangle"). Leibniz first writes A for the series of the inverse of natural numbers and B for the series of the inverse of the triangular, then treats them as if they were genuine numbers to get the "equation": $A - 1 + \frac{1}{2} B = A$ (hence $B = 2$).[32] What is striking for us is, of course, that a letter can designate in this context a diverging series and that Leibniz has no more trouble in treating it as a *totum* than the other series. So the letter here is not a way to designate a real quantity, but is really a fiction. We are acting *as if* the series could be gathered as a whole and designated by a letter like any other genuine quantity. The

---

[29] Note on Spinoza, cited earlier p. 35.

[30] For the exchange with Bernoulli, see (text [**32**]).

[31] Our emphasis. See also the famous passage from the *Essais de Théodicée*: "We embarrass ourselves in the same way with number series which go to infinity. We conceive a last term, an infinite number, or an infinitely small one; but this is all nothing but fictions" (text [**36**]). Note the fact that the introduction of a last term, of an infinite number and of an infinitely small number go hand in hand for Leibniz – a situation which corresponds to what is found in Bernoulli's position.

[32] See A VII, 3, 712-713 and Chap. 2, p. 36.

3.2 Mathematical Fictions in Leibniz

reliability of this procedure is due to the fact that what this manipulation depends upon, as explained in the text [**33**], is not a collective whole, but a distributive one. In other words, the letters here stand for generic variable terms, which can be handled in the same way.[33]

Now it is important to note that for Leibniz:

> ...in general it can be said that an infinite number, an infinite line, a series composed from infinitely many terms, or an aggregate of an infinite multitude of things, is in metaphysical rigour not one thing, *since they always involve the greatest number, which is impossible*. But in mathematical matters they are taken as one thing as an abbreviation of speech, since they have a foundation in reality. (text [**33**], §8)

We will come back to the question of the "foundation in reality" later, but note that in the same text, Leibniz says that we can also treat the universe as a whole *just as* we use "fictive quantities" in mathematics (with his usual example of imaginary roots).[34] There is no ambiguity here that Leibniz continues (at the end of the 1690s) to put on the same level the use of "fictive" entities in mathematics and contradictory objects such as infinite wholes exactly as he claimed 20 years earlier.[35] But it is also important to note that he states that such a fictive notion is *always* involved in the handling of infinite lines and infinite series as soon as one considers them as "one thing" (collective wholes).

Interestingly enough, Leibniz also mentions, as early as 1673, the *geometrical* interpretation of the infinitely small in the form of the difference between two "minimal ordinates" as fiction:

> But at infinity, where the quantity of the rectangles is unassignable, and the difference as well as the distance between two minimal ordinates is less than any straight line *which can be, I do not say thought, but feigned*.[36]

---

[33] See Chap. 2, p. 36 (Leibniz disposes of a formula involving a generic term in the simple cases of his harmonic triangle, but it is not clear how he could have expressed algebraically the rule for the inverse of figurate numbers of higher order).

[34] What he says in a previous passage of this text is equally interesting, since it repeats the strategy of using "incomparables" even in this case: "And we will keep in mind that the universe taken as one thing is similar to those fictitious quantities algebra uses, when it designates with marks the imaginary roots of impossible equations, which we know to be marvellously useful for reasoning and calculations. Meanwhile, however, I prefer to use 'unassignable' or even 'incomparable' for the imperfect infinity of aggregates of things" ("*Towards a Science of the Infinite*", chapter 9 in Arthur and Ottaviani, forthcoming).

[35] See also the discussion in the *Nouveaux essais* (text [**34**])).

[36] *At in infinito, ubi inassignabilis est rectangulorum quantitas, et differentia pariter ac distantia inter duas minimas applicatas minor est qualibet recta* **quae non dicam cogitari, sed fingi possit** (A VII, 4, 317, 1673; our emphasis). One should keep in mind that a contradictory notion corresponds to no idea and hence cannot *stricto sensu* be conceived (A VI, 3, 463, on the example of the *numerus maximus*). This may be the first occurrence of the vocabulary of fiction applied to infinitely small quantities in geometry in Leibniz.

All of these passages indicate, if needed, that the recourse to fictions is not a strategy created in the 1690s in order to respond to criticism raised against his differential calculus, but a practice existing even before the creation of the differential calculus (which was achieved only in the autumn of 1675). In fact, as Leibniz will recall to Bernoulli in 1698 in reference to the geometrical methods expounded in the *DQA*, these questions are *independent* of the differential calculus.[37] They concern more generally the recourse to infinitely large and infinitely small quantities in mathematics, which occurred at the time in various techniques (in Leibniz and in previous authors such as Cavalieri, Wallis or Mercator). This is why it is so important not to confuse the question of the existence of infinitely small quantities with that of the justification of the differential algorithm, which is a separate and independent issue.

On a larger scale, a closer attention to the sources indicates that all the types of "fiction" we encountered so far are to be found in Leibniz. First of all, he explicitly mentions on several occasions one of the most famous cases of fictions at the time: negative numbers.[38] In addition to the passages we already quoted, we can mention a dialogue he wrote in order to introduce algebra around 1679. In it, he presents the traditional interpretation of negative numbers as debts and recalls their usefulness as fictitious entities.[39] The interpretation of negative quantities as debts is very common in Leibniz[40] and, as we will see, belongs to a more general strategy in which a problem *male formatus* can make sense thanks to a proper reinterpretation of the data. As we have already emphasized, this was a common way to deal with *numeri ficti* in algebra since the sixteenth century. Very early, Leibniz also called the ratio of a negative quantity to a positive one "imaginary"[41] and this meaning persists in the *Observatio quod rationes* (1712).[42] In the dialogue from 1679, one also finds a mention of the use of "fictitious" numbers for the unknown in algebra.[43] In other texts devoted to the various kinds of analysis and synthesis, Leibniz points to the fact that algebra works with a

---

[37] *Ubi apparet objectionem non tantum nostrum calculum, sed et Geometriam jam antea receptam pari jure ferire* (A III, 7, 855).

[38] Amongst the first appearances is a commentary on Descartes's construction of equations from their roots, which can be, "although useful, nonetheless fictitious" (*etsi utiles tamen fictitias esse*) – the context does not permit discriminating between negative and imaginary roots here (A VII, 2, 242). See also A III, 1, 652.

[39] More specifically, the fiction is useful for representing the hope for a future acquisition (Knobloch 1976, p. 121: *Habet ergo usum fictio haec, in spem futurae acquisitionis*).

[40] See, for example, Leibniz (2018, pp. 149; 174).

[41] See A VII, 3, 40, quoted above note 28.

[42] GM V, 388; see text [41]. The general argument already occurs already in a letter to Mariotte from 1679 A III 2, 776 ("- 1 à part est une chose purement imaginaire").

[43] Knobloch (1976, p. 45): "*Nodus est vindice dignus, set ut melius rem intelligas et per te quaestionem resolvas loco c et a fictiorum adhibeamus numeros veros, exempli causa c significet 2 et a significet 3*)." On the use of fictitious entities in algebra since Vieta, see also the *De Analyseos historia* (1674–1675), A VII 8, 45. In the *Dialogus*, Leibniz also calls "fiction" the dimension greater than three, one of his examples of "imaginary" from 1672 ("**Quadrirectangulum** *m abcd (per*

3.2 Mathematical Fictions in Leibniz

kind of *synthesis fictitia*.[44] In all of these contexts, the fictitious character of the entities introduced is strongly related to the fact that one proceeds through hypotheses or *suppositiones* (which may be true or false).

This relationship is very apparent in another crucial context, that of the *numeri fictitii*, also called *suppositi*. These numbers were invented by Leibniz to designate the rank of coefficients in equations.[45] Hence the expression '32' was not supposed to designate the natural number 'thirty two' but a number occurring in the third line and the second position of an equation.[46] Indeed, it is maybe the most widespread use of term "fictitious" in Leibniz' mathematics and, strangely enough, is usually forgotten by commentators dealing with the role of fictions in his mathematical practice. Yet this use is perfectly in line with one we came across before: when an expression is used in order to refer to something different than its intended denotation, i.e. that its reference has to be suspended.[47] More generally, one would find in Leibniz's mathematics innumerable examples of the use of "*fingere*" in order to introduce an expression for which the denotation is not yet determined (typically a "fictitious" equation) and of "imaginary" (typically imaginary constructions in geometry). This is taken in an even broader context in which *fictio* and *fingere* were used in order to designate any kind of hypotheses, typically in natural sciences.

As should be clear from the preceding examples, "fiction" is not in Leibniz a way to designate a special kind of entity. It occurs in a variety of contexts and relates to a general methodology. It is a way to designate some expressions which seem to denote genuine entities although they may fail to do so (for various reasons). This is why Leibniz can sometimes claim that any mathematical entities are fictitious or imaginary, although he can distinguish some amongst them as "real".[48] Accordingly, one and the same "entity" can be called in the surface language sometimes "real" sometimes "fictitious", depending on the

---

fictionem scilicet quasi dimensiones lineares ultra solidum ascendere possent)" Knobloch 1976, p. 153).

[44] Leibniz (2018, p. 111); A VI, 4, 350-351 and 386; GM VII, 214.

[45] See GM VII, 5-8 for a text from 1678 in which these numbers are already used.

[46] See, for example, Leibniz (2018, p. 136).

[47] See E. Knobloch, « Déterminants et élimination chez Leibniz », *Revue d'histoire des sciences* 54 (2) (2001). Accordingly, Leibniz asserts in his first attempt at a logical calculus using numbers (*characteristici numeri*) that, not knowing the true numbers coding notions, we should employ "fictitious ones" (A VI, 4, 222).

[48] In early texts, at the time of the stay in Paris, Leibniz can designate geometrical objects such as the circle, the parabola or the angle as fictions, in the sense of not existing "in nature" (A VI, 3, 492; 498-499). In later texts, Leibniz would more likely talk of the "imaginary" nature of mathematical entities, in particular when dealing with spatial features (such as figure, size, position…), which should be so called because they cannot constitute the substance of a body (See letter to Arnauld, 30 April 1687) – although he still refers this imaginary character to a "fiction de l'esprit". On this sense of Leibniz's fictionalism, we refer the reader to Rabouin (2022).

interpretation to which the expression is attached.[49] In fact, this should be no surprise since Leibniz has the same contextual use of the term "real". According to his mature philosophy, only simple metaphysical substances, at some point called "monads", are "real". Yet one can speak of "real phenomena" (or "phenomena *bene fundata*") each time there is a way to ground a phenomenon in such a reality. By extension, a mathematical fact which can, under suitable interpretation, serve to describe a phenomenal regularity can be called "real" or even to exist "in nature".[50] When this interpretation is not a direct reference, the fictitious expression will be said to have a "foundation in reality".[51]

At this point, one could consider that the originality of Leibniz lies in extending this context of use for "fictions" to infinitesimals, by possibly inflecting it in an important manner. But we saw that this is certainly not the case. Firstly, he himself put the various kinds of fictitious entities on equal footing as early as 1672 when referring to the practice of other mathematicians such as Wallis and Cavalieri (and continues to do so in later texts such as the *Observatio quod rationes* from 1712). Even after the creation of differential calculus, he still talks at the end of 1680s of the "fiction" of infinitely small quantities not as a new strategy, but as a *common practice* amongst geometers of his time.[52] What would he have in mind when referring to such common practice? When discussing with Wallis in 1699 about the proper interpretation of infinitesimals (which Wallis took as "absolute nothings"), Leibniz refers very clearly to what he considers as its origin:

> But for the calculation, it is useful to feign [*fingere*] infinitely small quantities, or as Nicolaus Mercator called them, infinitieths [*infinitesimas*]: which are of such a kind that, since a ratio between them is sought that is certainly assignable, it is already illicit to hold them to be nothings....[53]

Indeed, although the term used by Mercator was *infinitissima*, and not *infinitesimas*, it remains the case that one of the first mentions of infinitely small quantities as "imaginaries" in Leibniz is made as early as 1673 in the context of a study of Mercator's work on the quadrature of the hyperbola.[54] More than 15 years later, Leibniz would repeat that in

---

[49] For example, in the exchange with Wallis, Leibniz calls "reals" any non-imaginary solutions, including the negative ones (A III, 8, 9), although in other places we saw that he calls negative numbers either fictive or imaginary.

[50] Leibniz (2018, p. 107).

[51] See Rabouin (2022).

[52] See *Specimen inventorum de admirandis naturae Generalis arcanis* (1688; A VI, 4, 1628/LLC 327-9), or *De abstracto et concreto* (1688; A VI, 4, 991).

[53] *Sed pro calculo utile est fingere quantitates infinite parvas, seu ut Nicolaus Mercator vocabat infinitesimas: quales cum ratio earum inter se utique assignabilis quaeritur, jam pro nihilis habere non licet* (Leibniz to Wallis, March 30/[April 9], 1699, text [**46**]).

[54] See text [**13**] (A VII, 4, 298): "A logarithmic figure is necessarily quadrilinear, that is, it consists of three straight lines, two of which are parallel to the ordinate of the height, and the curve. Since a ratio cannot be taken with respect to a point, a mean proportional is not intelligible unless with respect to a

3.2 Mathematical Fictions in Leibniz

logarithms, one introduces a fictitious *infinitesimum vel infinite parvum* in order to deal with all quantities at stake as having a ratio to one another:

> Meanwhile by a certain fiction we can conceive all homogeneous quantities to be as if commensurable with each other, namely by feigning some infinitieth or infinitely small element. The calculus of logarithms is founded on such a fiction, by the establishing of a certain logarithmic element. (text [26], GM VII, 39)[55]

On this occasion, he will make it clear that, according to him, this meaning is *the same* as the one used by geometers relying on infinitesimal techniques (which is concordant with Mercator's practice, as taking place in the context of a reinterpretation of the *geometrica indivisibilium*):

> A similar fiction takes place in geometry, by conceiving the matter just as if all lines are made up out of infinitely many infinitely small straight linelets, and so just as if curved lines were polygons having infinitely many sides, or just as if surfaces were made up of infinitely many planelets, that is, just as if concave or convex solids were all polyhedra with infinitely tiny faces. (text [26])[56]

As can be already seen in the texts of 1673, this kind of parallel was first motivated by the use of a fictitious infinite number and its inverse as lengths of the intervals in which the axis was subdivided, a technique introduced by Wallis in his *Arithmetica infinitorum* in order to "arithmetize" techniques from the Cavalerian *Geometria indivisibilium*—the two contexts explicitly mentioned at the end of 1672 in the discussion on infinite wholes.[57] Interestingly enough, Wallis recalled that the principle of Cavalieri's technique was to handle surfaces "as if" (*quasi*) they were composed of lines—an "as if" which will unleash the mockery of Hobbes,[58] and which is to be found again in the previous quote. Moreover, if Cavalieri himself did not talk about fictions, it should be remembered that he made an explicit

---

certain line that is infinitely smaller than a straight line, and infinitely smaller than a point, which kind of thing is imaginary". (Although Mercator is not mentioned by name, the use of the word "ratiuncula" in the text is a clear sign that Leibniz is reading his *Logarithmotechnia* here). On the importance of Mercator in Leibniz's first reflections on infinitely small during the Parisian Stay, see (Probst 2008).

[55] We will come back to the role of "fictitious" common measures in Leibniz's conception of numbers taken *lato sensu* in Chap. 5.

[56] On the interpretation of the use of these "quasi-minima", see Sect. 5.2.

[57] See A VII, 3, 69, quoted above. Here is the way Wallis introduced these objects at the beginning of his *Treatise on Conic Sections*: "*Suppono in limine (juxta Bonaventurae Cavallerii Geometriam Indivisibilium) Planum quodlibet quasi ex infinitis lineis parallelis conflari: Vel potius (quod ego mallem) ex infinitis Prallelogrammis aeque altis; quorum quidem singulorum altitudo sit totius altitudinis $1/\infty$, sive aliquota pars infinite parva; (esto enim ∞ nota numeri infiniti;) adeoq; omnium simul altitude aequalis altitudini figurae*" (Wallis, 1655, prop. 1).

[58] Wallis (2004), p. xxiii.

parallel between the introduction of these extraordinary "indivisibles" and the introduction by algebraists of "inexpressible" quantities.[59] Finally, we saw that Caramuel, following Maignan, interpreted all the methods of indivisibles as a typical case of the use of "fictions" in mathematics on a par with logarithms, *reductio ad absurdum* and *regula falsi*. All of this should make it clear, once again, that Leibniz is situating himself, as he claims, in a common practice of his time with regard to the use of infinitely small quantities as fictions in mathematics.[60] In fact, although commentators have put a lot of emphasis on this characterization of infinitesimals as "*fictions utiles*", they often fail to notice that this expression is pretty rare under Leibniz's pen, especially when it comes to *public* declarations.[61]

Now, there is a particular type of fictitious entity which was of special interest for Leibniz in the years 1675–1676 and which should be of particular interest to us too, since it constitutes an example that he regularly paralleled with the infinitely small in later texts: square roots of negative numbers. Indeed, it forced Leibniz to look more precisely at the question of the different forms of impossibility in mathematics and it will give us an occasion to clarify his subtle and complex position on this issue.

3.3 Fictions and the Question of Impossibility

As we have seen, a "fiction" is a way to handle an expression which, on the face of it, refers to a state of affairs for which we have no guarantee that it exists or is possible. The "suspension" of the referential regime can be related to various circumstances: we can ignore whether a certain situation holds or not, we can have doubts about its existence, we can know in advance that it cannot be possible, we can intend to prove this impossibility, we can suspect that some extra conditions should be added in order for the problem to be solvable, we can ask a reader not to interpret a sign in its ordinary meaning, etc. All of these uses existed in the mathematical literature before Leibniz and are to be found again in his texts.

---

[59] Cavalieri (1647), p. 202.

[60] See, in particular, A VI, 4, 991, where Leibniz compares the introduction of abstract predicates with the common practice of algebraists and geometers of his time (*Interea commodioris ratiocinationis causa adhiberi talia possunt, uti adhibemus in Algebra radices imaginarias, et in Geometria fortasse lineas infinitas et infinite parvas*).

[61] Although Leibniz uses the term fiction in his letter to Varignon from February 2, 1702 (text [**38a**]), this is only in a negative sense. The positive sense appears in the letter from April (text [**38c**]) and June (text [**38d**]), but they are private letters. As far as we can judge, the first public appearance of infinitesimals being fictions is in the Preface to the *Théodicée* (1710), a philosophical treatise which was not directed at mathematicians. Another public appearance is in the letter to Wolff (1713), text [**44**], already quoted above p. 56.

3.3 Fictions and the Question of Impossibility

Accordingly, when we stumble upon an impossibility in the process of handling a fiction, it can take several forms. It can be a sheer contradiction, which is typically what happens when we conduct a *reductio* proof. This meaning is familiar to Leibniz as appears, for example, in the following passage taken from the Correspondence with Clarke:

> If space and time were anything absolute, that is, if they were anything else besides certain orders of things, then indeed my assertion would be a contradiction. But since it is not so, the hypothesis (that space and time are anything absolute) is contradictory, that is, it is an impossible fiction.
>
> And the case is the same as in geometry, where by the very supposition that a figure is greater than it really is, we sometimes prove that it is not greater. This indeed is a contradiction, but it lies in the hypothesis, which appears to be false for that very reason. (Leibniz-Clarke Correspondence, Fourth Paper, §16–17, 374)

We saw that it was ranked amongst the common uses of "fictions" in mathematics by Caramuel and that it appears in Leibniz as early as 1666.[62] It is involved in many occurrences in which Leibniz says that we should check with care that a certain fiction is not "empty" (or, as stated in our passage in an interesting way, an "impossible fiction").

Another case, already mentioned by Pappus when describing the practice of "analysis", appears when the analysis stumbles upon a contradiction which is related to the conditions of the problem and can be fixed by restating the data. In this case, the impossibility is not consubstantial or essential to the problem and appears "accidentally" (i.e. because we did not specify certain conditions). This had been a typical way to justify negative numbers since Cardano, namely by showing that the problem could be reinterpreted so as to denote "real" quantities (typically when reinterpreting it as a way to ask for a debt or a position in the opposite direction). The same meaning is clearly endorsed by Leibniz in many texts such as the following one:

> Negative quantities, when it comes to subtracting a greater from a smaller, often occur in the calculation. It is permissible to consider that they do not answer the question, but in truth they answer it most perfectly. Not only do they indicate that the question was badly conceived (even if it had been accepted with indulgence, since one could not foresee [the answer]), but they also indicate how it should have been conceived, and what should be answered if it was correctly conceived (Leibniz 2018, p. 149)[63]

---

[62] See p. 55 above and note 17.

[63] In the continuation of the passage, Leibniz gives exactly the same kind of example as Cardano by mentioning the amount of money possessed by Titus in the case where the solution to the problem gives a negative quantity. He concludes that one should not have asked what amount of money he owed, but how much he would need to be free from debts. Then Leibniz gives a similar example in terms of geometrical lines by interpreting them as progression and regression of a traveler along her journey (Leibniz 2018, p. 149).

The idea that the introduction of a fictitious entity helps in seeing that a problem was "ill posed" appears in a number of texts concerning not only negative quantities, but imaginary roots and infinitely small quantities.[64]

Note that the idea that mathematical analysis stumbles upon various kinds of impossibilities for which we introduce expressions which work as *succedanea*,[65] or *expedients*[66] such as negative, fractional or irrational numbers, is presented explicitly by Wallis and Leibniz.[67] In conformity to what we saw in the previous section, the latter presents the introduction of infinite quantities in continuity with the other type of "impossible" entities:

> Sometimes also the operation which must be made in actuality is impossible, either for the time being or not possible at all, even if it could be exhibited at least by a construction in our characters, or is already exhibited in nature. Thus it is impossible to subtract when there is nothing, and yet this is represented in nature, for example when someone owes more than he has. In the same way it is impossible for one prime integer to be actually divided by another; hence there arises a fraction, which represents the division to be made—with the thing which is designated by this number being divided into parts more suitable for exhibiting this division. In the same way there arise incommensurable quantities, i.e., surd roots, where [root] extraction cannot take place. And some extractions are such that these surd roots do not even manifest themselves in nature. Then they are called imaginary, and the problem is then impossible.... (*Elementa nova matheseos universalis*; A VI, 4, 520)[68]

Now a particular case arose, as mentioned by Leibniz in the last sentence, when there was no concrete situation in which the fiction could be reinterpreted (and thus could appear as "real" or "existing in nature")—contrary to what happens for negative, fractional and irrational quantities, for which there exists a uniform geometrical interpretation. Although some expressions involving imaginary roots were in fact denoting real quantities (more on this below), there was no geometrical interpretation allowing this always to be the case. A typical example was provided in problems from algebraic geometry. Leibniz's favourite example was that of the finding of an intersection between a circle and a line moving in the plane. Although a single equation could be devised for all the relative positions of the circle and the line, parameters entering into the expression of the data rendered the problem of intersection either constructible or not. Impossibility (of construction) was therefore marked by the appearance of imaginary roots. This led to a typical case of diorism since one just had to fix some parameters to get rid of this problematic case. In the example, having devised $y = \sqrt{r^2 - x^2}$ as the geometrical constraint, Leibniz concludes:

---

[64] As early as 1675, see (A VII, 2, 752), and until the 1700s, see Leibniz (2018), p. 153.
[65] GM VII, 68.
[66] Wallis (1685), p. 316.
[67] See Rabouin (2024) for more detail.
[68] The continuation of this passage, to which we will come back later, deals with infinite quantities.

3.3 Fictions and the Question of Impossibility

> Here, then, is the way in which nature has been able to indicate that y, in the case where x is greater than r, is impossible. From this we learn that the question is not well posed, and that either the circle ABC, i.e. its radius r, must be assumed to be larger, or, the circle remaining the same, x or AF must be assumed to be smaller, in order that the question may be solved.[69]

Interestingly enough, Leibniz draws a parallel with some situations in which infinitely large and infinitely small quantities appear.

> In fact, the imaginary or impossible by accident, *which cannot be exhibited because what is necessary and sufficient for an intersection to occur is missing*, can be compared with infinite and infinitely small quantities, *which arise in the same way*. (our emphasis)

We will come back to the expression "impossible by accident", but let us follow Leibniz's analogy. In the continuation of the above passage from *Elementa nova matheseos universalis*, he then expounds a problem in which one studies the angle between two lines connected by a perpendicular segment CA, one of which, CB, is rotating from the position where it coincides with the perpendicular (when B coincides with A) to the position where it is parallel to the other line (see Fig. 3.1).

In the margin, Leibniz has indicated that there is, however, a difference between the two extreme cases, since in one case (when the point of intersection being at infinity becomes "imaginary"[70]) the problem is not properly soluble (the angle in B, between CB and AB, when B is at infinity is not a "real" angle[71])—and not in the other (the introduction of an infinitely small quantity as marking the coincidence between the two segments CA and CB corresponds to a null angle in B).[72] Notice that both cases are ranked amongst the *impossibilis per accidens*. This is something Leibniz often emphasizes: the recourse to fictions allows one to deal on a par with cases in which a problem is possible and cases in which it is not.[73] This is one of the reasons why it makes no sense to equate the use of "fiction" with a fixed kind of entities (or a fixed kind of impossibility).

The place where Leibniz encountered this variability of interpretation quite early was with roots of cubic equations when applying Cardano's formula.[74] Leibniz first thought,

---

[69] The development was introduced by the following comment: "How they in fact indicate that the question is ill-formed will become apparent in examples *where the imaginaries vanish once the required change in the data is made.*" (GM VII, 73)

[70] *Discrimen tamen inter haec duo imaginaria, quod uno modo problema solubile, altero non* (A VI, 4, 521).

[71] *hoc tamen commune, quod saltem per angulum solubile non est, parallelismus enim revera non est angulus*. Further on in the text, Leibniz calls it a "quasi angle" (*quasi angulus*), A VI, 4, 521.

[72] The case of parallel line intersecting at an imaginary point is taken up again in the *Cum prodiisset* as an example of application of the Law of continuity (see text [**50**]).

[73] See, for example, Leibniz (2018), p. 154.

[74] We would like to thank Shinji Ikeda, Osvaldo Ottaviani and Siegmund Probst for many useful insights during discussions on the nature of imaginary roots in the early Leibniz.

Fig. 3.1 From *Elementa nova matheseos universalis* (A VI, 4, 521)

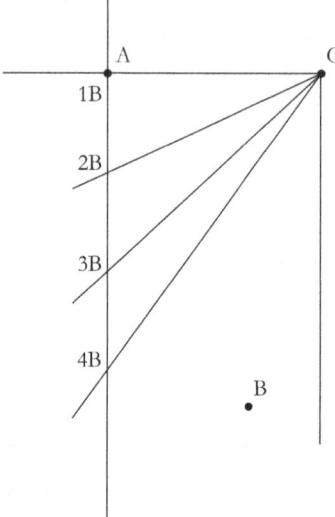

like many of his contemporaries, that the appearance of imaginary roots in the expression was an oddity which could be fixed, since in the so-called "irreducible case" (*casus irreducibilis*), it was known that the three solutions were real. But by trying to eliminate imaginary expressions, he progressively realized not only that it was not possible and that Cardano's formulae were truly general, but that it was not problematic since it could sometimes be shown, after some manipulation, that the imaginary expressions could cancel out (or, in Leibniz' terminology, be "virtually destroyed").[75]

This was a first step in which Leibniz realized, as he will put it later, that the expressions involving square roots of negative quantities are not contradictory in and of themselves since one can devise formulae involving them which are shown (by other means) to denote real quantities. His favourite example was: $\sqrt{(1+\sqrt{-3})} + \sqrt{(1-\sqrt{-3})}$ which denotes the real number $\sqrt{6}$.[76] In this case, Leibniz will say that imaginary quantities are only *in speciem* (or that they are "impossible *in speciem*"),[77] an expression difficult to translate since it means at the same time "in appearance" (but, as we saw, Leibniz also uses *apparens* in a slightly more general context)[78] and "in the symbolic expression" (*species*

---

[75] See Ottaviani (2021). Accordingly, Leibniz proposed to introduce a "sixth operation" consisting in reducing imaginary roots to real ones (see A VII 2, N. 51; A III 1, N. 96).

[76] A III, 1, 625-626 (Spring-Summer 1675). On the way in which such an equality can be proved, See A VII, 2, 682-683 (October 1675).

[77] "But although the reduction of imaginary quantities *in speciem* to real ones cannot always be rendered by a certain palpable expression, it nonetheless remains that these imaginaries *in speciem* must be held in truth to be real" (A VII, 2, 684).

[78] That this is part of the meaning is made clear from the fact that Leibniz translates *impossibilis in speciem* by "impossible en apparence" in French (A VII, 2, 797; 805).

3.3 Fictions and the Question of Impossibility

was the word used by Vieta to designate algebraic symbols, hence the name of widespread use in seventeenth century: *algebra speciosa*). If an imaginary expression can denote a real quantity, when entering in some regulated procedure, it shows already, as Leibniz objected to Prestet, that imaginary quantities cannot be said to be contradictory without qualification.[79]

Two things are, however, very important to note here: first, imaginary *in speciem* is a *kind* of imaginary which is *contrasted* with other appearances of imaginary roots where the impossibility cannot be eliminated, and which are hence said to be *truly* impossible.[80] This is crucial to note because it is common to find amongst commentators general talk about imaginaries in Leibniz which deals in fact only with the "*in speciem*" situation, but do so as if it applied to any kind of imaginary roots (or worse, of imaginary *numbers*, as if Leibniz had the idea of a field of numbers of that sort). When Leibniz refers to the fact that imaginary roots can have a *fundamentum in re*, for example, he does not speak of imaginary roots in general, but precisely of those which enter into the designation of real quantities.[81] Since this is the precise context to which he often refers in parallel with the infinitely small,[82] the comparison supports the view that the latter have a *fundamentum in re* because they can designate ordinary quantities. Second, it is important to note that these symbolic expressions raised particular difficulty, according to Leibniz, because they could not be translated into a geometrical representation, as was the case with ordinary algebraic manipulations after Descartes.[83] Accordingly, it is no surprise to read at the end of the text from October 1675: "Although, to go further, one must employ geometry to show the reality of the imaginary quantities *in speciem*, and a fortiori for the surd quantities, *in order not to hold them as empty fictions of the human mind*" (our emphasis).[84] We find here again the same need to check that a "fiction" is not "empty" by providing an interpretation in which it can express (possibly in an indirect way) a "real" entity. So it was important for Leibniz to check that the *in speciem imaginariae* also had a geometrical interpretation, in order to insure that they were not "empty fictions" (*inania figmenta*).

---

[79] A VII, 2, 795.

[80] "véritablement impossibles" (A VII, 2, 796) or "impossibles en effet" (A VII, 2, 805).

[81] A VI 4, 520: *Et has quantitates appello* **in speciem impossibiles***; cum reapse sint reales* ("and these quantities, I call them impossible *in speciem*, because they are in truth real").

[82] See the Letter to Varignon from 2 Feb. 1702 (where the example is the same as the one mentioned to Huygens in 1675), A III, 9, 14.

[83] A III, 1, 625: "From this it is clear, by the way, that there is something in the analysis that geometry cannot relate to in the same terms and the construction of this value: $\sqrt{(1+\sqrt{-3})} + \sqrt{(1-\sqrt{-3})}$ could not be constructed through geometry unless a previous analysis would have provided another value or another different relation".

[84] *Tametsi ultro largiar geometriam ad ostendandam realitatem quantitatum in speciem imaginarium, imo et surdarum adhiberi debere, ne scilicet pro figmentis inanibus humanae mentis habeantur* (A VII, 2, 687)

A geometrical interpretation of imaginary entities is something upon which Leibniz reflected as early as 1673.[85] He comes back to the question in the autumn of 1675, in particular in a text entitled "*Imaginariae usus ad comparationem circuli et hyperbola* ["The use of an imaginary for comparing a circle with a hyperbola]" which, as the title indicates, studies the link between the circle and an "imaginary hyperbola" derived from it.[86] Interestingly enough, he mentions on this occasion that this is a typical case of use of "fictitious" entities, as compared with incommensurable magnitudes, which can be useful in order to know real quantities (in that case the circle). More importantly, he then relates this use to what he calls elsewhere "blind thought":

> I am bringing these things up so that I can show with an example how we often reason about fictions on the example of true things. But whether the quantity $\sqrt{-1}$ is nothing at all, or whether it contains in truth I don't know what, should be discussed with great care. For although it cannot be carried out, it can be understood in a way, not in itself, but by means of characters and analogy, an example of the kind of thought I call 'blind'. And just as there are incommensurables which are commensurable in power, so there are imaginaries whose powers are real; that is to say, impossible ones whose squares are possible, such as $\sqrt{-1}$, whose square is -1, even if it is assumed that there is absolutely nothing in nature corresponding to such a quantity. It is sufficient, however, that its character be useful, because, when joined with others, it expresses real things.[87]

Note that the justification of the use of fictions here is that it acquires a non-contradictory meaning only when entering into some relations.[88] This relational aspect of fictions is something we will encounter on several occasions in the study of infinitely small quantities, and it plays a crucial role in understanding what Leibniz has in mind when talking of a "syncategorematic" interpretation of the infinite.

---

[85] A VII, 4, 424. We thank Siegmund Probst for having drawn our attention to this early text in which Leibniz talks about imaginary figures (*imaginariae figurae*) and in particular an "imaginary hyperbola".

[86] A VII, 7, 561 (November 1675). See also, the text *Curvarum dimensio evolutio expansio* (A VII, 5, 372), dated October 8th, 1675.

[87] "*Haec ideo adduco ut exemplo ostendam quomodo saepe de fictitiis ratiocinemur exemplo verorum. An autem quantitas $\sqrt{-1}$. sit omnino nihil, an vero contineat nescio quid discutiendum diligentius. Etsi enim non possit effici, potest tamen quodammodo intelligi, non seipsa sed ope characteris et analogiae, exemplo illius cogitationis quam caecam appello. Et vero quemadmodum sunt incommensurabiles quae sunt potentia commensurabiles, ita sunt imaginariae quarum potentiae sunt reales; seu impossibiles quarum quadrata sunt possibilia, ut $\sqrt{-1}$, cujus quadratum -1 etsi autem ponatur nihil omnino esse in natura tali quantitati respondens; sufficit tamen ejus characterem esse utilem, quia cum aliis junctus realia exprimit.*" (A VII, 7, 560. Corrected.)

[88] What Leibniz has in mind is the most simple example: the expression "$(\sqrt{-1})^2$" denotes a "real" quantity (in the sense of geometrically interpretable, like $\sqrt{2}$ is). Notice that mathematical signs indicating relations between quantities are called by Leibniz in later texts "syncategorematic" (Leibniz 2018, pp. 129–130).

3.3 Fictions and the Question of Impossibility

Fig. 3.2 The square root of $-x^2$ as a perpendicular (A VII, 2, 745)

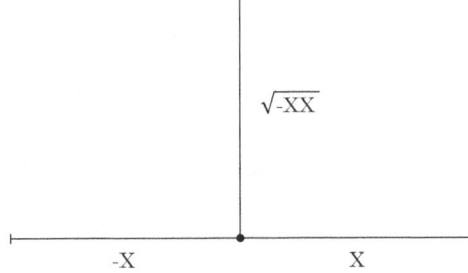

Leibniz develops the geometric interpretation of imaginary roots in a series of papers written under the influence of his encounter with Tschirnhaus in October 1675 (and possibly resulting from insight the latter got from his discussion with Wallis in London).[89] Indeed, Tschirnhaus realized that it was possible to give a consistent interpretation of $\sqrt{-x^2}$ by representing it as a mean proportional between x and $-x$, i.e. a perpendicular in the circle centred on zero with x as radius (See Fig. 3.2. A VII, 2, 745).[90] The conclusion Leibniz derives from this study is that in this case the impossibility does not come directly from the notion in and of itself, but from some conditions related to the laws of the *situs*.[91]

Just after the diagram, Leibniz has written the following comment:

> Two origins of impossibility, one from the nature of magnitude, e.g. when there is 1 ⊓ 3. Another from the nature of situation, when 1 and √−1 have to be combined into one quantity, in which there is a contradiction, because they are not taken in one line. (A VII, 2, 745).[92]

As can be seen, it is certainly not a question of accepting imaginary entities on the ground that they behave nicely in algebraic manipulations, as claimed by the "formalist" interpretation of imaginary entities as "ideal elements". On the contrary, the whole problem is, according to Leibniz, that it is not possible to define an addition between real and imaginary

---

[89] See A VII, 5, 372: *Haec a Tschirnhausio observata*.

[90] As indicated in A VII, 5, 372, this can be generalized to unequal segments, giving rise to the various positions in what we now call the unit circle.

[91] That *situs* provides some reality to symbolic expression is something Leibniz also emphasizes in his commentary on Prestet (A VII, 2, 804). This is in accordance with the way in which a mathematical entity can be said real in Leibniz's philosophy, see (Rabouin 2022).

[92] See also A VII, 2, 753: "*Ecce contradictionem non ex magnitudine sed situ ortam* [Here is a contradiction arising not from magnitude, but from situation". Interestingly enough, Leibniz explains in this passage that the problem of the addition of real and imaginary quantity is "ill posed" (*male formatus*)—the very same expression he uses in his later text about fictions. The reason he gives is that we are looking for a segment in the continuation of the one given, although it should be perpendicular to it.

quantities which would be consistent with the existing geometrical interpretation as the joining of two segments. This insight will be instrumental in his philosophical conceptions of impossibility, as attested by the following passage from the same period:

> The origin of impossibility is two-fold: one from essence, the other from existence or from position [*ab existentia seu positione*]. In the same way, there is a two-fold reason for impossible problems: one, when they are analysed into a contradictory equation, and the other, when there is an analysis into an imaginary quantity, for which no situation (*situs*) can be understood. This is an excellent image of those things which neither have been, nor are, nor will be. (*De mente, de universo, de Deo*, A VI 3, 464; transl. Ottaviani, modified)

The last sentence is an allusion to a comparison which Leibniz will make in some of his metaphysical texts in order to illustrate the fact that some entities may well be non-existing in our world without being contradictory—a claim central to his considered metaphysical views on modalities. In this last case, the entity may be called impossible *per accidens*.[93] In a text contemporary to the *Elementa nova*, he even uses the same example (the distinction between pure contradiction and an "impossible" intersection between a circle and a line) to illustrate this point (*De libertate et necessitate*, 1680–84, A VI, 4, 1448).

Since in *Elementa nova* Leibniz also opposes absolute contradiction with impossibility by accident, it is tempting to conflate the two contexts and claim that both in mathematics and in metaphysics the second type of impossibility does not entail contradiction. Accordingly, what would be at stake here would be an intermediary state of existence in mathematics, comparable with what happens for *possibilia* which are not realized in our world.[94] Yet, a too literal interpretation of these passages spawns a lot of difficulties. The most obvious one is that it blurs the very distinction which Leibniz wants to make, i.e. that between pure *possibilia*, such as mathematical objects (characterized by non-contradiction) and worldly existence (characterized by compossibility).[95] If all that is meant by "impossible by accident" is that these entities are not possible "in our world" (whatever interpretation we give to this expression), then the "solution" they provide should be perfectly acceptable when one stays inside of mathematics. There should be no particular problem, for example, in having an imaginary intersection between a circle and a point even when the moving line exceeds the size of the radius. But this not at all what Leibniz says.[96] The

---

[93] See Ottaviani (2021). Note that the expression "impossible par accident" is used in *the Elementa nova matheseos universalis* (A VI 4, 521; Leibniz 2018, p. 108).

[94] This is the position defended by Esquisabel and Raffo Quintana (2021).

[95] As Leibniz nicely puts it to Bernoulli: "*Possibilia sunt quae non implicant contradictionem; Actualia nihil aliud sunt quam possibilium (omnibus comparatis) optima. Itaque quae minus perfecta ideo non sunt impossibilia*" (A III, 8, 64).

[96] Esquisabel and Raffo Quintana (2021) try to overcome this difficulty by interpreting the case of incompossibility in mathematics as related to the absence of a geometrical representation (in the sense of a diagrammatic representation). But the problem is that Leibniz and Tschirnhaus precisely found a diagrammatic representation of imaginary roots and that this is what establishes their

3.3 Fictions and the Question of Impossibility

conclusion of the passage where he makes the comparison with the infinitely small, makes it clear that what he has in mind in both cases is the kind of "diorism" situation we already indicated:

> And yet imaginaries of this kind have a remarkable use, not only in Conics but also everywhere else, for the discovery of universal constructions; which is so true that it often happens that calculation necessarily leads us to them, at which point those who still lack experience in these things marvellously torment themselves and think they have stumbled upon some absurdity. *Those who understand them, however, are well aware that this apparent impossibility only means that instead of the straight line making the angle sought, it is drawn parallel. (Elementa nova matheseos universalis.*[97] (A VI, 4, 521/Leibniz 2018, pp. 108–09; our emphasis)

But there is a much greater difficulty involved in trying to incorporate "impossible by accident", taken in the metaphysical sense of *incompossibilia*, directly into the realm of mathematical *possibilia*. Indeed, in mathematics, the very reason why they appear as "impossible" is because, in the surface language they entail a conceptual incompatibility. Since Leibniz considers any mathematical truth to be reducible to a definition and an identical axiom, this superficial impossibility thus amounts to the positing of something which is incompatible with what is contained in the definitions. We saw that this is the very origin of the vocabulary of "fictions" in mathematics. For example, the very definition of a number in the Euclidean tradition prevents the existence of negative, rational and surd "numbers". If we want to overcome the "apparent impossibility" without relying on the "diorism" strategy, we hence have to provide an *alternative* definition of number or magnitude allowing us to overcome the apparent incompatibility. This definition would have to be general enough for it to make sense to talk about imaginary and/or infinitely small numbers and/or magnitudes. But although Leibniz certainly provided an extended definition of number applicable to rational, surd and even transcendental numbers, there is no such definition to be found for negative, imaginary or infinitesimal ones. We recount numerous texts from Leibniz in which he defines what a "number" or a "magnitude" is and they are simply *incompatible* with the various "fictitious" entities encountered so far.[98] The

---

non-contradictory character to their eyes, when we take them in isolation (in modern terms: there is a model). According to Leibniz, the problem of impossibility does not derive from the representation in and of itself, but from the fact that this representation is incompatible with the ordinary representation of addition amongst magnitudes. At the end of the day, as he expresses it very explicitly, it remains a kind of contradiction (with what he considers as being the laws of the *situs*—these laws being geometrical, and certainly not physical or related to the diagrammatic representation in and of itself).

[97]"*Et tamen hujusmodi imaginaria egregium usum habent tum in Conicis, tum alibi passim; ad constructiones universales inveniendas; quod adeo verum est ut saepe calculus ad eas necessario ducat, ubi harum rerum nondum satis periti mire torquentur et in absurdum se incidisse putant. Intelligentes vero sciunt apparentem illam impossibilitatem tantum significare, ut loco rectae angulum quaesitum facientis ducatur parallela; hunc parallelismum esse angulum illum seu quasi angulum quaesitum.*"

[98]We detail this view in Sect. 5.2.

burden of the proof is here on the shoulders of the supporters of the heterodox view of mathematical modalities: they should provide texts in which Leibniz formulates the alternative definition of number or magnitude they have in view.[99]

Fortunately, there is a much more straightforward interpretation: in mathematics, the distinction amongst impossible entities is not between contradictory notions and those which would not entail a contradiction, as is the case in metaphysics (this is why it is presented as a comparison or "an image"); it depends on whether the contradiction is immediate or derived (from the conditions of the problems, typically, in the case of imaginaries, conditions stated in terms of the geometry of the *situs*). In this last case, we can have various resolutions and this is why the impossibility is only "apparent": the problem might well become possible provided one can change the data or under a suitable a reinterpretation of the data which does not amount to changing them (what Leibniz calls *impossibilis in speciem*). Without this distinction, Leibniz' position would rapidly become inconsistent, since he can sometimes state that $\sqrt{-1}$ is not a contradictory notion,[100] whereas he can explain in other places that there is a contradiction entailed in the very the handling of $\sqrt{-1}$.[101] The situation is in fact the same with negative numbers, as Leibniz explains to Mariotte: they do not entail a contradiction in and of themselves, since one can manipulate them in problems in which all the quantities at stake at the end of the process are real. Yet, because of paradoxes such as the one related to the ratio between -1 and 1, a negative taken in isolation should be considered as a *purely imaginary* quantity.[102] They do not behave in a way consistent with the manipulation of ratios between ordinary quantities.

---

[99] Contrary to what is repeated in a number of papers by M. Katz and various co-authors, when Leibniz characterizes "incomparable magnitudes" as not satisfying the Archimedean axiom this cannot be taken as a definition of an alternative kind of magnitude. It describes a certain relation between magnitudes, which holds already in Euclidean geometry. For example, a segment and a surface are incomparable in this very sense. In order for this relation to characterize a particular type of magnitude, one would have to provide a general definition of magnitude in which could enter at the same time "ordinary" and "extraordinary" magnitudes—a definition which is never provided by Katz and al. and of which we can find no trace in Leibniz. In Sect. 5.2, we show that such an extended definition of magnitude would be incompatible with Leibniz's characterization.

[100] A VII 5, 372: *Itaque non est contradictio in quantitate $\sqrt{-1}$ [...]. Vera contradictio est v.g. aequatio inter 1 et 2.*

[101] See the text quoted above in note 92. In the same vein, it is interesting to see that Leibniz finds Prestet's argument inconclusive (From the fact that imaginary roots are neither positive nor negative, Prestet inferred that they are contradictory, cf. Prestet (1675, p. 355), although he uses the exact same argument in the *Specimen Geometriae Luciferae* to argue that they are impossible (GM VII, 272). This makes perfect sense if one realizes that the impossibility stems not from the notion, as was stated by Prestet, but from the geometrical constraints (as explained in the *SGL*).

[102] A III, 2, 776: "*je vous reponds que -1 ou -2 n'est rien en effect, a moins qu'on ne le prenne comme un nombre qu'on doit oster de quelque autre, de la maniere dont je me suis servi pour l'expliquer cy dessus: et cela estant: ce cas: (-1 est à $+1$ etc.) est une chose qui en elle meme et a part ne peut et ne doit jamais arriver. Car -1 a. part est une chose purement imaginaire*".

3.3 Fictions and the Question of Impossibility

All of this becomes straightforward as soon as one makes the distinction between absolute and conditional contradiction.

This is a crucial remark, since the same is true about "infinitely small" quantities. In some texts, Leibniz will insist that such a notion implies a contradiction.[103] But in other texts, Leibniz will explain that one can manipulate these entities without contradiction. As he puts it to Bernoulli, just after recalling that there cannot be an infinitieth term in a series, since the number of all numbers implies contradiction, "meanwhile, I confess that the maximum is something different from the infinite and the minimum from the infinitely small. From this [the proof of the impossibility of a maximum number], one cannot refute the possibility of our infinitely small".[104] Now, we should not conclude from this last sentence that the infinitely small are thus existing entities from a mathematical point of view. Neither should we conclude that they are impossible without qualification. Indeed, just after this passage, Leibniz immediately adds the following caveat: "when I said that if I thought the infinitely small to be possible, I would have to concede that they exist, I did not go as far as saying that they were impossible, but I left things in the middle".[105]

This is a very nice place to end our first journey in company of "fictions", since this is a claim which makes absolutely no sense if the "infinitely small" (or any other fiction) is a fixed entity. An entity is either possible (and hence a real mathematical being) or impossible (and hence an empty fiction). But, as we saw on many occasions in the preceding development, fictions are not entities but linguistic expressions, the possibility of which depends on the interpretation they receive. As recalled in the *Elementa nova matheseos universalis*, an imaginary expression can sometimes refer to a real entity, sometimes not, and we saw that Leibniz drew a direct parallel with infinitely small and large quantities by distinguishing these two cases using a simple example.

Let us recapitulate some of the results obtained so far. Very early on, Leibniz realized that one should be careful in the handling of infinite quantities in mathematics, because they entail contradiction. Yet one could use them as "fictions", on the model of a very widespread strategy employed at the time with a variety of mathematical expressions such as negative numbers, surd numbers, geometrical dimensions greater than three, logarithmic elements, "indivisibles" or imaginary roots. Indeed, a fiction is something which does not have to be taken literally and hence does not commit its user to the possibility of what it designates in the surface language. Moreover, it allows one sometimes to neutralize the apparent impossibility provided we find a way to interpret the fiction in a non-literal manner and provided that this interpretation can be couched in a demonstrative setting. At

---

[103] "It would be a mistake to try to suppose an absolute space which is an infinite whole made up of parts. There is no such thing: *it is a notion which implies a contradiction*; and these infinite wholes, *and their opposites the infinitesimals*, have no place except in geometrical calculations, just like the imaginary roots in algebra." (*New Essays*, A VI, 6, 158, text [**34**]; our emphasis).

[104] *Interim fateor cum aliud sit maximum ab infinito, et minimum ab infinite parvo; non hinc statim refutari possibilitatem nostrorum infinite parvorum* (A III, 7, 884).

[105] We come back to the interpretation of this intriguing passage in Sect. 5.3 below.

that point, we did not enter much into this last aspect (what kind of proof can support the use of infinitesimals as fictions in mathematics?). We saw in Chap. 2 that Leibniz devised a general Difference Principle, which he applied to finite and infinite differences, in association with an interpretation of the "last term" of series in terms of variable quantities. This led in 1676 to a rigorous definition of what the "sum" of an infinite series consists in. But it still remains for us to show how to insert this general strategy into proofs, especially when no numerical series are at play. Leibniz worked on this issue in the context of his quadrature of the circle and intended in 1676 to provide what he presented as "the most solid foundations for the Method of indivisibles" (*fundamenta totius Methodi indivisibilium firmissime jacienda*).[106] An attempt to describe this approach will be the object of the next chapter devoted to the *De Quadratura arithmetica circuli, ellipseos et hyperbola*.

But before undertaking this study, it should be noted that the talking in terms of literal/ nonliteral meaning and the emphasis on the linguistic nature of "fiction" is not an observer's category here. In the *Observatio quod rationes* (text [41]), Leibniz is very clear about the fact that formulae involving negative and imaginary quantities are only "true within a tolerance" (an expression he borrows from Jungius and uses again in the letter to Wolff (text [44])). He compares this situation with that of Euclid talking of the angle of contact being "smaller" than any rectilinear one and concludes that many of these modes of speaking (*genus loquendi*) used in geometry contain "something figurative and cryptic". He then continues by mentioning infinite and infinitely small numbers and lines— and here it should be kept in mind not only that this text was published in the *Acta eruditorum*, but that it was supposed to deliver, as indicated by its title the "true meaning of the infinitesimal method" (*De vero sensu methodi infinitesimalis*). Later in the text, he explains that the infinite can never be a quantity, but that when we take it as such, it is just a "way of speaking" (*modus loquendi*), which does not hold properly (*proprie*), and that we attribute a quantity to the infinite "by analogy". When coming back to the "truth within a tolerance" at the end of the text, he also makes it clear that it is not rigorous and cannot work without an explanation (*explicatio*).

In this connection, it is worth taking note of an interesting unpublished text in which Leibniz discusses an apparent "amazing exception" to his Law of Continuity, the *Regula de Transitu per saltum non admittendo* (text [43]). The example concerns taking powers of 0. For all $x < a$, $0^{a-x} = 0$, but for $x = a$, $0^{a-x} = 0^0 = 1$. So as x goes from indefinitely close to a to a itself, the value of the function 0^{a-x} leaps from 0 to 1, skipping all the values between. There is then an immediate leap in the value of 0^{a-x} from 1 to infinity if x continues to increase beyond a, since "$0^{a-x} = 0^{-y} = \frac{1}{0^y} = \frac{1}{0} = $ infinity." Thus "Nothing, Unity, and Infinity will follow one another immediately, with nothing between," in flagrant violation of the Law of Continuity, if it is taken to apply to variations in any mathematical expression without qualification.

---

[106] A VII, 6, 521.

3.3 Fictions and the Question of Impossibility

This counterinstance arises, so Leibniz tells us, because we are using fictitious expressions, which are "true within a tolerance". But this case contrasts with that of the asymptote of a hyperbola, where an infinite ordinate occurs. For in that case there is an accompanying *explicatio*: the ordinate may be understood as a finite line that "grows to a width greater than any given" by passing through "all straight lines smaller than an infinite one", and if the infinitely small distance from the axis is similarly understood as a variable finite distance shrinking proportionally to a length smaller than any given, calculations can be performed. We are not treating infinity itself as a quantity. Likewise, as he explains to Jenisch, "when we calculate by our method, we ought not to use nothings, but infinitely small quantities" (text [**48**]). There he gives the following example: if in the differential equation $adx - dxdx = dydy$, instead of the variable differentials dx and dy, one were to take, $dx = 0$ and $dy = \underline{0}$, that is, "nothings" then we would have "a0 − 00 = $\underline{0}.\underline{0}$, that is, a.0. = $\underline{0}.\underline{0}$., that is to say $\underline{0}$ would be the mean proportional between a and 0". He concludes

> But it is absurd to talk in this way about nothings, and so instead of nothings, infinitely small quantities should be used, and whatever is said concerning these, will always be verified by a rigorous method, by means of the infinitely small, that is, smaller than any given.

This suggests that the problem with the counterexamples provided by the calculation with powers of zero is that it is not permissible to treat 0^{a-x} as if "nothing" is a quantity, something varying with x. In the absence of an accompanying explanation in terms of finite quantities, it remains a mere fiction.

On many occasions, Leibniz will refer to the framework of infinitesimals as "a way of speaking" (*modus loquendi*) or "abbreviation of speech" (*compendia loquendi*).[107] Defending his Differential Calculus in 1701, he will write that in it "one does not differ from Archimedes' style *but for the expressions* which in our method are more direct and more in accordance with the art of discovery."[108] More importantly, he emphasizes the "figurative" nature of these modes in other places: besides the *Observatio*, we should mention here the letter to Wolff in which the law of continuity according to which we can consider vanishing quantities, although they entail "something fictitious and imaginary" is presented as a "Philosophical-rhetorical Figure". In a letter to Bodenhausen, Leibniz compares the way one deals with some letters occurring in computation sometimes as finite and sometimes as infinitely small quantities with the reading of the Holy scripture,

---

[107] As early as the *DQA* (A VII, 6, 585). Amongst other places, see H&O, 43; GP II, 305, GM V, 389.

[108] In one of the versions of the Letter to Gallois from October 1705 (not included in our translation, text [**40**]), Leibniz intends to exemplify his views by a figure taken from Galileo in which one can illustrate the velocity and the acceleration on a single diagram. He then concludes: "All the extraordinary magnitudes (in appearance) are therefore similar to those that Galilei was allowed to use; it is obvious that all this infinitesimal Analysis consists only in the calculation adapted to these continuous changes of magnitudes, to pass as is necessary from the estimation of the magnitudes to that of their infinitesimal increases or differences (…). *We also see that the innovations were not made in the things, but in the abbreviations of the expressions*" (our translation).

which can receive a literal and a "figurative" interpretation.[109] The linguistic nature of the use of infinitesimal as fiction also appears in the way Leibniz emphasizes regularly how we handle them. In the *Observatio*, he states that when we talk (*dicimus*) about an infinitely small error, what we understand (*intelligi*) is a magnitude which can be rendered smaller than any given one. In the exchange with Jenisch (text **[48]**), he even claims:

> Whenever we show with the help of the infinitesimal calculus, or method of differences, two quantities to be equal to each other, then we *tacitly* apply a rigorous method, as if *we were to say*: Let there be a difference between them, from which very thing we will prove that there is none, ôr that no error can be assigned.

The tacit understanding of the infinitesimal techniques rests heavily on the fact that we substitute the meaning of the terms at play, an operation on which Leibniz will put a lot of emphasis in his description. Moreover, it supposes that we are always able to support the "ways of speaking" by rigorous proof, as indicated at the beginning of text **[39]** (*Quaestio de jure negligendi*):

> one needs to allow for pure mathematics to be preserved intact from metaphysical controversies. We can do this if, caring not at all whether the infinite and infinitely small in quantities, numbers and lines are real, we use the infinite and infinitely small as apt expressions for condensing thought. Thus it is granted that they are imaginary. These quantities can nevertheless be employed like the imaginary roots in algebra.
>
> Hence it follows that this abbreviation is never to be used without an explanation of what is substituted, by which the matter reduces to a rigorous demonstration in the style of Euclid or Archimedes.[110]

All of these elements played a crucial role in Leibniz' uses of infinitesimals as fictions and will be expounded and explained in the following sections.

---

[109] A III, 4, 524. On the fact that this makes Leibniz's version of fictionalism very close to what Yablo has coined "figuralism", see Rabouin (2022).

[110] See also in a letter to Varignon from 1713: "But one must not rely on reasonings about infinite series, unless one can demonstrate their truth with finite quantities by the methods of Archimedes. [*Mais il ne faut se fier aux raisonnemens sur les series infinies, que lorsqu'on en peut demontrer la vérité par les finis à la façons d'Archimède*]." (GM IV 191).

The De Quadratura Arithmetica (DQA)

4.1 Quadratures as Riemannian Sums

Leibniz was a quick learner, and the alacrity with which he absorbed and improved upon the methods of his contemporaries is often remarkable. An obvious case in point is his mastery of the method of indivisibles, due initially to Cavalieri. On his arrival in Paris in the Fall of 1672, his knowledge of this method seems to have been gleaned almost entirely from his reading of Hobbes.[1] But under Huygens' felicitous guidance he immediately began a study of the mature version of the method of indivisibles that he found in Pascal's *Lettres de Dettonville*, as well as John Wallis's pioneering arithmetical methods in his *Arithmetica infinitorum*.[2] We have already seen in Chap. 2 how in a little over a year he was able to combine these methods to derive his infinite series expression for the quadrature of the hyperbola. In his derivation he employed Wallis's idea that an infinite series is completed, suppositionally, by the inclusion of an ultimate term or *terminatio*, his own results on the summing of infinite series by the Difference Principle, and also Pascal's notion of the infinitesimal as a difference of successive values of a variable standing as a "unit", combining these ideas with the infinite series expansion of $1/(1-x)$ that he had learned from Wallis and Mercator. By the end of his stay in Paris he was already contributing original results in almost all the areas of higher mathematics of his time. In

---

[1] Hobbes's idea that a line is a surface whose depth is "smaller than considered" in the reasoning in which it occurs, a point a line whose length is smaller than considered, etc., is not without impact on Leibniz's later idea of incomparables. See Jesseph (2008) for an enlightening analysis.

[2] Also evident is what he learnt in Paris from his studies of Mercator and Grégoire de Saint Vincent, and later from the controversy between Huygens and James Gregory.

particular, he was able, using similar techniques, to derive his famous series for the quadrature of the circle.[3]

As a culmination of his studies of quadrature, during his last full year in Paris (1675–1676), Leibniz composed a substantial treatise, the *De Quadratura Arithmetica*, and left it in Paris with his friend Soubry in the hope of its being published there, as part of his efforts to gain admittance to the *Académie Royale des Sciences*. After Soubry's untimely death and the subsequent loss of this fair copy, and the failure of his later efforts to publish it, the original lay undisturbed in the library in Hanover, apart from Scholz's publishing of a fragment of it in 1934.[4] It is not to be found, for instance, in Carl Gerhardt's 7-volume edition of Leibniz's mathematical writings,[5] and although the historian of mathematics J. E. Hofmann recognized its importance, it did not otherwise feature in discussions of the foundations of the Leibnizian calculus, such as the excellent two-part analysis given by Henk Bos (1975–1976). It had therefore virtually disappeared from view until its publication in the 1990s, in a critical edition with commentary by Eberhard Knobloch (Leibniz 1993). The treatise is notable for the derivation of Leibniz's famous quadrature of the circle through the infinite series expansion of the ratio of the area of the circle to the square on the diameter (which we would write as $\pi/4$) as $\frac{1}{1} - \frac{1}{3} + \frac{1}{5} - \frac{1}{7} + \frac{1}{9} - \frac{1}{11}$ etc. to 1, but also contained several other seminal innovations of Leibniz's own invention.

Not the least of these innovations, as Knobloch pointed out, was the theorem derived as Proposition 6, which consisted in a methodical approach to quadrature from first principles that was tantamount to providing a rigorous foundation (given the standards of rigour for the time) for what is now termed Riemannian integration. Leibniz describes that proposition in his *Index notabiliorum* as a "very thorny" [*spinosissima*] one,

> in which it is demonstrated in fastidious detail how certain step-spaces, and likewise certain polygonal spaces, can be increased continuously, to the point where they differ from each other or from curves by a quantity smaller than any given, which is something that is most often [simply] assumed by other authors. Although one can skip over it at first reading, it serves to lay the foundations for the whole Method of Indivisibles in the soundest possible way. (Leibniz 1993, p. 24).

As has since been shown in articles by Knobloch (2002), Arthur (2008, 2013), Levey (2008) and Rabouin (2011, 2015), the nub of the proof is an exploitation of the Archimedean property to prove that quantities whose difference can be reduced to a

---

[3] See Siegmund Probst (2006).

[4] See Knobloch's *Einleitung* of (Leibniz 1993) for details.

[5] Gerhardt did, however, publish both a letter in French, presumably to the editor of the *Journal des sçavans*, seeking the publication of the *DQA* (GM V, 88–92), and a Latin Preface to the work, *Praefatio opusculi de quadratura circuli arithmetica* (GM V, 93–98), as well as the *Compendium Quadraturae Arithmeticae* (GM V, 999–112; text [**17**]), all of which he had found among Leibniz's manuscripts in Hanover.

4.1 Quadratures as Riemannian Sums

quantity smaller than any given quantity are equal. Leibniz did not claim originality for the latter principle, which we have styled the Principle of Unassignable Difference, or *PUD*, as explained in the introduction above. Versions of it can be found in the work of Torricelli, Pascal and Wallis, and even in the work in analysis that Newton was engaged in contemporarily with the *DQA*.[6] But the particular way in which Leibniz put it to use was certainly innovative and powerful, and it is possible to regard it as an early version of his Principle of Continuity.[7]

Now, in order to appreciate the originality of Leibniz's contribution to quadrature, we should compare it with the methods for deriving quadratures extant at that time. Of these there were two main ones, the traditional method of exhaustion and the modern proofs by indivisibles. The first method, that of Archimedes, relied on inscribed and circumscribed polygons and concluded by a double *reductio ad absurdum* that the area could not be greater or smaller than the intended result without contradiction. The second worked with infinite collections[8] and concluded by transferring some ratios holding for the finite case to the infinite case.

Both entailed well known limitations: the first method presupposed the construction of inscribed and circumscribed polygons to the curve; it also presupposed that one knew in advance the result that was to be obtained (and, because of the use of *reductio*, it had no heuristic or explanatory power); the second one presupposed that an area can be identified with an infinite collection (typically "all the lines" under the curves), and that some ratio remains when passing from the finite to the infinite, a fact that remained in need of a rigorous justification[9]; finally, even when people understood "all the lines" as a convenient way of talking about small rectangles under the curve, they usually presupposed that their widths were equal.[10]

Leibniz's first major innovation in the *DQA* is to propose the basic technique of "transmutation" (*transmutatio*) of a given curve into another one (known as the Quadratrix). One first considers the lengths A_1T, A_2T intercepted on the ordinate axis

---

[6] For a comparison of Leibniz's use of the principle with Newton's, see Arthur (2008).

[7] This is how Knobloch regards it in his (2002) article.

[8] "The Method of Indivisibles" is usually attributed to Bonaventura Cavalieri (1635), although the work itself was rare and little known, and what most mathematicians learned of the method was acquired through Torricelli's or Pascal's rather different understandings of "indivisibles". Cavalieri himself proposed two methods, a collective one (depending on his concept of "all the lines") and a distributive one designed to avoid the paradoxes of the infinite consequent on interpreting "all the lines" as composing a plane. See Andersen (1985) for a pellucid exposition.

[9] The best that could be done, and was done by mathematicians such as Pascal and Torricelli, was to show through examples that the same results were obtained either by using indivisibles or by relying on the "method of the Ancients".

[10] This is typically the case in Pascal and Wallis, who both believed "indivisibles" to stand for infinitely small, equal, homogeneous quantities (Pascal is explicit about the fact that one takes "des petites portions égales").

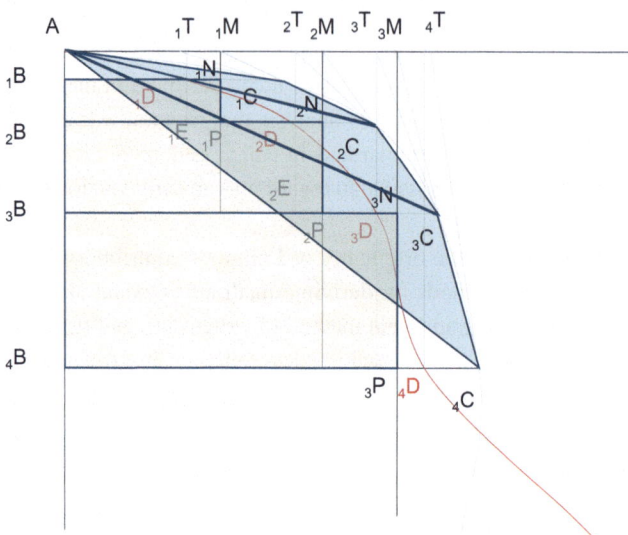

Fig. 4.1 The "transmutation" theorem

(horizontal in the figure) by tangents at points $_1C$, $_2C$, etc. on the curve and reports them as abscissas $_1T_1D$, $_2T_2D$, etc., for the new curve. One then reports in the same way the intersections of the chords passing through the points $_1C$, $_2C$, etc. into points on the ordinates of the *Quadratrix*: $_1N$, $_1P$, $_2N$, $_2P$, etc. This gives birth to a step-space (*spatium gradiformis*) with each of the triangles A_1C_2C, A_2C_3C, etc., being a half of each of the corresponding rectangles $_1B_1N_1P_2B$, etc. (see Fig. 4.1).

Leibniz then has to demonstrate: first, that the areas under the two curves (not taken from the origin) can be estimated, respectively, by the polygon $A_1C_2C_3C\ldots_nCA$ and by the step-space $_1B_1N_1P_2N_2P\ldots_nN_nP_nB$; second, that the relationship given term by term between triangles and rectangles holds for these two areas.

As noted, the two extant techniques available to Leibniz for establishing this were the traditional proof by exhaustion and the modern proof by indivisibles. Leibniz's proof is an original mixture of (and improvement on) both of them. It is "in the manner of the Ancients"[11] in the sense that it does not appeal to infinite collections (or infinitely small quantities, or "indivisibles", etc.). But it shares with the method of indivisibles the advantage that one does not need a method for constructing a polygon under the curve,[12]

---

[11] As we will see below, this qualification was explicit when Leibniz presented a variant of the proof to Bodenhausen in 1690.

[12] One just has to fix a point as origin and take triangles under the first curve from this point, each triangle being transformed by the general construction into a rectangle of which one can estimate the difference with the second curve. This is crucial in the demonstration for two reasons: first, it enables one to give a proof which works in general (there are, of course, some conditions on the curve, the first one being that it has a tangent at every point, but Leibniz makes these constraints explicit, see

4.1 Quadratures as Riemannian Sums

and that one can proceed directly by identifying the area under the *Quadratrix* with a collection of rectangles. To obtain this identification without inscribed and circumscribed figures, Leibniz uses rectangles passing *through* the curve at arbitrary points. Another originality is the systematic use of upper bounds for the estimation of the error, which enables one in particular to have decompositions of the area under the *Quadratrix* into unequal portions. This is the sense in which Leibniz can be regarded as providing a foundation for Riemannian integration.[13]

As Leibniz proceeds to show, the usual methods of indivisibles are special cases of this general construction, and he emphasizes the fact that he has provided in this way a "rigorous" foundation for them.[14] Here by "rigorous" Leibniz clearly means the way in which he has shown how the method of indivisibles—usually used without any justification—could be rephrased, as was usual at the time, "in the manner of the Ancients". However, it should be emphasized that his method is *direct* and does not rely on a *reductio*. It should also be emphasized that it does not rely on "infinitely small" quantities either, even in the language of description. It is based, like the proofs of the Ancients, on finite quantities, and the idea is that one can translate the usual demonstrations using "indivisibles" into ones involving only these finite quantities. The fundamental insight is to work directly on the "difference" (or "error") between the approximating quantity and the area under the curve, taken as a variable quantity, which may be rendered as small as one wishes.[15] This corresponds well to a kind of "reductionist" strategy to which Leibniz will continue to appeal late in his career:

> For in place of the infinite or the infinitely small one takes quantities as great and as small as is necessary in order for the error to be less than the error given; which differs from the style of Archimedes only in the expressions, which are more direct in our method and better adapted to the art of discovery (to Pinsson, 29 August 1701, text [**37**])

This explains why previous commentaries on the significance of the *DQA* have concentrated on Proposition 6, regarding it as the foundation stone for Leibniz's interpretation of the infinitesimal methods used in his calculus. But the focus on Prop. 6, though

---

Rabouin (2015, pp. 355–356)); second, it allows one to transform the "difference" between the triangulation and the curve comparable with a well known figure, of which one can estimate the magnitude (in this case a rectangle, one brilliant idea of Leibniz being the use of an upper bound for the width of this rectangle).

[13] See Knobloch (2002, p. 65) and Rabouin (2015, p. 356) for details.

[14] Leibniz mentions the need for "rigour" in several places in the *DQA*. Thus at the beginning of Prop. 6 he writes "The reading of this proposition can be omitted if in demonstrating Prop. 7 one does not desire the utmost rigour [*Hujus propositionis lectio omitti potest, si quis in demonstranda prop. 7. summum rigorem non desideret*]." Elsewhere he writes of *severas demonstrationes,* and *severe demonstrare.*

[15] This idea already appears in Torricelli and in Pascal. For the latter see (Whiteside 1961, p. 341) and (Cortese and Rabouin 2019).

justified, had the unfortunate consequence of concealing other very important aspects of this treatise, to which we now turn.

4.2 The *DQA* and the "Direct Method"

The first point to note in this regard is that the following proposition, Prop. 7, is *not* direct; and the second point is that the language of infinitesimals is not introduced before Prop. 8. Let us dwell a little bit upon these two features.

Prop. 6 gives the general process on which the quadrature of the given curve relies, but it does not give the actual quadrature. Indeed, there is something missing in the proof: nothing ensures, in the given framework, that the relationship between each triangle under the first curve and the corresponding rectangle of the step-space will "hold in the limit", as we would say nowadays (although the advantage of the "direct" method is to give a heuristic justification that this is the relationship we ought to demonstrate properly). To prove this fact, in Prop. 7 Leibniz relies on a *reductio*. He first assumes a difference Z between twice the *Trilineum* (T)—the area under the curve—and the *Quadrilineum* (Q)—the area under the *Quadratrix*. Then by reducing the distance between the points taken on the curve, he shows that the difference between the *Trilineum* and the Polygon P can be made smaller than one fourth of Z and the same for the difference between the *Quadrilineum* and the *Spatium Gradiformis*. But since this *Spatium Gradiformis* G is twice the area contained by the Polygon P (this is a *finite* sum), we finally get the following inequalities:

$$T - P < Z/4, \text{ hence } 2P - 2T < 2Z/4$$
$$Q - G < Z/4 \text{ and } G = 2P, \text{ hence } Q - 2P < Z/4$$

From this, it follows that $Z = Q - 2T < |Q - 2P| + |2P - 2T|^{16} < Z/4 + 2Z/4 = 3Z/4 < Z$. *Quod est absurdum.*

The fact that Leibniz uses a *reductio* here is very striking and should not be underestimated.[17] In particular, in order to render the difference between the two quantities smaller than $3Z/4$, it should have sufficed to conclude directly by using the same reasoning as in Prop. 6. But this is not what Leibniz did, which indicates that to produce a direct proof was not a central issue here. Indeed, it is easy to see from the preceding reasoning that each direct proof in terms of unassignable differences is equivalent to a *reductio* concluding

---

[16] The validity of this general result on sum of "differences" had been demonstrated by Leibniz in Prop. 5.

[17] Leibniz still insists on this in the *Compendium* of ca. 1690, when commenting on the *DQA* and warning about the danger of reasoning with the infinite: "In the hyperbolic conic, since a zone is equal to a conjugate zone, this makes the whole equal to a part. Whence it is clear that the matter should be reduced to apagogical demonstrations." (text [**17**]; GM V, 106).

4.2 The *DQA* and the "Direct Method"

from the fact that a quantity is rendered smaller than itself—and vice versa.[18] As we shall see, however, this kind of conversion is not immediate for other examples of "direct" proofs provided by Leibniz. We shall not enter here into an explanation of the choice of an indirect proof,[19] but emphasize that it gives a specific meaning to the famous scholium inserted just after this proposition:

> For my part I confess that there is no way that I know of up till now by which even a single quadrature can be perfectly demonstrated without an inference *ad absurdum*. Indeed, I have reasons for doubting that this would be possible through natural means without assuming fictitious quantities, namely, infinite and infinitely small ones; but of all inferences *ad absurdum* I believe none to be simpler and more natural, and more proper for a direct demonstration, than that which not only simply shows that the difference between two quantities is nothing, so that they are then equal (whereas otherwise it is usually proved by a double *reductio* that one is neither greater nor smaller than the other), but which also uses only one middle term, namely either inscribed or circumscribed, rather than both together; and so brings it about that we have clearer comprehensions of these matters. (Scholium to Prop. 7, text [**16**]; Leibniz 1993, p. 35)

First, one should keep in mind that Leibniz did not use the "fiction" of "infinitely small" quantities either in Prop. 6 or in Prop. 7. So the horizon of the reference to a direct proof in the first sentence cannot be—at least, cannot solely be—the method used in these propositions (this will be confirmed by a study of Prop. 8). Second, one should notice that Leibniz praises here a certain method *amongst* all the *reductio* proofs (and not as an alternative to them), which is "more proper for a direct demonstration". In sum, we have a *first* dictionary which allows the translating of descriptions in terms of indivisibles into ones in terms of finite quantities, and of *reductios* into (pseudo)-direct proofs, in propositions 6 and 7. But this first dictionary is *not* the one relying on fictions yet, and is not direct in this sense. Moreover, such a method had already been sketched by some authors such as Pascal.[20] Rather than dwelling on existing methods, Leibniz (rightly) emphasized other aspects of his results: that he does not have to resort to inscribed and circumscribed polygons (and, accordingly, not to a double *reductio* either), that he has

---

[18] Interestingly enough, when Leibniz rewrites the proof of Prop. 7 for the *Compendium*, as we shall see in Sect. 4.3, he formulates it in a mixed way: starting out as in a *reductio* by supposing that the intended result does not hold (i.e. that the *Quadrilineum* is not twice the *Trilineum*), but then, rather than refuting the original supposition after deriving an absurdity, instead concluding with a direct argument appealing to the *PUD* to show that the difference is "smaller than any given quantity whatever" and thus null.

[19] For some hypotheses in this direction see (Rabouin 2015).

[20] According to Whiteside (see note 15 above), Pascal is the first author to have the idea of reasoning directly on the difference (which can be made smaller than any given quantity). In the *Lettres d'A. Dettonville* he applies this reasoning to the case of the division of the basis into a "sum of lines" (i.e. the basic technique of the "method of indivisibles"), (Pascal 1659, pp. 10–11).

provided a general proof,[21] and that he can rely on quadratures with rectangles and with triangles.[22]

All of this is crucial for a proper reading of Prop. 8, the one in which the fiction of "infinitely small" entities will be used for the first time. The purpose of proposition 8 is to generalize the result to the case in which one begins the quadrature at the origin of the curve. But why is there a particular problem here? The answer is simple: in this case, the geometric figures produced in the sub-divisions degenerate (*degenerare* is the verb used by Leibniz). The triangle under the first curve degenerates into a sector (which Leibniz calls a "segment" here) and the rectangle of the step space becomes an "orthogonal *trilineum*".[23] This is a crucial remark because in this case, there is, at first sight, no way one can apply the relationship between the basic triangle under the first curve and the corresponding rectangle of the step-space. Still, Leibniz maintains that the proof could consist "in one word" (*uno verbo confici potest*) since the relations used in proposition 7 were considered between arbitrary small segments and hence also holds in the infinitely small case (*Quae propositione 7 demonstravimus generalia sunt, et locum habent, utcunque parvae sint rectae*). They hold in particular when A, $_1C$ and $_1B$ coincide, that is to say, when their distance is infinitely small (*ac proinde sint infinite parvae, sive etsi puncta coincidat*).

This kind of reasoning was already used by Leibniz as early as 1674 in a series of texts to which we shall return later, since they provide a first form of what would later be called the "Law of Continuity".[24] The context was the elaboration of a general formula for conic sections (one of Leibniz's favourite examples when mentioning his law of continuity later) and the fact that it implies the consideration of some parameters as either going to infinity in some cases, or as vanishing in others.[25] Thus, Leibniz remarks that in algebra one often needs some equation to hold even in degenerate cases, typically when points come to

---

[21] This aspect should not be underestimated. According to Whiteside, it was one of the major obstacles which prevented early modern authors from correctly assessing the power of exhaustion methods, when taken in their logical form (Whiteside 1968, p. 331).

[22] See the Scholium to Prop. 7 in text [**16**]. As can be seen, Leibniz insists on the fact that he can use triangles or rectangles to perform quadratures.

[23] In the *Compendium Quadraturae Arithmeticae* Leibniz defines a *segment* as "the space comprised between two lines, a curve, and another straight line", and a *sector* as "a *trilineum* comprised between two straight lines and a curve" (GM V, 101). See text [**17**].

[24] On this genealogy, see (Grosholz 2007, chap. 8.2, "Leibniz on Transcendental Curves").

[25] "To understand the Parabola and the straight line it is necessary to make use of infinite and infinitely small lines. Thus, if we suppose that the line q, or the *latus transversum* of the Parabola, is of an infinite length, it is evident that the equation $2axq \pm ax^2 = qy^2$ will be equal to this one: $2axq = qy^2$, or $2ax = y^2$ (which is that of the parabola), because the term of the equation ax^2 is infinitely small with respect to the others, $2axq$ and qy^2, for since there are as many letters or dimensions of one term as the other, those one of which one letter is infinite will be infinitely greater than those whose letters are only ordinary: which, consequently, can be neglected, since the error which will result from them will be infinitely small, or less than any given error, that is to say, null." (chap. XLIV. Notations modernized. A VII 7, 103–104).

4.2 The *DQA* and the "Direct Method"

coincidence. The simplest example is a segment AB and a point C taken at random on the same straight line. If we want the equality $AC = AB \pm BC$ to be general, "In this case of the coincidence of B and C we must conceive the line BC as infinitely small, so that the equation does not contradict the equality between AC and AB".[26]

The important point to notice in the first "proof" of Prop. 8 is the following: in this case, the proof is "direct", and in fact immediate, in a different sense than in Prop. 6. By the introduction of the "fiction" of infinitely small quantities, we do indeed have the degenerate case entering into the ordinary one, even if the geometrical shape of the objects under consideration is not preserved.[27] This directly contradicts a common view according to which Leibniz would have developed this kind of argument, based on continuity, as an *alternative* to the methods presented in the *Quadratura*. In fact, the introduction of "fiction" corresponds precisely to this argument (and not to Proposition 6). This is made explicit in the Scholium to Prop. 23:

> What we have said up to this point about infinities and infinitely small quantities will appear obscure to certain people, as does everything new—although we have said nothing that cannot be easily understood by each of them after a little reflection: indeed, whoever has understood them will recognize their fecundity. It does not matter whether there are such quantities in nature, for it suffices that they be introduced by a fiction, since they provide abbreviations of speaking and thinking, and thereby of discovery as well as of demonstration, so that *it is not always necessary to use inscribed or circumscribed figures and to infer* ad absurdum, *and to show that the error is smaller than any assignable*. Nevertheless it is evident that the latter can easily be done by means of what we have said in Props. 6, 7, and 8. (text [**16**]; our emphasis)

Although we cannot presume at this stage that the Law of Continuity acts in the same way here as in later texts, it is crucial to keep in mind that the *DQA* already presents *two* strategies of proof, one direct (Prop. 8) and one indirect (Prop. 7), the former being the one based on fictions. Moreover, Leibniz then proceeds to show that this direct proof in Prop. 8 is equivalent to the other one in Prop. 7 because it can be translated into it.

Leibniz therefore supposes a difference Z between the two figures and sets about showing that it leads to a contradiction. But he cannot use the relationship between the triangles and the rectangles directly, so here is how he proceeds: first he takes the rectangle $A_1B_1C_2T$ smaller than $1/4 Z$ and concludes that its constituents, the small "segment" A_1CA and the small *trilineum* A_1B_1DA are also smaller than $1/4 Z$. Then he considers these two figures as parts of bigger figures in which one part is not degenerate. Typically the small segment A_1CA is the difference between a greater "segment" $A_3C\ _2CA$ and the sector

---

[26] "In the case of the coincidence of the points B and C it is necessary to conceive the line BC as infinitely small, so that the equation does not contradict the equality between AC and AB." (A VII 7, 88). The example of the coincidence taken as infinitely small distance is mentioned in the letter to Varignon from Feb. 2 1702 as a typical case of application of the Law of continuity (A III 9, 14). See text [**38**].

[27] We'll see that this point was at the core of the discussion with Wallis in 1699; see text [**46**].

$_1CA_3C\ _2C_1C$ (a non-degenerate figure). By the same reasoning, the small *trilineum* is the difference between the large *trilineum* $A_3B_3D_2B_2DA$ (the figure we are interested in) and the *quadrilinum* $_1D_1B_3B_3D_2D_1D$ (a non-degenerate figure). Since the intended relation holds between the non-degenerate figures, he can then conclude that the difference between the large *trilineum* and twice the small segment (seen as the first difference above) is smaller than $^3/_4Z$.

Notice that in this case, the "translation" between the two proofs does not amount to simply taking, instead of the infinitely small, finite quantities which approach them more and more closely, because this would not solve our first problem (they don't have the same shape as the non-degenerate figures). The reasoning is hence slightly more complicated and necessitates the introduction of a second type of differences (on both sides of the inequality).

Here we have a clear example of a translation between a "direct" proof relying on a continuity argument together with the fiction of the infinitely small, and a *reductio* proof in the manner of the Ancients. The feature we would like to emphasize is that this meaning of "direct" is different from the one relating the *reductio* in Prop. 7 to the direct proof in Prop. 6. This is a subtle issue since Prop. 8 proceeds by translating the direct proof into the method of Prop. 7, which itself can be rendered in terms of the direct construction of Prop. 6. This means that, at the end of the day, all of these methods rely on the *PUD*. But, this being said, they don't proceed in the same way, in particular when it comes to their use of fictions. Only the third one relies constitutively on fictions and only this one concludes by using the continuity argument.

Now Prop. 8 is not the only proposition that has been neglected by scholars dealing with infinitesimals in *DQA*—which is surprising, considering the fact that this is where the concept is first introduced. Prop. 11 is also neglected. This is a no less surprising fact, considering that this is where Leibniz proposes a clarification of his views on the infinite.

Let us recall the content of this proposition. Prop. 11 is intended to show the following (somewhat paradoxical) result: from any curvilinear figure, no matter how small, one can extract a part, which is twice the magnitude of a figure *with infinite length* (see Fig. 4.2).[28]

In order to show this result, Leibniz starts again from the given curve $_1C_2C_3C..._nC$, but, instead of its origin A, considers an arbitrary point μ, where he draws a tangent to the curve $\mu\lambda$. He then draws a perpendicular to this tangent as an axis of coordinates, chooses a point A on this straight line where it meets the curve again[29] and considers a perpendicular to this perpendicular as a second axis (this line being, by construction, parallel to $\mu\lambda$). By recourse to the previous construction, Leibniz can then perform exactly the same reasoning as in Prop. 6, with the only difference that the *Quadratrix* now has an asymptote $\mu\lambda$. Relying on

---

[28] This proposition is crucial for dealing with the quadrature of the simple hyperbola, a case which Leibniz deals with in Prop. 12.

[29] This presupposes that the curve is, at least piece-wise, convex—a condition which is stated by Leibniz when describing the curve in Prop. 6.

4.2 The *DQA* and the "Direct Method"

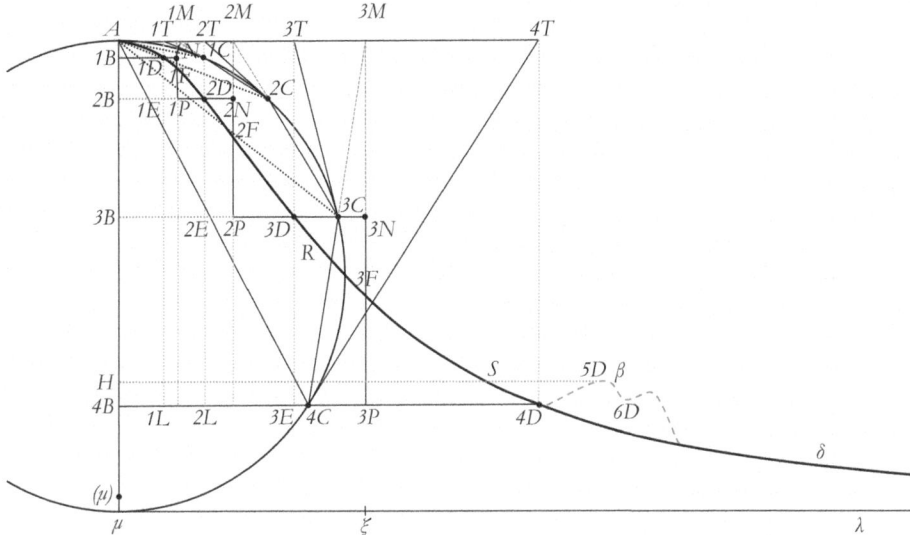

Fig. 4.2 The construction of an asymptote to the *Quadratrix*

his previous results, he can show that the area comprised between the *Quadratrix* (starting at $_2D$) and the asymptote—a figure which is infinite in length—is twice the portion of the area under the curve delimited by the points $_2C_2B\mu_3C_2C$.

Now the conceptual difficulty is of course to establish a finite ratio between a space infinite in length and a finite area. Here Leibniz comments that he knows no other way (*non aliter fieri potest*) than to introduce a point (μ) at infinitely small distance from the axis. In this case, indeed, the straight line $(\mu)\lambda$ will still be infinite in length in the sense than it can be made greater than any given quantity (*major qualibet assignabili*), yet will be bounded (*terminata*). Accordingly the line "$(\mu)\lambda$ will not be an asymptote to the curve $D\delta$, but will cut it in a point λ, this point being distant by an infinite interval". This kind of proof will allow the introduction of a kind of "arithmetic of the infinite" in which Leibniz explains the ratios between finite and infinite quantities (Prop. 20, then used in 21).[30]

Following this proof, Leibniz comments on some paradoxical results occurring with the development of the "method of indivisibles", and in particular on the famous construction by Torricelli of an "infinitely long solid" with a finite volume.[31] According to him, this kind of result no longer seems mysterious as soon as one realizes that it is based on the fiction of a line (or a surface) at the same time *terminata* and *infinita*. We find here again the same strategy as the one he had devised for infinite series: the addition of a fictitious *terminatio* allows one to calculate with the infinite. It should be noticed that another way of describing

---

[30] Knobloch (2002) has given other examples, which can be generalized from Prop. 20.

[31] See Mancosu and Vailati (1990). Leibniz also mentions other examples of the same kind in Grégoire de Saint Vincent and Huygens.

the introduction of this *terminatio* would be the addition of a point at infinity—again one of Leibniz's favourite examples when mentioning his Law of Continuity.[32] *Caeteris paribus*, it resembles what is done in modern analysis by recourse to "compactification": one transforms a space which cannot be covered by finite means into one which can by adding a point at infinity, and this allows one to calculate more easily—for example, some integrals.

It is here that Leibniz then comments about the difference between "indivisibles" and "infinitely small" quantities, in parallel with *interminatum* and *infinitum*. The crucial issue, as was already noted by several authors at the time, is that of homogeneity, an indivisible and a non-terminated object being related in a heterogeneous manner to their whole and parts respectively. This is why, Leibniz explains, a rectangle with an infinitely small width and an infinite length can be equal to a finite rectangle, as happens when one performs the quadrature of the hyperbola. This leads to the following characterization in a cancelled version of the Scholium to Prop. 11:

> I call unbounded that in which no ultimate point can be assumed, at least on one side. On the other hand, I call a quantity, infinite, whether bounded or unbounded, whenever we conceive it as greater than any quantity that is assignable by us, or expressible in numbers. Now, whether the nature of things tolerates such quantities is for the metaphysician to discuss; for the geometer, it suffices to demonstrate what results from supposing them. (text [**6**])

We find here another very important feature of Leibniz's strategy when dealing with infinite and infinitely small quantities: nowhere does he express his conviction that they are contradictory notions, although this was the way—as we have shown—by which he was led to his own conception of infinite wholes and their reciprocals as fictions. Everything is done so that the surface language of the mathematician remains *neutral*, leaving the philosophers to disentangle the difficult question of the *Labyrinthus continui*. Since mathematicians that Leibniz held in high esteem, such as Galileo or Grégoire de Saint-Vincent, did not share his views about the non-existence of non-standard quantities, it was of crucial importance to show that this question does not belong to mathematics proper because the mathematician can develop a language which can be interpreted in both ways.[33] This is the strategy which Leibniz will follow in all of his public declarations from then on, and it is all the more surprising that some commentators maintain that one

---

[32] See the beginning of the *Cum Prodiisset*, (text [**50**]). The parallel between the introduction of point at infinite distance and infinitesimals, which appears in the *DQA* and was already present in Desargues, plays also a prominent role in the *Elementa nova matheseos universalis* (circa 1683; text [**15**]) and the *Matheseos Universalis pars prior* (1699; GM VII, 75–76).

[33] This consideration will become even more significant when Leibniz realizes that the main supporters of the Differential Calculus in the Académie Royale (the Bernoullis, L'Hôpital and Fontenelle) all believe in the existence of infinitesimal quantities.

4.2 The *DQA* and the "Direct Method"

can infer from the surface language to the existence of entities populating the mathematical continuum.[34]

Although there are many other very interesting propositions in the *DQA*, Props. 8 and 11 will be sufficient to help us clarify the connections between Leibniz's views on the infinite and the conduct of proofs, which is not necessarily a matter of simple and direct paraphrase. Leibniz's "syncategorematic" view, to which we will return in greater detail in Chap. 6, says that whenever we seem to make reference to an infinite quantity in mathematics, be it an infinite number, an infinitely long distance, an infinite bounded line or an infinitely small quantity, what we are doing in fact is pointing to some relations between finite quantities. As we have seen, Leibniz calls a quantity infinite, "whether it is bounded or unbounded, whenever we conceive it as greater than any quantity that is assignable by us, that is to say, numerically expressible" (Leibniz 1993, p. 68). Taken as such, this view entails the possibility of expressing any use of the infinite through sentences in which it does not occur (as was already the case in Ancient Geometry).

Now, it should be stressed that this does not tell us anything about the way to conduct the proofs.[35] As the previous developments should make clear, this conception is compatible with (at least) two different kinds of proofs: a proof "in the manner of the Ancients" in which one could replace the difference which can be taken *quantumlibet parva* by an "infinitely small" error[36]; and a direct proof relying on a continuity argument such as the "one sentence proof" at the beginning of Prop. 8. It is crucial to note that in this second case, the complete justification is *not* given, according to Leibniz, by the paraphrase in and of itself, because it is clear that those who do not accept "infinitely small" quantities will not be convinced. The reason is profound: even if we translate the "one sentence proof" into a longer proof involving finite quantities that can be taken as small as we wish, the difficulty will not be solved since it relies on the fact that the shape of the objects under consideration degenerates. If we want to translate the direct proof into a "rigorous" one, we need to do more than just rephrase it without recourse to infinite quantities—as Leibniz does in the remaining part of Prop. 8.

---

[34] Especially when it comes to interpreting the "incomparables", since what Leibniz expressly says is the following: "This is what makes me speak on other occasions of incomparables, because what I said there holds *whether one means infinitely small magnitudes or one employs magnitudes of a smallness that is inconsiderable and sufficient to make the error less than that which is given*", (*Defense du calcul*, text [**49**]; our emphasis).

[35] A similar point was made by Tzuchien Tho in his (2012), where he distinguished the question of the elimination of reference to infinitary terms by "syncategorematic" paraphrase from that of how Leibniz conducted his proofs in the *DQA*.

[36] Although Leibniz did not use this language in props. 6 and 7, this is what he alludes to when introducing the fiction of "infinitely small" at the beginning of Prop. 8: "*Hoc uno verbo confici potest, ex eo quod quae p r o p o s i t i o n e 7. Demonstravimus generalia sunt, et locum habent, utcunque parvae sint rectae, ac proinde etsi sint infinite parvae.* [This can be gathered from what was in Prop. 7. We have demonstrated general things, and they hold however small the straight lines may be, and therefore even if they are infinitely small.]"

4.3 The Posterity of the *DQA*

But what is the status of the *DQA* in the long term? Some critics have downplayed its significance for Leibniz's later defence of the calculus, even while acknowledging its importance for understanding the genesis of his views. It is true that, in his published reference to it in 1691 (text [**19**]), Leibniz acknowledges that "it did not seem worth the trouble to expound at length in the common fashion what our new analysis exhibits in a few lines." Indeed, the treatise does not contain any Differential Calculus, and although it indicates a general strategy that could be useful in justifying the Leibnizian algorithm, the move toward this justification is not immediate.[37] Moreover, the *DQA* appears in many aspects as a piece of juvenilia that Leibniz overestimated for political reasons (Leibniz wrote it for submission to the *Académie des Sciences* in the hope of securing a place in that illustrious assembly).[38] It has been charged that it deals only with "well behaved curves",[39] and that it relies on a main result, which, from the point of view of the differential and integral calculus, is still limited.[40] Finally, it is alleged that if this treatise had the significance that some commentators have given it, then it is striking that Leibniz did not rely on it when justifying the calculus in later periods. "What about the fact," Viktor Blåsjö objects, "that he didn't submit it, nor publish it later, nor reproduce its key results in his extensive subsequent correspondence on the foundations of the calculus?" (Blåsjö 2017, p. 136).[41] In

---

[37] Amongst other problems, one difficulty is that it is not easy to find an equivalent of a generic case on which to conduct such a justification.

[38] Victor Blåsjö writes: "it is well known that Leibniz was desperate to fashion a career for himself in intellectual circles at this time. The fact that he wanted to submit his work to the French Academy could very well be a reflection of this desire more than an assessment of the quality of the work, so this in itself proves nothing." (2017, p. 136). Douglas Jesseph "suspect[s] that he set aside the *Arithmetical Quadrature* without publishing it because he had turned his attention to more powerful methods that he would introduce in the 1680s…" (Jesseph 2015, p. 200). This echoes the judgement of Gerhardt in 1859: "Since the subject treated in this treatise increased in size from day to day as a result of the algorithm of higher analysis discovered by Leibniz, but especially because the original structure of the entire treatise and the treatment on which it was based were still based on the old methods that had been eliminated by the discovery of the algorithm of higher analysis, after a few years Leibniz no longer considered it time to publish the treatise in question in print" (GM V, 84–85; our translation).

[39] Thus Jesseph writes that Leibniz's procedure in the *DQA*, in its dependence on the construction of auxiliary curves, "requires that we have a tangent construction that will apply to the original curve", and that although "this is readily available in the case of the circle, and tangents to conic sections and other well-behaved curves are also constructible with classical methods" (Jesseph 2015, p. 199), this would not extend to more general curves for which no geometrical tangent construction was available.

[40] Blåsjö (2017, p. 136). We will return to this below.

[41] Many of these remarks denigrating the treatise depend on the bias we noted above concerning the concentration on Proposition 6 to the exclusion of what else is contained in it. When considered in its entirety, the *DQA* contains many beautiful results which Leibniz had not published at the time and

4.3 The Posterity of the DQA

the present section, we shall answer this question and more generally provide a reply to these various objections.

Beginning with the easiest argument, it is a historical fact that the last claim is false, as is more generally the idea that Leibniz gave up the *DQA* because it was obsolete and superseded by the techniques of the Differential Calculus. This fact has been known since the end of the nineteenth century when Gerhardt published a *Compendium* of the *DQA* (text [**17**]), which Leibniz had prepared for publication. Gerhardt speculated that the *Compendium* might date from 1678–1679,[42] but we now know (thanks to philological tools not available to him) that it was prepared at the beginning of the 1690s at the earliest.[43] Although this is predominantly a faithful abbreviation of some of the main results of the treatise, two points are of interest. The first is the version of the proof of Prop. 7 given there. As mentioned above, this begins as a *reductio* by supposing that the intended result does not hold (i.e. that the *Quadrilineum* is not twice the *Trilineum*). Then, supposing that there is a difference Z between the two quantities, he demonstrates that the difference can be made smaller than Z. But instead of directly concluding *quod est absurdum*, he continues: "by Prop. 5 the difference between Q and T will be smaller than $^3/_4 Z$ and so smaller than Z, and so smaller than any given quantity whatever, and so this difference will be zero." (GM V, 101). From this we can see very clearly the way in which any *reductio* of this type can be transformed into a direct argument using the *PUD* and vice versa. The second point of note about the *Compendium* is that in it Leibniz had no trouble employing the differential algorithm—although this is even clearer in the next piece to be considered, the appendix that he sent Bodenhausen in October of 1690 (text [**18**]), where he translates

---

will still praise in later writings: an easy way to find the quadrature of the cycloid, his famous series for the quadrature of the circle, a unified and analytic treatment for trigonometric functions, a study of the logarithmic curve, the presentation of his "harmonic triangle", and a proof of the impossibility of an algebraic quadrature of the circle. By the time he wrote to Bernoulli, it is true, these would not have carried the same weight, since all of this was then published in other places. But as we shall see, this did not derogate from the importance of Proposition 6 for Leibniz.

[42] Gerhardt writes: "For a moment, he seems to have had the idea of publishing the whole work in a shorter version containing only the theorems, partly with the use of higher analysis, as a '*Compendium Quadraturae Arithmeticae*'; however, this plan also remained unrealized. ... The time at which this extract was written is not given; it may have been in the years 1678 or 1679, in which the development of the algorithm of higher analysis was already significantly advanced." (GM V, 85–86; our translations).

[43] GM V, 99–113. This piece has not yet been published in the *Akademie* edition. Knobloch (Leibniz 1993, p. 12) writes "*Um diese Zeit dürfte Leibniz das Compendium quadraturae arithmeticae angefergit und die Randbermerkung zum Beweis von Satz 47 ergäntz haben, in der er sich auf das Compendium bezieht*. [Around this time, Leibniz probably produced the *Compendium quadraturae arithmeticae* and added the marginal note to the proof of Theorem 47, in which he refers to the Compendium.]"—"this time" being when he is first engaged in trying to get it published by Rieuwertsz (around 1682). The letter to the editor of the *Journal des sçavans* is reprinted at GM V, 88–92; the *Praefatio opusculi de quadratura circuli arithmetica* at GM V, 93–98.

the proof of Prop. 8 into the differential calculus. Thus the two techniques, that of the original *DQA* and that of the calculus, are certainly not exclusive of one another.

Moreover, Leibniz certainly sought to publish the treatise itself. In an exchange of letters in 1682–1683 he discusses the project of publishing the *DQA* in Holland. Ferguson found a publisher: Jan Rieuwertsz (Spinoza's publisher), but Leibniz finally decided that the *DQA* should be inserted in a more general discourse on *Ars Inveniendi*—and, as usual, he did not find the time to complete this project.[44] More interestingly, though, 10 years later Leibniz reopened the possibility of publishing the *DQA*. In a letter to Henri Basnage de Bauval of late August 1692, he wrote:

> A certain Mons. Rieuwerts, a bookseller in Amsterdam who has published some writings of Spinoza—to whom the late M. Ferguson sent me—was disposed at one time to publish certain geometrical speculations that I had. But the distractions that I then had did not permit me to lay it out in full, and I contented myself by giving certain abstracts in the *Acts* of Leipzig. Perhaps Mons. Rieuwerts would again take charge of it, since it appears more plausible and more accessible to the majority of readers. One could add a preamble containing some curious particulars on what Mons. Descartes invented or took from elsewhere. (A II, 2, 559)

This plan to relaunch publication (in which connection he might have written the *Compendium*) could well have been stimulated by Leibniz's exchanges with Bodenhausen. For when Leibniz initiated Bodenhausen into the new algorithm (which he intended to publish as an appendix to the *Dynamica*, mimicking Newton's strategy in his *Principia*), Bodenhausen raised a few *dubia*, to which Leibniz answered methodically.[45] After a few exchanges, as Bodenhausen came to understand the calculus better, he signalled to Leibniz that it would be very useful to have at one's disposal a gentle introduction to shut the mouths of the Euclidean "Pied Pipers" (*Rattenfänger*), who were hostile to the new method (A III 4, 564). Leibniz responded positively to the demand and sent as an appendix to his letter from October 26, 1690, a presentation of the calculus for those who were trained in the "manner of the Ancients" (text [**18**]). And what did he provide on this occasion? A presentation of Prop. 6 of the *DQA* accompanied by a translation into the differential calculus corresponding to Prop. 8.

An interesting feature is the variant proof of Prop. 6, which proceeds by considering not the chords to the curve $_1C_2C$, but small triangles obtained by the intersection of the line passing through the point $_1C$ and the tangent to the curve passing through the next point $_2C$. This allows Leibniz to recover a technique by inscribed and circumscribed figures without having the defect of relying on an explicit construction of polygons, which he presents as

---

[44] See A III, 3, 645, 800 and 813.

[45] See in particular A III 4, 520 which Leibniz titles: *Responsio mea ad dubium hujus Epistolae*. Among the difficulties Bodenhausen raises is a case where Leibniz concludes from the equation $eg = m$ (where e is extension and m is mass) that $d(e)g = d(m)$, even though g is not a constant. In a lengthy and intriguing reply, Leibniz writes that while e and m are sometimes *assignabiles* and sometimes *inassignabiles*, g is introduced by "a kind of fiction" (A III, 4, 523).

4.3 The Posterity of the *DQA*

"Archimedean" (A III 4, 635). To be sure, by 1690 all of these results from the *DQA* have been superseded by the many researches in which Leibniz had been engaged since then, and he does warn his correspondent that the same results are almost immediate with the new calculus.[46] But precisely, it is all the more striking that when it came to giving a translation of this calculus into the language of the Ancients, the example he chose to provide in 1690 was still Prop. 6 of the *DQA*. This confirms something that he already said when presenting the proof of Prop. 6 in detail: this kind of proof is boring and painful, satisfying only those who are scrupulous. However, one should perform such proofs as the occasion demands to confirm that the general strategy is well grounded.

This is precisely the strategy Leibniz adopts in 1695 in responding to Johann Christoph Sturm, who had found difficulty in understanding his quadrature. Leibniz replies:

> Concerning my Quadrature, I am pleased to remove the difficulty that the distinguished gentleman has encountered. But since he seems to desire a demonstration, I attach here some foundations from which he will easily be able to resolve it, for there is no time to explain the whole thing now at length. I have, indeed, several ways of demonstrating the same thing, but this one seems most elegant. (text [20]; A II, 3, 102)

What Leibniz gives immediately following this is an outline of the main argument of the *DQA*, stretching over three pages (text [20]; A II, 3, 102–104). His First Lemma is Prop. 1 of the *DQA*, that showing the equivalence of the triangle construction with that of rectangles twice their size (Leibniz 1993, pp. 24–25; see Rabouin 2015, pp. 15–16). On this basis he constructs the transformed curve, his *Quadratrix*, so that the area under the first curve (decomposed into triangles) is transformed into an area under the second curve that is decomposed into rectangles twice their size. He then gives an epitome of the method of the *DQA* encapsulated in Props. 6 and 7, which allows him to calculate the quadrature. "From this theorem," he writes,

> there follows almost at once the quadrature of all the paraboloids and hyperboloids which Fermat, Wallis and others have squared by sums of numbers, by assuming a kind of induction. ... And from this theorem I also found the absolute quadrature—hitherto unknown—of that segment of a cycloid which is cut off if you draw from the vertex two straight lines to a point on the curve, on which a parallel to the base drawn through the centre of the generating circle meets the curve. (text [20]; A II, 3, 103)

This describes Proposition 14 of the *DQA* (Leibniz 1993, p. 45; 2016, p. 66). He also reports his result in Prop. 32, that "the circle is to the circumscribed square, that is, the arc of the quadrant is to the diameter, as $1/1 - 1/3 + 1/5 - 1/7 + 1/9 - 1/11$ etc. is to unity" (Leibniz 1993, p. 79; A II 3, 104). Here Leibniz adds that when he was in Paris, Jean Prestet, "author of the *Elements of a Universal Mathematics* (Prestet 1675), had persuaded

---

[46] Notice however that Leibniz adds that he is still glad to employ this method, because it gives some representation in the imagination corresponding to the operations of the calculus.

himself that the cycloid is an arc of this circle—which is impossible, as I easily showed him" on the basis of the foregoing arguments.

In sum, when Leibniz is pressed to explain his method of quadrature to someone having difficulty with it in 1695, the "most elegant" way he can conceive of demonstrating it is not in terms of the more powerful methods he has developed since his youth, but by reference to the very presentation in the *DQA* which those methods had, according to Jesseph and Blåsjö, rendered inadequate and obsolete. So it is certainly not the case that in the 1690s, flush with the success of his calculus, Leibniz regarded the *DQA* as containing nothing of interest to colleagues familiar with the new methods, or that he never saw fit to reproduce its key results.

Moreover, contrary to Blåsjö's claim that he never quoted any of its results in foundational discussions, we have not only the example of the exchange with Bodenhausen of 1690, but also what Leibniz wrote to Johann Bernoulli in 1698. In a letter dated July 29 (text [**32e**]), he returns to Proposition 22 of the *DQA* in connection with the notorious case mentioned above of Torricelli's construction of an "infinitely long solid" with a finite volume, and how the air of paradox can be dispersed by recourse to the fiction of a line (or a surface) that is at the same time *terminata* and *infinita*. The context is an objection raised by De Volder "against the infinitesimal calculus", which Leibniz credits Bernoulli with having "correctly solved". It concerns a hyperbola whose equation is $y = a^3/x^2$, and the fact that if x is supposed infinitely small, then y is an infinity of second degree. Leibniz comments:

> I myself once formulated a cognate objection to this very one in the Scholium of Proposition 22 of the unpublished treatise which I composed in France on my Arithmetical Quadrature, shortly after its discovery. There the objection appeared to apply not only to our calculus, but also with equal right to the geometry already accepted up to that time. [...]
>
> But between ourselves I would also add this, that I also wrote in the said unpublished manuscript that it is possible to doubt whether there could be infinitely long straight lines that were nevertheless also in fact bounded. I wrote moreover that it suffices for the calculus that they be taken as fictions [*fingantur*], like imaginary roots in algebra. For it is always the case that what is concluded by means of the infinite and infinitely small can be evinced by a *reductio ad absurdum* by my method of incomparables (the Lemmas for which I gave in the *Acta*). So you also shouldn't wonder that I doubt whether there is an infinitely small thing, or an infinitely great one bounded on both sides. For even though I concede that there is no portion of matter that is not actually cut, one does not on that account come to uncuttable elements, or minimum portions, *nor indeed to infinitely small things, but only to ones perpetually smaller, and yet ordinary; similarly, in increasing one comes to perpetually greater ones.* (text [**32e**]; our emphasis)

Here the connection between the language of fictions from the 1690s and that of the *DQA* is made fully explicit. Indeed this last document is also crucial, since this is the place where Leibniz first publicly divulged to Bernoulli his idea that infinitesimals are imaginary

4.3 The Posterity of the DQA

entities that can be introduced by a kind of fiction. This is the very passage to which he will later refer when talking to Varignon of his *"fictions utiles"*[47]:

> Moreover, I wrote some years ago to Mons. Bernoulli of Groningen that the infinite and infinitely small could be taken as fictions, similarly to imaginary roots, without that having to do a wrong to our calculus, these fictions being useful and founded in realities. (14 April 1702, text [**38c**])

Moreover, it is the beginning of the very long exchange in which Leibniz makes it clear that he does not share Bernoulli's view on the continuum, as comprising actually infinitely small elements.

This is not the place to enter into a detailed commentary of all of these fascinating pieces. Our first aim is to emphasize that they exist: there are in fact many documents in which Leibniz refers to the *DQA*, most of the time very explicitly, as the place to go to find a justification for the use of infinitesimals. These documents have been so far completely neglected, if not ignored.

---

[47] This is also the context in which he mentions his syncategorematic view on the infinite as grounding his conception of "fictions".

Infinitesimals and Their Existence After 1676

5.1 First Publications on Differential Calculus, and Foundational Reflections

Publication was a complicated affair in the seventeenth century. Leibniz had managed to publish five articles in the *Journal des sçavans* by 1678,[1] but, as we have seen, his hopes that the same journal might publish the *DQA* (which he had left in the hands of his friend Soudry after leaving France) did not materialize. In 1682 a new journal was established in his home town of Leipzig, the *Acta eruditorum*, and when the editor Otto Mencke requested a lead article from him for its earliest issues, Leibniz sent him *De vera proportione circuli*.[2] In it Leibniz presented the infinite series expansions he had devised for various curves in a succinct format, including the alternating infinite series expansion for $\pi/4$. It also contained the first appearances in print of the term 'transcendent' and of an equation with an unknown in the exponent, x^x (i.e. x to the power of x).[3]

These were mostly results that Leibniz had derived as much as nine years earlier. Yet he was already in command of a far more powerful algorithm, by whose means he could solve any tangent problem, and a whole range of quadratures, even of what he had termed transcendent curves. This he had achieved by October 1675, as is testified by his

---

[1] These included an article on time-keeping in portable watches in 1675, one describing the report of a strange array of antlers on a deer and another expounding Crafft's discovery of phosphorus, both appearing in 1677, and two papers on mathematics in 1678.

[2] Leibniz (1682). In English, the full title translates as "On the True Proportion of the Circle to the Circumscribed Square Expressed in Rational Numbers"; there is a French translation by Marc Parmentier in (Leibniz 1995, pp. 61–81).

[3] This article was published two months later in English translation in the *Philosophical Collections* of the Royal Society (April 1682).

manuscripts, although he did not appreciate the full significance of his new algorithm until just over a year later.[4] One of the most illuminating of these is the *Pars Secunda* of his *Analyseos Tetragonisticae*,[5] in which, while reviewing his new analytic method of quadratures by transforming that problem into one of finding tangents, Leibniz invents his new symbolism for the calculus. The piece begins:

> I believe we can finally give a method by which the analytic quadratrix may be found of any analytic figure, when this is possible, or, when it cannot be done, it will nevertheless always be possible for an analytic figure to be described that performs the function of the quadratrix as nearly as possible. (A VII, 5, 288)[6]

He then describes a first approach, based on a generalization of the method of indeterminate coefficients. One first sets a general algebraic equation for a curve, derives an algebraic expression of the ordinates to the tangents and then relies on the general relation $\frac{t}{y} = \frac{a}{v}$ between the subtangent t to the curve, its coordinates and the ordinate of a new curve v expressed in terms of the characteristic triangle $v = a \frac{dy}{dx}$ (A VII, 5, 288). A standard process of elimination among the various unknown then allows one to express the ordinate of the quadratrix v in terms of x.

A few pages further on, Leibniz gives a more recognizable presentation. Indeed, this presentation is fascinating, for in it we can see the very origin of the calculus, stated in Leibniz's own words, and recognized by him as a powerful new algorithm. For in his calculations up to this point he had expressed what we now call the integral of y with respect to x either (in words) as "the sum of all y applied to x [*summa omnium y, ad x applicatarum*]" or "the sum of all y in relation to x [*summa omnium y ad x*]", or (in symbols) as "*omn y ad x*"—or, when there was no ambiguity in leaving the independent variable x implicit, simply "*summa omnium y*" or "*omn y*". But in this text, dated October 29, 1675, Leibniz now comes up with a new symbolism:[7]

---

[4] It is in the *Calculus tangentium differentialis* of 26 November 1676 that Leibniz finally explicitly recognizes that his formulas for taking differentials of the first order do not depend on any particular progression of the variables, and that one can take as constant the differences of any one of the variables in the problem.

[5] This ms., dated "29. Oktober 1675", may be found in A VII 4, 288–295. Leibniz notes that it is Part II in relation to Part I of 25 October 1675, "*Analysis Tetragonistica ex Centrobarycis*" (A VII 5, 263–69).

[6] "*Credo nos tandem dare posse methodum, qua cujuslibet figurae Analyticae figura analytica quadratrix inveniri potest, quando id possibile, aut quando id fieri non potest, poterit tamen semper figura describi analytica, fungens vice quadratricis, quam proxime.*"

[7] The significance of these passages and the notational changes is well explained by A. Weil in his review article of Hofmann's *Leibniz in Paris 1672-1676* (Weil 1975, pp. 678–679).

5.1 First Publications on Differential Calculus, and Foundational Reflections

> The following is even a Theorem: $omn. (xl) \sqcap x (omn. l) \sqcap omn. (omn. l)$,[8] taking l to be a term of the progression and x to be the number that expresses the place or the order of l corresponding to it, ôr x to be an ordinal number, and l to be the thing ordered.
>
> Note that in these calculations the law of homogeneity can be observed. For if *omn.* is prefixed to a number or ratio to the infinitely small, it makes a line, if to a line, it makes a surface, and if to a surface, it makes a body, and so on in dimensions to infinity.
>
> It will be useful to write $\int$. for *omn*, for instance, $\int l$ for *omn. l.* that is, the sum of all l. And so we will have $\frac{\{\int l\}^2}{2} \sqcap \int\{\int l \frac{l}{a}\}$ and $(\int(xl) \sqcap x \int l - \int \int l$. And so the law of homogeneity is always observed, which is useful in order to avoid errors in the calculation.

Then, after a further page of restating his results in this notation,

> This is new and remarkable enough, since it indicates a new kind of calculus.
>
> I propose we retrace our steps. When l is given in relation to x, one asks for $\int l$. This is now achieved by the contrary calculation; that is, if $\int l \sqcap ya$, we put $l \sqcap ya/d$ ...[9] (A VII 5, 292, 293)

Here Leibniz is representing the finding of the ratio of the differences of ya in relation to x by the notation ya/d. In these calculations a is a constant for preserving dimensional homogeneity: as evidenced by the second paragraph, at this time Leibniz had supposed that $\int$ increased the dimensions by 1, and that $/d$ reduced them by 1. Realizing that in fact $y = \int l$ is a sum, not of lines, but of elementary areas ($y = \int l\, dx$), he sees that neither $\int$ nor $/d$ changes the dimension, and on November 11, 1675, introduces the notation dy for ya/d.[10] Thus if in the above passage we set $y = \int l$, i.e. $y = \int l\, dx$ with x as independent variable, the first equation can be rewritten as $\frac{y^2}{2} = \int y \frac{l}{d} = \int y\, dy$, and the second as $\int x \frac{dy}{dx} dx = xy - \int y\, dx$, or $\int x\, dy = xy - \int y\, dx$. This is a neat encapsulation of his analytic method for finding integrals by inverting the problem: $\int l\, dx$ is found by setting $y = \int l\, dx$ and solving the inverse problem, $l = dy/dx$.[11]

---

[8] Here Leibniz is using '⊓' for equality; later he will use the '=' familiar today. We have been unable to reproduce the overbar that he uses for grouping variables, however—for instance, under a square root or integral sign—and have used parentheses instead.

[9] *Analysis Tetragonistica ex Centrobarycis* (A VII 5A, 292-; 25 October 1675).

[10] *Methodus tangentium inversae exempla*, A VII 5, 321–331. The change of notation from x/d to dx is introduced on p. 324.

[11] As Weil reminds us, however, "Gradually and after no little fumbling, notations like $\int y\, dy$ creep in when y is not the independent variable. But there is no hint yet of Leibniz' decisive discovery of the invariance of the differential form $y\, dx$ with respect to *all* changes of variable. To us this is an all-important fact, more so perhaps than the so-called 'fundamental theorem of the calculus', about which so much ink has been spilled in vain." (Weil 1975, p. 683).

This general method of finding quadratures analytically is something Leibniz had shown Ehrenfried Walther von Tschirnhaus after the latter had arrived in Paris in the Fall of 1675. Tschirnhaus had come to Paris with letters of introduction to Huygens and Leibniz from Henry Oldenburg, and he and Leibniz had immediately set about collaborating, studying extant mathematical literature and working on algebraic problems. A year later they both had to leave Paris, Leibniz to Hanover via his second trip to England and on to the Low Countries, and Tschirnhaus to Rome and Milan. But they continued collaborating on mathematics. In fact, Tschirnhaus wrote to Leibniz several times about quadrature, and in a letter of December 1679 (A III, 2, 921–930), Leibniz explained his differential calculus to Tschirnhaus again. "But," he writes, "since you attest that it is difficult for you to see how to find all squarable figures, I will show how easy it is",[12] and he proceeds to present the statement of the method in almost exactly the same terms as it was given in the beginning of Part 2 of the *Analyseos Tetragonisticae* as reported above, prior to the invention of the new symbolism.[13] Having carefully reprised that statement of the method, he reprimands his correspondent: "And thus by the calculation of several of these we will have a universal rule for the general algebraic quadrature of any algebraic figure whatever. And I remind you that I already told you these things in Paris, but I see that you did not pay attention."[14]

Thus it was with considerable annoyance that Leibniz discovered Tschirnhaus's publication in the 1683 *Acta eruditorum*, "A Method for determining either the Quadrature of a given figure, bounded by straight lines and a geometric curve, or the impossibility of the same quadrature" (Tschirnhaus 1683). For its subject was the very "easier" explanation of his method that he had sent to Tschirnhaus in 1678/79, although admittedly now elaborated in greater detail, and stated without any accompanying demonstration. Moreover, Tschirnhaus had failed to ask Leibniz's permission to use his work, or even to notify him of his intention to publish it. The article begins by presenting a "geometric curve", i.e. one that can be expressed by an algebraic equation, and proceeds to state that the problem is that of determining a second such curve with ordinate z which is such that for every x, the area of the mixtilinear figure under the curve can be equated with the rectilinear area xy, i.e. such that the sum of all the products of z and the differences of successive abscissae (between 0 and x) is equal to xy. But instead of determining y for a given z, the method consisted in inverting the problem, and seeking the curve z whose integral would be y. Tschirnhaus then stated theorems (given without accompanying proofs) relating the coefficients of the second curve to the first. This method of finding quadratures by inverting the problem to finding tangents is, of course, the gist of Leibniz's method described above,

---

[12]"*Quoniam autem testaris Tibi difficile videri, invenire omnes figuras quadrabiles, ideo ostendam facilitatem*" (A III, 2, 927).

[13] For an informative account, see Kracht and Kreyszig (1990).

[14]"*Atque ita calculo aliquot horarum habebimus universalem regulam pro quadratura generali algebraica figurae algebraicae cujuscunque. Et memini me Tibi jam haec Parisiis dicere, sed ut video non attendisti.*" (A III, 2, 928).

5.1 First Publications on Differential Calculus, and Foundational Reflections

and although expressed geometrically, and without demonstrations, comes close to revealing Leibniz's new calculus. Nevertheless, Leibniz found Tschirnhaus' presentation of the method insufficiently rigorous, as he subsequently reported in a publication of his own in the *Acta eruditorum* in May of the same year.[15] The chief deficiency is that it fails to distinguish the impossibility of a general algebraic quadrature from one of a determinate figure. Leibniz gives Hippocrates' lunule as an example of a figure whose particular quadrature was known, without anyone being able to obtain the general algebraic quadrature.

Although this episode put something of a dent in the erstwhile close friendship of the two men, it did have the positive consequence of forcing Leibniz's hand, persuading him of the need to publish his own method. Thus it was that in the October 1684 issue of the *Acta Eruditorum* Leibniz's differential calculus made its first public appearance in the *Nova methodus* (Leibniz 1684b; see text [21] for substantial extracts from the article). It contains clear statements of the rules for finding differences (or differentials) of sums, of differences, of products, of quotients, of powers, and of roots, e.g. for *multiplication*: $d(xy) = xdy + ydx$, *division*: $d\frac{v}{y}$ or, if one posits $z = \frac{v}{y}$, $dz = \frac{\pm vdy \mp ydv}{yy}$ and so forth, although there are no demonstrations of these rules. It explains how these rules can be applied to form differential equations, which can then be used to find maxima and minima of curves (as announced in the article's title). Given this application, as well as the later references in the article to infinitely small distances and infinitangular polygons, readers naturally supposed that the differentials were infinitely small elements of curves. Leibniz, however, does not say this, and studiously avoids defining differentials beyond the bare idea that they are differences in values of the abscissa, ordinates, tangents, arc lengths, etc. In fact, the differences are introduced as follows: "Now let an arbitrarily assumed straight line be called dx, and let the straight lines which are to dx as v ... is to XB ... be called dv ... ôr the differences of v ...."

Leibniz's demurral to define his differences as infinitesimals later caused some consternation among his supporters, especially when it was coupled with formulations in his *Lemmas on Incomparables* (from his *Tentamen*, published in 1689, text [23]), from a published response to Nieuwentijt (text [45]), and from a letter to Pinsson (text [37]), that gave finite examples for the differentials. It also puzzled historians of mathematics when they discovered a manuscript written in preparation for his publication of the *Nova Methodus*, the *Elementa calculi novi* (*Elements of a New Calculus*), in which the differentials are treated as infinitely small in the derivation of the rules for the differential of a product, etc.[16] There Leibniz writes: "If now these dx and dy are taken to be infinitely

---

[15] Leibniz (1684a), "On finding the measures of figures"; Leibniz also published an *Additio* in December of the same year addressing the efforts of Johann Christoph Sturm (GM V 123–26, 126–27); see Parmentier, (Leibniz 1995), 82–92, 92-92, for French translations.

[16] *Elementa calculi novi pro differentiis et summis, tangentibus et quadraturis, maximis et minimis, dimensionibus linearum, superficierum, solidorum, aliusque communem calculum transcendentibus*, first published by C. I. Gerhardt in his edition of the *Historia et Origo* (Leibniz 1846, pp. 32–38); translated into English by J. M. Child (1920, pp. 136–144).

small, or when the two points on the curve are understood to be a distance apart that is smaller than any given, ...", then the line joining the two points on the curve, when produced, will be the tangent (Leibniz 1846, p. 33; Child 1920, p. 137).

But in fact the treatments given in the two texts are not significantly different. The above conditional proposition from the *Elementa calculi novi* is matched by an equivalent statement in the *Nova Methodus*, that if one wants to determine tangents using the method, then the differences must be interpreted as infinitely small. When in the *Elementa* Leibniz says that (in calculating $d(xy)$, $dxdy$ can be ignored because "it is infinitely small in comparison with the rest, for it is supposed that dx and dy are infinitely small (because the lines are understood to be continuously increasing or decreasing by very small increments throughout the series of terms)" (Leibniz 1846, p. 36; Child 1920, p. 143), this is eminently understandable on the reading of the infinitely small that he had established in the *De Quadratura Arithmetica*. To say that dx is infinitely small is to say that it can be taken as a straight line that is small enough that the error in the resulting expression is made smaller than any preassigned quantity. This conception can apply equally well to the "arbitrarily assumed straight line" dx of the *Nova Methodus*.

Some commentators have suggested that Leibniz chose to define dx in terms of a finite line in the *Nova Methodus* in order not to upset those traditionalists who disapproved of the use of indivisibles and infinities as not well-founded.[17] But this would have been a poor strategy, since in order for the calculus to apply to the finding of tangents, the differentials *had* to be interpreted as infinitely small, as Leibniz explicitly maintained in the *Nova Methodus*: "to find a tangent is to draw a line which joins two points on a curve an infinitely small distance apart, that is, the produced side of an infinitangular polygon, which to us is equivalent to a curve. That infinitely small distance, however, can always be expressed by a known differential, such as dv..." (see Text [21]). If Leibniz had meant differentials to be defined as exclusively finite, that would have undermined such an application. And on the other hand, in the *Nova Methodus* he mentions the possibility that dv "is infinite with respect to dx", which makes sense for dx as a finite difference and dv as a (fictionally) infinite one in comparison with it, but not for dv and dx as infinitesimals of the same order.

Thus it is not that Leibniz has two differing ways of defining his differentials, one in terms of finite lines, the other in terms of infinitesimals. Rather, dx is defined neutrally in terms of a line arbitrarily taken, precisely so that the differentials can be interpreted either as finite or as infinitesimal, depending on the problem to be solved. On Leibniz's conception of his calculus, it applies to differences of values of a variable term, and is valid irrespective

---

[17] For example, Bos, in his justly lauded article on Leibniz's calculus, writes "Leibniz did not give reasons for choosing this definition for the differential, but it seems most likely that he chose it to avoid controversies on infinitesimals. That it was a conscious choice may be inferred from a manuscript which Gerhardt identified as an alternative draft for the first publication of the rules of the calculus, in which the differentials are introduced as infinitesimals" (Bos 1974–75, p. 63).

5.1 First Publications on Differential Calculus, and Foundational Reflections

of whether the differences are finite or infinitely small.[18] This twofold interpretation is clearly put to the fore at the beginning of the text Leibniz wrote in defence of his originality in 1714, the *Historia et origo calculi differentialis*. Thus as Marc Parmentier has observed in introducing his French translation of the *Nova Methodus*, even in the manuscripts where Leibniz is developing the calculus, he is content to write "dx, that is, the difference between two neighbouring x", without attempting any greater precision. For example, in the third part of the *Analysis Tetragonistica*, dated 1 November 1675, Leibniz writes: "Now this is true of every progression, whether of numbers or of lines—that is, even if we do not use curvilinear figures, but ordinated polygons—in other words where the differences between the terms are not infinitely small" (A VII 5, 311; Child 1920, p. 84). What is important is only that the dx are differences of what can be taken as "neighbouring" values, not whether they are finite or infinite. They stand for lines of arbitrary length, and so may be taken as small as needed either for the problem at hand, or for the error in the calculation to be less than any assignable error, and thus zero, as explained above.[19] And this remains the case throughout his career. In a published article from 1702 Leibniz expresses the sum of an infinite series by an integral with x ranging over the integers 2, 3, 4, etc. and $dx = 1$:

> For example, $1/3 + 1/8 + 1/15$ etc. ôr $\int \frac{dx}{xx-1}$ with x equal to 2, 3, 4, etc., is a series which taken as a whole to infinity can be summed, and dx is here 1. For in numerical series the numbers are assignable.[20]

In the *Nova Methodus* the plasticity of the calculus so conceived is evident when Leibniz comes to show how to solve the problem de Beaune had set for Descartes. The problem was

---

[18] "*Dans le cas des séries de nombres, dx a une valeur finie et constante, or dans son esprit, le calcul vaudra toujours indifféremment pour les différences finies et les infiniment petites. ... L'assignation à dx d'une valeur infinitésimale pourra dès lors se présenter comme cas particulier d'un calcul plus général, irréductible en tout cas à ses applications géométriques.*" (Leibniz 1995, p. 100). Cf. Weil (1975), p. 686: "In the *Historia et origo* and elsewhere, Leibniz never tires of explaining how finite differences were at the source of his calculus. The analogies between the two topics, or rather (in view of his way of looking at them) their substantial identity, appear to have been at all times at the very center of his thoughts on this subject."

[19] Thus we take issue with Bos's claim in the previously mentioned article, that "the choice of (11) [sc. the definition in terms of an arbitrary line in the *Nova Methodus*] as definition in Leibniz 1684a was an anomalous and rather unfortunate one (indeed, the term *differentia* in relation with this definition is a misnomer)." (Bos 1974–75, p. 64). On the contrary, *differentiæ* are precisely the subject of Leibniz's differential calculus, differences that may be finite or infinitesimal. We will return to this topic later in connection with Leibniz's *Cum prodiisset...*, the importance of which was stressed by Bos in this article.

[20] "*Exempli gratia $1/3 + 1/8 + 1/15$ etc. seu $\int \frac{dx}{xx-1}$ posito x esse 2 vel 3 vel 4, etc. est series quae tota in infinitum sumta summari potest, et dx quidem hoc loco est 1.*" (*Specimen novum analyseos pro scientia infiniti circa summas et quadraturas, Acta eruditorum*, 1702; GM V, 356–57). Parmentier calls Leibniz's setting of $dx = 1$ "*étonnante*" (astonishing) (Leibniz 1995, p. 100, n.22).

to find the curve such that, in drawing the tangent *WC* at a point on the curve to the axis *AX*, *XC* would always be equal to the same constant segment *a*. Leibniz writes

> Now, *XW* ôr *w* is to *XC* ôr *a* as *dw* to *dx*; therefore, if one takes the *dx* (which one can take as one wants) constant, that is to say invariable, namely, equal to *b*, that is, if *x* ôr *AX* increases uniformly, we will have $w = \frac{a}{b}dw$, and the ordinates *w* will be proportional to their increases or differences *dw*; that is, if the *x* are in arithmetic progression, the *w* will be in geometrical progression, and if the *w* are numbers, the *x* will be their logarithms; so the line *WW* is logarithmic.[21]

This is reminiscent of Leibniz's solution of the quadrature of the symmetric hyperbola (text [5]), where the area under the continuous curve comes out as $1 - y + y^2 - y^3 + y^4$ etc. for values of the ordinate between $y = 0$ and $y = 1$, yielding the infinite series of discrete terms $1 - 1/2 + 1/3 - 1/4 + 1/5 - 1/6$ etc.

The novel status of the *dx* is one of the reasons why Leibniz's contemporaries had such difficulty understanding the "*nova methodus*" of Leibniz's 1684 article—especially in the absence of any supplementary explanation. But to this must be added the astoundingly complicated and forbidding form of the first example that Leibniz chose to give of his method in the 1684 paper. In order to illustrate how his rules could be used "even in a difficult calculation", he began with the equation

$$\frac{x}{y} + \frac{(a+bx)(c-xx)}{(ex+fxx)^2} + ax\sqrt{gg+yy} + \frac{yy}{\sqrt{hh+lx+mxx}} = 0$$

(This equation appeared even more intimidating in the original, in which division was expressed as a ratio, the square of *y* in the third term was written as "quad *y*", and overbars were used instead of parentheses.) It was then differentiated four times, with numerous changes of variables, and with Leibniz's ambiguous signs (designed to keep all the abscissas and ordinates positive) giving the formulas even more complexity. He then gave three "more immediately intelligible" examples: a proof of the law of refraction, whose result he had quoted in his 1682 paper on optics; a curve that is "such that the sum of six segments drawn from any one of its points to six fixed points on an axis is equal to a given line *g*"; and finally de Beaune's problem mentioned above.

One of the huge advantages of his new approach over existing methods, as Leibniz was at pains to emphasize in his publications of 1684 and 1686, was that it could be applied to problems that were intractable by the existing methods of indivisibles. In particular, it could

---

[21] "*Jam XW seu w ad XC seu a, ut dw ad dx; ergo si dx (quae assumi potest pro arbitrio) assumatur constans sive semper eadem, nempe b, seu si ipsae x sive AX crescant uniformiter, fiet w aequ. $\frac{a}{b}dw$, quae erunt ipsae w ordinatae ipsis dw, suis incrementis sive differentiis, proportionales, hoc est si x sint progressions arithmeticae, erunt w progressions Geometricae, seu si w sint numeri, x erunt logarithmi: linea ergo WW logarithmica est.*" (GM V, 226).

5.1 First Publications on Differential Calculus, and Foundational Reflections

be applied to finding quadratures of what he had termed *transcendent* curves. It was not just that it could determine that the particular curve to be squared does not admit an algebraic quadratrix, as Tschirnhaus had claimed on behalf of "his" method: "By the same method I can find what has a quadratrix, if not an algebraic one, then at least a transcendent one, that is, one presupposing the quadrature of the circle or hyperbola, or at least that we reduce the remaining measures to these simpler ones."[22] Consequently, as he stressed, such transcendent curves as the cycloid and the logarithmic curve

> can be expressed by a calculus and even by finite equations, albeit not algebraic ones or equations of a determinate degree, but equations of an indefinite degree, that is, transcendent ones; and thus can be subjected to a calculus by the same method as the rest, even though that calculus is of a different nature than that which is commonly employed.[23]

An illustration of the difficulty his contemporaries experienced is provided by the fact that his mentor Christian Huygens, despite his appreciation of Leibniz's mathematical talents, could not make much of his calculus until very late in life. It was not until 1693, in a letter of 17 September, that Huygens was finally able to write of "your wonderful calculus, which I can now handle with moderate ease"—an appreciation that would have been of immense value to Leibniz in his priority dispute had it been stated publicly before Huygens' unfortunate demise in 1695.

By that time, of course, Leibniz's calculus had been taken up by others, after lying fallow for the first few years. In particular, the Bernoulli brothers had taught themselves how to use it, with Jacob Bernoulli seeking Leibniz's advice. Both Johann and Jacob had subsequently competed with one another in applying it to challenge problems. In the May 1690 issue of the *Acta eruditorum*, Leibniz had been delighted to discover an article by Jacob Bernoulli using his calculus to provide a neat solution to the problem of finding the curve along which a descending heavy body recedes from or approaches a given point uniformly—the problem Leibniz had set as a challenge for the Cartesians in the course of his exchange with the Abbé de Catelan in 1689. There Bernoulli had proposed the problem of the catenary or hanging chain as another problem suitable for treatment, and Leibniz had immediately solved it, and proposed it as a challenge problem, to be solved within a year. Leibniz's own comprehensive and very elegant solution was published in the June 1691 *Acta eruditorum*, together with the solutions of Huygens, Johann and Jacob Bernoulli.

---

[22] "*Eadem methodo invenire possum, quam habeat quadratricem, si non Algebraicum, saltem transcendentem, hoc est Circuli aut Hyperbolae aut alterius figurae quadraturam supponentem, ut scilicet saltem dimensiones reliquas ad has simpliciores reducamus*" (GM V, 124).

[23] "*Verum sciendum est, istas ipsas quoque, ut Cycloidem, Logarithmicam, aliasque id genus, quae maxime habent usus, posse calculo et aequationibus etiam finitis exprimi, at non Algebraicis seu certi gradus, sed gradus indefiniti sive transcendentis, et ita eodem modo posse calculo subjici ac reliquas, licet ille calculus sit alterius naturae, quam qui vulgo usurpatur.*" (GM V, 124).

Meanwhile, however, there had been another momentous event in the history of the calculus, namely the publication of Newton's *Philosophiæ Naturalis Principia Mathematica*, which Leibniz was able to borrow briefly and make notes on when he was in Vienna in 1688, on his way to Rome. For Newton's book included his Method of First and Ultimate Ratios, outlining the principles of the fluxional geometry that underlay his approach to cosmology and to motion in resisting media. Leibniz had already applied his calculus to certain problems about the paths of rays in optics, and to motion in resisting media, but he had not published the results. So once again it was the publication by a rival of results similar to his own unpublished work that provided the stimulus for him to publish, which he did in three papers submitted to the *Acta eruditorum* of February 1689: on optical lines, on motion in resisting media, and on celestial motions. The third of these, the *Tentamen de motuum coelestium causis,* was the most important, an application of his calculus to cosmology, which Leibniz unwisely pretended to have elaborated prior to seeing the *Principia*.[24] Nonetheless, it contained an elegant derivation of the inverse square law through a differential equation containing second order differentials. In explanation of his method, Leibniz devoted a section to how his calculations with infinitesimals should be understood (text [**23**]). But where Newton had laid out his Method of First and Ultimate Ratios *more geometrico*, Leibniz's corresponding statement of his own method in the *Tentamen* was surprisingly laconic. It consisted in a single, rather informal paragraph, which ever after he would refer to as "my lemmas on incomparables". But there is no statement of individual lemmas (and their corollaries), complete with demonstrations, as there is in Newton's comparable foundations. Instead we are presented with something more akin to Newton's Scholium to his Method, in which Newton gives a defence of his use of line-elements, ultimate velocities and "ultimate ratios of vanishing quantities", and an explanation of how these are to be understood.

The main novelty is Leibniz's use of the term "incomparably small" to describe the relationship of his differentials to "ordinary" quantities, for instance, of dx to x. That is, he again declines to define his differentials as infinitely small, advising that "if someone does not wish to employ *infinitely small* quantities, one can assume them as small as one judges sufficient for them to become incomparable, and to produce an error of no importance, indeed, an error smaller than any given" (text [**23**]). The first examples he then gives are of taking the Earth as a point, or its diameter as infinitely small with respect to the heavens; these are examples where the error is finite but of no importance, typical in the application of mathematics to physics. The further examples are of *unassignable* differences: for example, the angle of a triangle whose base dr is incomparably smaller than either side r, will be incomparably smaller than a right angle. Likewise, the sine of such an angle will

---

[24]"An Essay on the Causes of the Celestial Motions" (Leibniz 1689). English translations of all three of these papers can be found in (Leibniz 2023, pp. 81–84, 85–94, 95–111).

5.1 First Publications on Differential Calculus, and Foundational Reflections

also be incomparably smaller than the sine of a finite angle, and its versed sine will be incomparably smaller than it.[25]

Here the notion of comparability is in fact the traditional one: a point cannot be compared with a line because they are not homogeneous: they cannot be put into a finite ratio to one another. This depends what is now called the Archimedean axiom, as Leibniz would explain to Nieuwentijt in 1695 (text [**45**]): "For only those homogeneous quantities are comparable, I hold with Euclid Book 5, Definition 5, one of which can be made greater than the other when multiplied by a finite number" (GM V, 322). Thus if we suppose that the difference between the sine and the versed sine is the least assignable difference Δ comparable with the sine, then (by the Archimedean property and the relation between sine and versed sine) we will be able to choose some value v of the sine which gives a difference smaller than Δ, which had been supposed as the least assignable. So the difference is smaller than any assignable difference, and is null compared with the sine. As Leibniz says in the same letter to Nieuwentijt, such an unassignable difference between two quantities "is the very thing which is said to be a difference smaller than any given", and its being null by comparison with that of which it is supposed to be the difference is something that can always be confirmed "by a process that is indeed Archimedean" by means of a *reductio ad absurdum*, like the one just given. This describes what we have called above the *Principle of Unassignable Difference*.

Between the publication of the *Nova Methodus* and the *Tentamen*, Leibniz had published another important article in the *Nouvelles de la République des Lettres* of July 1687, the *Extrait d'une Lettre de M. L. sur un Principe Générale* (text [**22**]). It was in this article that he announced his Principle of Continuity, which he expressed as follows:

> When the difference of two cases can be diminished below any given magnitude *in datis* or in what is asked, it must also be diminished below any given magnitude *in quaesitis* or in what results from it. Or, to speak more familiarly: When the cases (or what is given) continually approach and finally disappear one into the other, it is necessary that the sequences or outcomes (or what is required) do so also. (text [**22**])

As explained earlier, this principle can be seen as a generalization of the Principle of Unassignable Difference, now applied not just to geometric magnitudes, but also to the laws of motion in physics. The point of Leibniz's article is to show Malebranche that the laws of motion he gave in his *Recherche*, although modified from the Cartesian laws he had endorsed earlier, were still not compatible with the Principle of Continuity in this wider acceptation.

---

[25]This may be justified as follows: $\sin d\phi = d\phi - d\phi^3/6 + $ (smaller terms) $= d\phi$, since $d\phi^3$ is incomparable with respect to $d\phi$. Again, the versed sine of an angle ϕ is $r(1 - \cos \phi)$, so the versed sine of $d\phi$ is as $1 - \left(1 - \frac{d\phi^2}{2} + \frac{d\phi^4}{24} - \text{(smaller terms)}\right) = \frac{d\phi^2}{2} -$ incomparably smaller terms, so that the versed sine of an infinitesimal angle is incomparably smaller than it.

Henk Bos understood Leibniz's appeal to the Principle of Continuity to constitute a different approach to the foundations of his calculus than the one deriving from his improvement of traditional methods based on the Method of Exhaustion.[26] As we shall argue in Chap. 6, there is some truth to this, but the method involving an appeal to the Principle of Continuity is not based on any assumptions about the existence of infinitesimals or infinities. Indeed, when Leibniz asks us to consider a parabola "as an ellipse one of whose foci is infinitely distant", he is careful to add "or (to avoid this expression) as a figure which differs from a certain ellipse by less than any given difference" (see text [**22**]), thus putting in evidence once again his belief that the calculus is neutral with respect to existence assumptions.

This is related to misplaced conceptions according to which Leibniz ought to have consistently defined differentials as infinitesimals. As we have seen, however, their status is neutral between their being differences of finite quantities and infinitesimal differences. Rather, as Bos has perceptively observed, they are *variables*, like the quantities of which they are the differences, and differentiation is for Leibniz an operation on variables. Moreover, "the concept of variable differs from the [modern] concept of function in that it is not necessary to specify on which 'independent' variable a given variable depends." (Bos 1974–75, p. 17). Variables are indefinite, and do not necessarily stand for given geometrical quantities specific to a given curve (what Leibniz called the "functions" of the curve). Nor do they depend on the choice of a "progression of the variables", that is, on which variable is taken as the "independent variable", as the parameter or unit of measure.

Supporting Bos on this point, Parmentier goes further, claiming that Leibniz's differentials "*dx, dy* are not quantities but variables".[27] But as a matter of fact, that contradicts Bos, who declares that "The differentials and sums, introduced by the operators d and $\int$, are quantities, and therefore they have a dimension" (22). To mediate this difference of opinion, it would probably be preferable to say that the differentials *stand for* quantities—as dx stands for an arbitrarily small line. When interpreted, the variables stand for geometrical quantities or magnitudes—and even non-geometrical ones, such as *conatus*, as we shall see—but quantities having a dimension. The variables themselves are indeterminate magnitudes, and when operated on by differentiation, what are produced are further indeterminate magnitudes of the same dimension: the differences are "affections" of these magnitudes, analogous to their powers, such as x^3.

---

[26] "Leibniz considered two different approaches to the foundations of the calculus; one connected with the classical methods of proof by 'exhaustion', the other in connection with a law of continuity." (Bos 1974–75, p. 55).

[27] "*Enfin, c'est parce que* dx, dy *ne sont des quantités, mais des variables qu'elles peuvent être non seulement différentiées à leur tour, mais surout par la même opération, le même transitus automatique, qui faisait passer la première variable à sa différentielle.* [Finally, it is because *dx, dy* are not quantities but variables, which can not only be differentiated in their turn, but above all by the same operation, the same automatic *transitus*, which achieves the passage of the first variable to its differential]" (Leibniz 1995, p. 101).

Thus when Huygens suggested (text [24]) that he had a method that differed from Leibniz's only in the expressions they used, Leibniz replied that Huygens was free to express matters in his own way. However, certain things would not come as easily as with his own expressions:

> It is almost as if instead of roots or powers one would always want to substitute letters, and instead of xx or x^3, to take m or n, after having declared that these must be the powers of some magnitude. See, Sir, how cumbersome this would be. It is the same for dx or ddx, and the differences are no less affections of magnitudes, indeterminate in their places, than the powers are affections of a magnitude taken apart.

What this discussion indicates, however, is that we should examine more closely Leibniz's own philosophy of mathematics, rather than merely speculating about it. For Leibniz, naturally, had his own definite views on the nature of quantity, magnitude, number, and the status of infinitesimals. This is the topic of the following section.

5.2 Mathematical Foundations: Quantity, Magnitude and Quasi-minima

As we saw in the previous section, although Leibniz was cautious not to have his algorithm depend on the acceptance of infinitesimals, he had no particular problem in talking about infinitesimal quantities when describing it. When he envisaged publishing the *Nova Methodus* as an appendix to his *Dynamica* in 1690, here is the introduction he wrote in order to present his work:

> The incremental or differential calculus takes place wherever there is some change. Every change in nature is continuous, and happens gradually, not by leaps. Accordingly, transitions happen through unassignable ôr indefinitely small quantities, an infinite multitude of which is needed to constitute ordinary quantities.[28]

Declarations of this kind, which are widespread in Leibniz, could give the impression that there are quantities of a special kind, which Leibniz called *inassignabiles* and which he thought of as genuine mathematical entities grounding his calculus. The "Lemmas on Incomparables", to which Leibniz systematically refers his interlocutors holding the latter view, show that this was not the case. There, Leibniz emphasizes that "if someone does not wish to employ *infinitely small* quantities, one can assume them as small as one judges sufficient for them to become incomparable" (text [23]). Accordingly, one can use the

---

[28]"*Calculus incrementalis vel differentialis locum habet ubicunque mutationi locus est. Omnis autem mutatio in natura continua est, ac per gradus fit non per saltum. Adeoque transitus fiunt per quantitates inassignabiles seu indefinite parvas, quarum infinita multitudine opus est ad quantitates communes constituendas.*" (A III, 4, 488).

differential algorithm, and more generally infinitesimal quantities, without being committed to their existence. At the beginning of the "*Défense du calcul des differences*", which Leibniz wrote in the early 1700s in order to tackle various criticisms raised against his algorithm, this will be noted as the first misunderstanding: "I learn that talented people are opposed to the Calculus of Differences, because it seems that in it one necessarily proceeds by infinitesimals"; and his answer will be exactly the same: "One can always show them that everything that is concluded by this calculus can be proved by a *reductio ad absurdum* in the style of Archimedes, and by using the Lemmas on Incomparables proposed in the Leipzig *Acta*" (text [**49**]).

This is Leibniz's most repeated motto: the differential calculus is neutral as regards the question of the existence of infinitesimal quantities, and mathematicians should not enter into these questions, which, at the end of the day, involve metaphysical issues such as the composition of the continuum. Everything that is formulated in terms of infinitesimals can always be rephrased without them, in the style of the Ancients and with recourse to the *Lemmata Incomparabilium*.

But precisely because the practice is neutral with regard to the question of existence, this paraphrase leaves open the possibility of interpreting the same practice in terms of infinitesimals (seen as genuine mathematical entities). Moreover, from the observer's point of view, it could seem that these two interpretations were put on an equal footing, in the sense that Leibniz did not find the need to decide between them. But this would be a confusion between what Leibniz advocated as a mathematician and what he thought as a philosopher. If he repeats that mathematicians do not have to enter into philosophical questions, that certainly does not mean that he did not have his own philosophical view on the question, as we already saw in Chap. 2. The neutrality of the mathematical practice is not a veil drawn in order to mask some indecision, but a clear strategy in order to have his calculus acceptable by all parties in the game.[29]

But what were Leibniz' philosophical views on the question of the existence of infinitesimals? Did the development of differential calculus modify in any substantial way his reflections elaborated at the end of the Parisian stay? From what we saw in Chap. 2, Leibniz considered that an infinite number was a contradictory notion. By the same token, infinitesimals interpreted as inverses of infinite numbers, as was proposed by Wallis, were also contradictory. These positions are repeated from 1672 to 1716 without any modification. But, we also saw that it was only in 1676 that Leibniz took a clear stance against infinitesimal lines as being absurd. In between, one finds various attempts to consider "thick" points lacking extension. This leaves us with two different questions to answer: first, can we accept some mathematical quantities which are immune from the

---

[29] It is important to keep in mind that, as Leibniz recalled in his letter to Dangicourt from 1716 (Dutens III, 500–501), his view on fictions were not shared by supporters of the differential calculus in the *Académie des Sciences*. His exchange with Johann Bernoulli at the end of the 1690s (text [**32**]) and his comments expressed to Varignon about Fontenelle's approach (text [**38d**]) largely confirm this statement.

5.2 Mathematical Foundations: Quantity, Magnitude and Quasi-minima

contradiction involved in the concept of an infinite number (be it infinitely large or infinitely small)? For example, could we have some infinitesimal magnitudes not entailing the existence of infinite numbers? We will study this question in the present section and show that it is not the case. Leibniz's definition of quantity or magnitude forbids such a possibility. Then in the following section we will turn to a second question: granted that there cannot be infinitesimal quantities, couldn't we have mathematical entities that are not reducible to quantities, yet which could correspond to infinitesimals (two obvious candidates being physical "endeavours" [*conatus*] represented by infinitely small lines, and horn angles)? In this case too, we will see that Leibniz showed that this was not the case—although the demonstration of this fact cost him more effort than answering the first question.

Leibniz gave different accounts of the nature of quantity, on the one hand, and number on the other. The manuscript *Initia Mathematica* (text [**25**]) gives a good summary of his views on the topic. In the first section, *De quantitate*, he defines quantity, number, and other foundational concepts. Thus

> *Quantity* is that which belongs to a thing insofar as it has all its parts, ôr, on account of which it is said to be equal to, greater than, or less than, another thing (with each homogeneous with the other), ôr can be compared with it. (GM VII, 30)

According to this definition, only things to which the part-whole axiom can be applied can count as quantities *stricto sensu*. (For Leibniz 'part' always means 'proper part'.) This characterization is repeated consistently by him at various periods, from the beginning[30] to the end of his work.[31] In this context, the order relation (greater or less than) between quantities is defined in terms of whole and part and this is why Leibniz takes the axiom "the whole is greater than the part" as an analytic truth which suffers no exception in mathematics.[32] As we saw in Chap. 2, Leibniz used Galileo's paradox to conclude that if an

---

[30] From *De Arte Combinatoria* (1666) A VI, 1, 170 (*Quantitas igitur est numerum partium*). See also "*Quantity* is the same as the totality itself of a thing, that is to say, that by which it is called a whole" A VI, 2, 507 (1671–1672)). In the *De magnitudine* from 1676, Leibniz explains that he used to define quantity (or magnitude) as a number of parts, but that he was not satisfied by this definition since it implicitly relied on the equality of parts and, therefore sounds circular (A VI, 3, 482). After having proposed to define quantity by congruence, he nonetheless came back to his initial definition and accepted that quantity and equality should be defined in association; see, for example, A VI, 4, 417 (dated from the beginning of the 1680s): "We can define *Quantity* as the affection of the whole of a thing in as much as it possesses all of its parts".

[31] See, amongst other places: GM VII, 53: "every quantity can be determined by the number of parts congruent with each other, or by the repetition of a measure"; Leibniz (2018), p. 206: "Nor can quantity be judged to obtain except with a whole inside which there are many homogeneous parts of that whole".

[32] The same proof in terms of whole and parts is found, for instance, in the *Demontrationum propositionum primarum* of 1671–1672 (A VI, 2, 482) and in the *Initia rerum mathematicarum Metaphysica* of 1715 (GM VII, 20).

infinite aggregate is interpreted as a whole, then it can be proven equal to its proper part, contrary to the part-whole axiom. It then follows that infinite aggregates, and therefore an infinite number, cannot be wholes, *and are therefore not quantities*.[33] Likewise he showed in October 1674 that the infinite area under the Apollonian hyperbola could be proven equal to its proper part, with the same applying to the rendering of its value as an infinite series. In that case, the infinite aggregate is not a cardinal number, but the measure of an area. And if infinite aggregates are not wholes, then the infinitely small areas, an infinite aggregate of which constitutes the finite area in a quadrature, cannot properly be their parts.

One might, however, suspect that such arguments only apply to quantities interpreted as discrete aggregates of parts, and might not apply to truly continuous quantities as magnitudes. However, such a distinction between quantity and magnitude has no textual basis in Leibniz. Quite the contrary, they are systematically identified.[34] Accordingly, Leibniz always defines continuous magnitudes in terms of whole and parts: "*Magnitude is that which is designated by a number of congruent parts*" (Leibniz 2018, p. 169). If infinite wholes are contradictory, infinite magnitudes and quantities are contradictory notions from their very definition, and so are infinitely small parts. Accordingly, to talk about infinitely small or infinitely large magnitudes is just a way of speaking.[35]

This is expressed most clearly in the *Observatio quod rationes*, which was published by Leibniz in 1712 as an explanation of the "true meaning" of the infinitesimal method (*de vero sensu Methodi infinitesimalis*):

> Moreover, just as I deny the reality of a ratio one of whose terms is a quantity less than nothing, I also deny that properly speaking there exists an infinite or infinitely small number, or an infinite or infinitely small line — even though Euclid often speaks of an infinite line, but in a sound sense. The *infinite*, whether continuous or discrete, is properly neither one, nor a whole, nor a quantity [*quantum*], and if we take it as such by a certain analogy, it is only, so to speak, a way of speaking; when, namely, there are more things than can be comprised by any number, yet by analogy we attribute to these things a number, which we call infinite. (text [**41**])

Here it is clear that Leibniz makes no distinction between discrete and continuous infinites and is explicit about the fact that neither of them can be taken as a quantity, except by

---

[33] We are indebted to Filippo Costantini (draft of *Leibniz's Mereology: a Logical Reconstruction*) for developing the above argument, based on Leibniz's definition of quantity in *De quantitate* from the *Initia Mathematica*.

[34] From *De magnitudine* (A VI 3, 482) of 1676 to *Initia rerum mathematicarum Metaphysica* (GM VII, 18) of 1715.

[35] See text [**35b**]: "It is therefore an abbreviation of speech when we say 'one' when there are more things than can be comprised in an assignable whole, *and treat as a magnitude something that does not have its properties*. For just as it cannot be said of an infinite number whether it is even or odd, so it cannot be said of an infinite straight line whether it is commensurable with a given straight line or otherwise; so that these are only improper ways of speaking of infinity, as though of one magnitude, based on some analogy, but which, if you examine them more carefully, cannot be sustained".

5.2 Mathematical Foundations: Quantity, Magnitude and Quasi-minima

analogy. In this latter case, it is just a "way of speaking" of a variable quantity which can be made greater (or less) than any given quantity. This is what Leibniz called a "syncategorematic" interpretation of the infinite, to which we will come back in Sect. 6.1.

Now, it is true that the general definition of magnitude or quantity rests on a definition of number in which it cannot be characterized as a discrete collection of units (as was the case in the Euclidean tradition). This is what Leibniz calls the notion of number *lato sensu*.[36] To assess this claim, we therefore need to investigate Leibniz's philosophy of number. In *De quantitate* he defines it as follows:

> *Number* is homogeneous with unity, and so it can be compared with unity by adding to it or subtracting from it. And it is either an aggregate of unities which is called an *integer*, ... or an aggregate of aliquot parts of unity, which is called a *fraction*... (text [**25**]; GM VII, 30)

As Kyle Sereda has argued (Sereda 2015), according to this definition (and others Leibniz gives elsewhere) number has an essentially *relational* aspect. It is not just a multitude of ones, as the Euclidean definition had it, but something that can stand in a ratio to 1. In his discussion of the above definition Leibniz argues that it follows from it that even a surd like $\sqrt{2}$ is a number. This can be shown, Leibniz argues, by taking a square *AEFG* whose side *AG* equal to 1. Its diagonal *AF* can be shown to be the side of a square whose area is 2.

> Therefore, since we have defined number to be homogeneous to unity, there must certainly be a number whose relation to unity is that of the straight line *AF* to the straight line *AG*, that is to say, assuming *AG* is 1, there must be a number which expresses the quantity of *AF*, which is called $\sqrt{2}$. (text [**25**])

Moreover, Leibniz intended this definition of number to encompass not only integers and surds, but also rational numbers, and transcendent ones. As he wrote in the *Initia rerum mathematicarum Metaphysica* of 1715,

> From these things it is evident that *Number* of whatever kind—integer, fraction, rational, surd, ordinal, transcendent—can be defined by a general notion, as that which is homogenous to unity, ôr what is related to unity as one straight line is to another. It is also evident that if the ratio of *a* to *b* is considered as a number which is to unity as one straight line is to another, ratio itself will be homogeneous with unity, whereas unity itself represents the ratio of equality. (GM VII, 24)

Leibniz evidently thought he could show that transcendent numbers were consistent with this definition by employing his definition of homogeneity:

> Those things are *homogeneous* which are either similar or can be rendered similar by a transformation. Two straight lines are homogeneous, because similar; but a straight line and the arc of a circle are homogeneous things, because a circle can be extended into a straight line. (text [**25**])

---

[36] *Nouveaux Essais sur l'entendement humain* II, 16, §4 (A VI, 6, 156).

Thus a semicircular line has a ratio to the radius of the circle; but since it can be transformed into a straight line of the same length, there must be a number expressing that ratio, namely the one we denote π, even though this number cannot be expressed as the root of an algebraic equation. That is, π is a "transcendent number"—a term Leibniz invented. The transformation in question here is what was called at the time "the rectification of a curve"—turning it into a straight line of the same length—that is, finding an expression for the arc length. Leibniz (along with Jacob and Johann Bernoulli) had been at the forefront of successful applications of his calculus to determine rectifications of curves. Such rectifications (integrals yielding the arc length) require an extension of the definition of transformation in order to cover the fiction of treating the line as an infinite sum (i.e. integral) of infinitely small parts. We will return to this matter below. Granting its success, we see that Leibniz's definition supports his claim that integers, fractions, rationals, surds, ordinals, and transcendents are all numbers, since they are all comparable with 1, since they are homogeneous with 1. But this definition of number is incompatible with treating an infinitesimal as a number in relation to 1 and "infinitesimals" are never included by Leibniz in his lists of numbers taken *lato sensu*. *Infinitesimals cannot be numbers, according to Leibniz, since they are not homogeneous with 1*. They cannot be "compared with unity".

Now, the inclusion of the idea of *comparability* in the definitions of number and quantity is deliberate: an infinitesimal is not a number because it is *incomparable* with 1. This means that Leibniz's definition of number coheres with his definition of quantity. If you take two different homogeneous quantities, one of them will be equal to a part of the other (or to a part of it after they have been rendered congruent by a continuous transformation). They will therefore have a ratio to one another, and this will be expressible by a number with a certain ratio to 1. Thus "there are only ratios between homogeneous things" (GM VII, 34), while homogeneous things are "those things whose magnitudes can be expressed by numbers by assuming the same measure as unity for all of them" (GM VII, 36). But of course, by assuming different measures as unity, different numbers can be assigned to the same quantity; "therefore quantity is not a definite number, but the material basis for a number [*materiale numeri*], ôr an indefinite number that is to be defined by assuming a certain measure" (text [**25**]).

It would be a mistake, therefore, to equate 'quantity' with 'number' in treating Leibniz's philosophy of mathematics. Nevertheless, such is the connection between his definitions of the two notions that one quantity is comparable with another one homogeneous with it if and only if their ratio is expressible as a number comparable with 1; therefore, if there are no infinitesimal numbers because they would be incomparable with 1, then accordingly there will be no infinitesimal quantities either, and vice versa: a purported infinitesimal quantity dx is not comparable with a finite quantity x, it has no ratio to it, there is no number that could express such a ratio.

One might object here that this heterogeneity (between infinitesimal and finite quantities) does not prevent infinitesimals from constituting a domain of homogeneous quantities of their own. But two things are worth noting: firstly, in order for these entities to

5.2 Mathematical Foundations: Quantity, Magnitude and Quasi-minima

be quantities and to be called "homogeneous", they would have to be expressible "by numbers by assuming the same measure as unity for all of them" (text [**26**]). We will come back to this condition, since it is closely related to the introduction of fictitious "quasi-minima", which allow us to proceed as if we were dealing with quantities. Secondly, even if it were possible to really (as opposed to: "by a fiction") introduce a unit of measure amongst infinitesimals, this would still not allow us to handle them in computations on a par with finite quantities—as is required in infinitesimal techniques such as differential and integral calculus. To be more precise, such computations would not be possible if they involved the acceptance of multiplication and division by an "infinite number"—a notion which, we recall, implies contradiction according to Leibniz. So there is no question of Leibniz rejecting infinitesimals as supposed reciprocals of infinite numbers, and yet upholding infinite and infinitesimal "quantities".[37] This is impossible to reconcile with his foundations of mathematics.

Even though infinitesimals are not quantities, however, it is possible to treat them *as if* they are quantities in certain well defined circumstances. That is, one may, *by a fiction*, treat them as if they are infinitely small parts of a continuous quantity, while treating the latter as a "sum" of infinitely many of them. This, of course, is what Leibniz does in calculating, for instance, the rectification of a curve. How is this consistent with his definitions?

The first thing to observe is the relational aspect of quantity in Leibniz's definition of it, as noted above. Thus although an infinitesimal dx has no ratio to a finite x, it is comparable with an infinitesimal ds (where s and x are functionally related). It follows, for instance, that under the fiction that these dx and ds are parts of the abscissa x and the trajectory of the curve s, respectively, they can be compared. Indeed, in his rectifications (as in his quadratures), Leibniz assumes an infinitesimal such as dx as the common measure, so that in performing his calculations with infinite sums, dx is taken as "unity". As Leibniz explains in his important essay *Specimen Geometriae luciferae*, the comparison of such quantities involves a resolution into infinitesimals understood as fictional common measures, justifiable by the usual reasoning involving the Principle of Unassignable Difference:

> Also the method of indivisibles and infinites, ôr rather of the infinitely small or infinitely large, ôr of infinitesimals and infinituples, is of the utmost use. For it contains a certain resolution as if (*quasi*) into a common measure, albeit one that is smaller than any given quantity, ôr a means by which it is shown that by neglecting other things which make an error smaller than any given, and thus null, one of two things that are comparable can be transformed into the other by a transposition. (text [**27**])[38]

---

[37] As suggested in a recent article (Katz et al. 2021).

[38] In the continuation of this passage Leibniz makes clear its agreement with his Lemmas on Incomparables (GM VII, 273; see (text [**27**])). On the introduction of a fictitious infinitely small common measure see also GM VII, 40 ([text [**26**]).

These fictional common measures Leibniz calls "*quasi-minima*"; and their use in the transformation of a curve into a straight line involves what he calls a "*quasi-transformation*". This is necessary because on Leibniz's understanding of transformation, the figure to be transformed is analysed into parts, the parts are transposed into congruent parts, and then reassembled into the transformed figure. But, as he writes in a text from 1685, when a curve is transposed into a straight line, "we recognize that the same thing persists as before, when nothing really does remain, for no part of the straight line is left in the curve; and if someone says that all the points are left, he says something very obscure, for there is no certain and definite number of points in the continuum, indeed points are merely modes" (A VI 4, 628/ LLC 273). Here Leibniz concludes that in order to bypass this difficulty, homogeneous things should be defined without appealing to transformation, namely as "those things which have a common measure, either an exact one, or one as exact as desired, namely such that the remainder is less than any given quantity" (628/ 273). In the *Specimen Geometriae luciferae*, he expresses the latter idea in terms of a quasi-transformation becoming a true transformation when the error can be made smaller than any given:

> On the other hand, there is a kind of transformation that does not conserve the parts, as when a straight line is transformed into a curve, a gibbous surface into a plane, and anything rectilinear into something curvilinear, and conversely. In that case, therefore, only the minima are conserved, and the transformation is when one thing is made from another with at least the same minima remaining, and so is conserved in a perfect real transformation through something flexible or liquid. But in a mental transformation instead of minima we can employ quasi-minima, that is, indefinitely small parts, in order to make a quasi-transformation, since instead of something curvilinear we can employ something quasi-curvilinear, namely a rectilinear polygon having as great a number of sides as you wish; and if then the quasi-transformation we seek in this way should succeed, that is, the error ôr difference between a quasi-transformation and a true one will always come out smaller and smaller, so that finally it becomes smaller than any given, then it can be inferred to be a true transformation. (text [**27**])

This is the justification of the passage from finite to infinite: one acts *as if* there were minima such as atoms or metaphysical infinitely small entities as common measures, although the principle of sufficient reason forbids the existence of such things in reality. As Leibniz wrote to Varignon, "the rules of the infinite succeed in the finite, as if there were metaphysical infinitely small things, even though there is absolutely no need for them, and the division of matter never does proceed to infinitely small particles" (text [**38a**])

In this way, Leibniz is able to defend the use of infinitesimals in, for instance, setting the ratio of dy to dx in Pascal's infinitesimal triangle equal to the ratio between the finite straight lines in determining the tangent to a curve, without having to settle the question of whether such infinitesimals actually exist as parts of the continuum. It is sufficient that they be taken as fictional parts that can be smaller than, greater than or equal to one another, for then they can bear finite relations to one another, and thus stand to one another in the same ratio as finite lines. On the fiction that they are infinitely small parts on the point of vanishing, they have a ratio to one another, even while, relative to the whole of which they are supposed to be parts, they are null. As Leibniz explains to Guido Grandi in 1713:

Furthermore, my opinion, very often expounded, is that infinitely small as well as infinite quantities are indeed fictions, although useful for reasoning compendiously and at the same time safely. And it suffices that they are understood to be truly as small as is necessary in order for the error to be smaller than any given; from which it is shown that there is no error. I have indubitable arguments for this opinion, but which would be too prolix to expound now. Meanwhile we conceive infinitely small [quantities] not as simply and absolutely nothings, but as *relative nothings* (as is well known), that is, as indeed vanishing into nothing, yet retaining the character of that which is vanishing. (text [**42**])

5.3 Further Thoughts on the Existence of Infinitesimals

The "quasi-minima" employed by Leibniz are not to be confused with true minima in the sense of indivisibles. Leibniz was already clear about this in 1676. In a cancelled scholium to Proposition XI of his *De quadratura arithmetica* (text [**6**]), he wrote: "The Geometry of Indivisibles is fallacious unless it is explained using the infinitely small; for neither points nor truly indivisible things may be safely used, but instead one must use lines that are indeed infinitely small, yet still lines, and therefore divisible" (A VII 6, 549). These infinitely small lines have in common with Cavalieri's "indivisible" lines that they may stand in finite ratios to one another, but while a finite line may fictionally be regarded as the sum of infinitely many infinitely small lines, a point as the minimum of a line has no magnitude. Thus no number of such points can constitute a finite line, just as no number of finite, bounded lines can constitute or exhaust an unbounded line. It is otherwise, though, with the fiction of a bounded infinite line. We will come back to this notion later. And just as such an infinite bounded line can be conceived as composed out of a multiplicity of finite ones "even if this multitude exceeds every number", so in quadratures a finite line can be conceived as composed of infinitely many infinitely small yet divisible ones. That is, once we allow that a finite line can actually be composed of infinitely small lines taken as units, we have allowed that there exists an "infinite ratio" between such a unit and the finite line. Given the theory of proportions, then, the same ratio could exist between the finite line and an infinite but bounded line. Thus we have a proportion: an infinite line bounded at both ends (which we will designate b_∞) is to a finite line x as that finite line x is to an infinitesimal line dx. Thus the finite line x "is the mean proportional, not in a certain way, but truly and exactly, between some infinitely small line and some infinite one": in symbols, $dx : x = x : b_\infty$. As we saw in Chap. 2, this is precisely the place where Leibniz stresses: "whether the nature of things tolerates quantities of this kind is for the metaphysician to discuss; for the geometer it suffices to demonstrate what follows from supposing them" (A VII, 6, 549; text [**6**]).[39]

---

[39] Leibniz refers to this passage in his correspondence with Bernoulli in his letter of 29 July/8 August, 1698: "I would also add something that I also once wrote in the said unpublished manuscript, that it is possible to doubt whether there could be infinitely long straight lines that were nonetheless also in fact

However, as we have seen in the previous Sect. 5.2, it is not possible to define a ratio between an infinitesimal and an ordinary quantity *stricto sensu*, since a ratio is only defined between homogeneous quantities, and homogeneous quantities are expressible "by numbers by assuming the same measure as unity for all of them". An "infinite ratio" therefore involves an "infinite number" which is a contradictory notion. So how is it that Leibniz can introduce a mean proportional between an infinitely small line and some infinite bounded line "truly and exactly"? In the framework of the geometry of the indivisibles, as in the framework of differential calculus, the answer is very clear: an infinitesimal dx stands not for a fixed infinitely small quantity but for a finite variable quantity, which one can take as small as one wishes (and the same will be true for infinitely large quantities considered as variable finite quantities taken as great as one wishes). An "infinite ratio" is just a way of speaking of a relation between variable quantities, just as Newton did with his "first and ultimate ratios". And in this sense, it is "true and exact". However, this answer does not settle the question of existence. What it shows is that infinite small and large lines cannot be taken as genuine quantities, not that they are contradictory in and of themselves (if not taken as quantities). The metaphysician still has to study whether nature tolerates this type of entity (a typical example of which was for the young Leibniz physical endeavours or *conatus*).

In order to tackle this question, one has thus to dwell a little bit on the notion of quantity in Leibniz. Because "infinitely small" involves a notion of smallness, it seems that the previous distinction (whether or not "infinitely small" lines are taken as quantities) makes no sense. As soon as something is said to be "small", it has to be a quantity and since we showed that these entities cannot be quantities, we also showed that an "infinitely small" entity is a *contradictio in adjecto* (hence cannot exist, either in mathematics or in nature). But it should be emphasized that Leibniz allows for what he presents as an "imperfect" estimation of quantity. In other texts, he will also talk of quantities taken *lato sensu* similarly to how he also allows for numbers taken *lato sensu*. But one should be careful not to conflate these two situations, since a quantity *stricto sensu* can have number attached to it, whereas a quantity *lato sensu* can be measured only indirectly.

A typical example of such quantities taken *lato sensu* are angles, which are not measured by a direct estimation, but by recourse to the arcs intercepted on the circle and thus by the ratio between the sides of the subtended arc (for example, by a sine).[40] More

---

bounded." (text [**32e**], A III 7, 857). In other letters in the exchange (text [**32**]), he also repeats that questions about the existence of the infinite and infinitely small are "μεταφυσιχώτερα".

[40] On the difficulty of directly defining whole/part (and hence quantitative) relations between rectilinear angles, see Leibniz (1995), pp. 92–93 and 110–115. Already in the *Numeri infiniti* from 1676 (text [**10**]), Leibniz criticizes his earlier view according to which the angle can be considered as the quantity of a point. He then introduces the indirect measure by the sine (*Erit ergo anguli quantitas nihil aliud quam quantitas sinus proportionalis*). Since this measure is invariant whatever is the size of the lines forming the angle, he calls it a "fiction".

5.3 Further Thoughts on the Existence of Infinitesimals

generally, any object not satisfying the *partes extra partes* condition,[41] such as "intensive magnitudes" (typically *conatus*), would require such an indirect approach and can be called quantities only *lato sensu*.[42] Leibniz seems to have adopted such a position in 1672, when he defines the "beginning" of a body or an infinitely small line as "the beginning of motion itself, i.e. an endeavour, since otherwise the beginning of the body would turn out to be an indivisible" (text [**3**]). In so doing he explicitly analogized them to angles, which enclose a space "infinitely smaller than any space that can be sensed", and are consequently points.[43]

This conception is explained in full detail in the *Scientia mathematica generalis* (circa 1700),[44] but the distinction between perfect and imperfect estimation of magnitudes is also expounded in the *In Euclidis πρῶτα*, and in the *Initia rerum mathematicarum Metaphysica*,[45] and goes back to manuscripts on the angle of contact written in the 1680s.[46] Indeed, a typical example of the latter is given by the fact that the angle of contact is said by Euclid to be "smaller" than any rectilinear angle although this meaning of "small" cannot be the same as the one used in perfect estimation. We will come back to this example later. In this broad sense, claims Leibniz, a line can be called "greater" than a point and a surface "greater" than a line. So the first thing to notice is that the main distinction is not between number and continuous magnitude (a magnitude is always characterized by Leibniz as a certain *number* of (congruent) parts), but between magnitudes taken *stricto* and *lato sensu*. In the second sense, nothing prevents the possibility of infinitely small lines, which would not be magnitudes *stricto sensu*, but geometrical entities possibly captured by indirect measure such as is the case for angles.[47]

---

[41] On the fact that angles are not quantities, because they do not satisfy the *"partes extra partes"* condition, see for example the *In Euclidis πρῶτα*: "Magnitude must be defined in such a way that it includes line, surface, and solid, but not angle, which I believe can be made to follow in this way: *Magnitude* is a continuum which has situation. But an angle is not a continuum. For a continuum two things are required: first, that any two of its parts equalling the whole must have something in common which, however, is not itself a part; second, that in a continuum there are parts exterior to one another [*partes extra partes*], as is commonly said, that is, it must be possible to take any two of its parts (but ones not equalling the whole) which have nothing in common, not even a minimum." (GM V, 184).

[42] See in particular Leibniz (2018, p. 189): "But even where the parts are not external to one another, if they cannot be known when discerned in themselves, they can perhaps be known in their effects, as for instance degrees of force are nevertheless estimated …".

[43] Fifty years later, in the *In Euclidis πρῶτα*, he repeats the analogy between angles and (instantaneous) velocity (GM V, 184).

[44] Leibniz (2018, pp. 188–209).

[45] GM VII, 22.

[46] See, in particular, the *De Angulo Contactus et curvedine et de natura quantitatis* (LH 35, 1, 19, fol. 7–8). A transcription of this text can be found at: https://eman-archives.org/philiumm/angles-of-contact/de-angulo-contactus-et-curvedine-et-de-natura-quantitatis.

[47] The fact that mathematics in general, and geometry in particular, include a quantitative orientation (associated to the notion of magnitude) and a "qualitative" one (associated to the notion of form), is

This might explain facts which could otherwise appear as oddities in Leibniz's thought, the first one being that he stumbles upon the impossibility of infinite wholes as early as 1672, although he explicitly rejects actual infinitesimals only in 1676. When one studies the various stances endorsed by Leibniz in between, as we did in Chap. 2, it is clear that infinitely small entities were precisely first conceived as geometrical entities, "thick points", which he sometimes called "metaphysical points".[48] In this context, an angle could defined as *quantitas puncti* and accordingly one point can be called "greater" than another one.[49] As we have seen, as late as February 1676, in *De arcanis sublimium*, Leibniz was still conjecturing that matter could be comprised of such "metaphysical points", making it a discrete, infinitely divided entity, in distinction from space, as a truly continuous one.[50] We should add that even in December 1676, after apparently having convinced himself of the absurdity of supposing that either atoms or infinitesimals exist, he still cites the division of matter into "points" as an argument for the existence of atoms, showing a lingering belief that profound metaphysical consequences might after all follow from the existence of points taken *latu senso*.[51]

Another intriguing fact is his repeating to Bernoulli in June 1698 that accepting infinitely small lines would be tantamount to accepting bounded lines that would be to ordinary ones as infinite to finite. Indeed, instead of ruling out this possibility by logical arguments, he then lists some paradoxical results of a physical nature which would derive from this and concludes that he would not dare to admit them, unless forced by indubitable proofs.[52] This reflection seems independent of the arguments based on the part-whole axiom. This is clear from the development of the exchange with Bernoulli in which Leibniz recalls the latter arguments in order to exclude the possibility of infinitesimal terms in numerical series, but still concludes: "Therefore it must be seen by what reasoning it could be demonstrated that it is possible (for example) for there to exist a straight line that is infinite and yet bounded at both ends." (A III 7, 943; text [**32k**])

Now, the fact that one cannot exclude a possible interpretation of the infinitely small in terms of geometrical entities does not mean that Leibniz accepted their existence as in the

---

something on which Leibniz insists in the 1680s (see Leibniz 2018, p. 99). On the fact that Angles should not be comprised in the realm of magnitudes *stricto sensu*, see note 41 above.

[48] See the discussion in Chap. 2 above.

[49] In particular, *De minimo et maximo* (1672–1673), (A VI, 3, 98–99; Leibniz 2001,pp. 10–15).

[50] A VI, 3, 473–74; Leibniz (2001), pp. 46–49.

[51] See his *Catena mirabilium demonstrationum de summa rerum* of 12 December 1676: "Supposing plenitude, *atoms* are demonstrated. In fact, they can also be demonstrated without plenitude from the sole consideration that everything flexible is divided into points [*Posita plenitudine demonstrantur atomi. Imo et sine plenitudine, ex ea sola consideration omne flexile dividetur in puncta*]." (A VI 3, 585/ DSR 108-09).

[52] (A III, 7, 796; text [**32a**]). See also the letter from 12/22 July 1698 in which Leibniz repeats that accepting infinitesimals in nature would amount to accepting *linea infinita terminata* (A III, 7, 828; text [**32c**]).

5.3 Further Thoughts on the Existence of Infinitesimals

modern nonstandard setting, where they must be non-contradictory entities. First of all, as recalled by Leibniz against Bernoulli, there will be no way to use *directly* these nonstandard entities in numerical computations (and in particular in the handling of infinite series). But, at any rate, what Leibniz repeats again and again is not that these geometrical entities would be non-contradictory, but that he entertained serious *doubts* about their existence as early as 1676. We will see how these doubts guided a very clear strategy developed in order to *banish* them from the mathematical realm. This strategy finally resulted in what Leibniz presented in the mid-1690s as a *proof* of their absurdity.

The first text to consider in this regard is the passage from the *Pacidius Philalethi* (1676), which we encountered already in Chap. 2.[53] Whereas in February of the same year, Leibniz could declare that the splendid success of the hypothesis of infinitely small in Geometry increases the likelihood that they really exist,[54] he now claims, under Pacidius' pen:

> I would indeed admit these infinitely small spaces and times in geometry for the sake of discovery, although they were imaginary. But I am deliberating whether they can be admitted in nature. For there seem to arise from this infinite straight lines bounded at both ends, as I will show elsewhere; which is absurd. (text [**11**])

This first argument[55] is remarkable since it concludes that infinitely small entities cannot exist in nature *because* they entail the existence of infinite bounded lines, *quod est absurdum*.[56] Leibniz announces that he will show this elsewhere. It might be a reference to his argument in the cancelled scholium to Prop. XI of the *De Quadratura* (text [**6**]), based on the mean proportional, which he intended to develop. At any rate, we do not know of any other text bearing on this issue from this period. However, there exists a text dating from the beginning of the Hanover period in which Leibniz intends to provide such a demonstration by *reductio ad absurdum* (text [**28**]).[57]

Leibniz's demonstration goes as follows. Suppose we accept the existence of an infinitesimal line *ab* (see Fig. 5.1. below), standing in an "infinite ratio" to a line *cd*, which is finite. We first look for a mean proportional between *ab* and *cd* and call it *ef*. *Ef*

---

[53] Leibniz enclosed the passage in square brackets. The meaning of this indication is not clear, but compared with other places in the manuscript, it is unlikely that it marks a deletion (which Leibniz usually indicates by crossing out the passage he wants to suppress).

[54] A VI, 3, 475.

[55] It is followed by another argument ("*Praeterea*") based on the principle of sufficient reason.

[56] This sense of the implication is always valid for Leibniz: if something is absurd, it cannot exist in nature—although the contrary sense of the implication (from non-existence in nature to absurdity) is not valid.

[57] LH 35, 8, 12, fol. 37. The date is not known, but the paper used by Leibniz suggests that it was written before 1682. This manuscript was already transcribed by Enrico Pasini (Pasini 1985–1986, App. 35–39) and by André Robinet (Robinet 1986, p. 292), and O. Bradley Bassler provided an English translation and commentary (Bassler 2008).

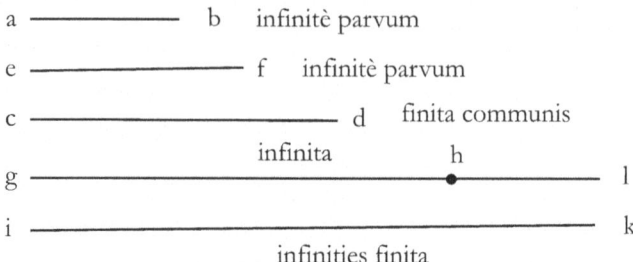

Fig. 5.1 The scale of infinites in LH 35, 8, 12

could be either in an infinite ratio to *ab* or a finite one. But the latter is not possible or else, by the definition of a mean proportional, it will also be in finite ratio to *cd* and thus be a finite ordinary line. Next we seek a line *gh* such that *cd* is the mean proportional between it and *ef*, which, for the same reason, would have to be infinite with regard to *cd*. Again, we construct another mean proportional such that *cd* to *gh* is as *gh* to *ik* (where *ik* will be "infinitely infinite").

Since we are dealing with lines, we can now transport *ik* onto *gh* and have its excess *hl*. Then Leibniz concludes that *gh*, being terminated in *h*, is finite. But this is absurd since it was constructed as infinite.

The "proof" is certainly not fully convincing, to say the least, and Leibniz might have realized this since he crossed it out. In the first place, his argument that *ef* must be infinitesimal (because otherwise *ab*, as the third proportional to *ef* and *cd*, would be finite, contrary to hypothesis) could equally well be applied to prove that it must be finite (otherwise *cd*, as a third proportional between *ab* and *ef* would have to be infinitesimal). A possible answer to this difficulty is suggested by Bassler (2008, pp. 139–142): perhaps *ab* could be interpreted as a second order differential, *ddx*. Then $ab:ef (= ddx:dx) = ef:cd$ ($= dx:x$), which is possible for the logarithmic curve $y = a \log\left(\frac{x}{b}\right)$. But, secondly, it is in any case hard to see why Leibniz needs the mean proportional *ef* at all, since he could have established that *gh* is infinite by taking *cd* as the mean proportional between itself and *ab*. Then, if *gh* is taken as the mean proportional between *cd* and *ik*, it follows that *ik* is infinitely infinite, as desired. To these objections one might add that it is difficult to understand why Leibniz thought *gh* being infinite and bounded is a contradiction, when all the infinite lines in his diagram appear to be bounded. Perhaps his intention was to prove that if one denies that infinite lines must be bounded, but insists that infinitely small lines can stand in a ratio to finite lines of 1 to infinity, then one is inexorably led (by this argument) to lines that are both bounded and infinite—which is a contradiction if one denies infinite yet bounded lines.[58]

---

[58]Cf. Leibniz's argument in his Annotations on White's *Euclides Physicus* [text [29]], in which he argues that if there are infinite lines one greater than another standing in ratios, then, since "every

5.3 Further Thoughts on the Existence of Infinitesimals

At any rate, it is interesting to see that at the beginning of the Hanover period, Leibniz intended to show that infinitely small lines are fictions *because* they entail a contradiction.[59] This already goes directly against an interpretation of "fictions" seen as "ideal elements" à la Hilbert, which can be introduced in a theory without contradiction, and more generally against the claim according to which Leibniz saw a crucial difference between fiction in the sense of infinite whole (contradictory notions) and fiction in the sense of infinitely small and large lines (which would be non-contradictory notions).[60] The second point to notice is that the proof is based on the fact that ordinary lines would be mean proportionals between infinitely small and infinitely large ones, as was argued in the *De Quadratura Arithmetica* (text [6]). In *Quicunque attentius ea considerabit* (text [28]), Leibniz tries to show (by adjoining progressively various third proportionals) that this would entail one of the lines constructed being at the same time bounded and infinite. The argument is therefore that if one takes a ratio between an infinitesimal line and an ordinary line to be something other than a "way of speaking" (of ordinary variable quantities), this would imply the existence of an infinite bounded line.

This contradictory conclusion is reminiscent of the argument Leibniz gives in "On Secrets of the Sublime" from February 1676, that "someone who lives for a number of years that is greater than any finite number can at some time die" (text [7]). This is in the context of imagining "another world that is of infinite magnitude, and yet bounded", in comparison to which we would be infinitely small. But there he takes this possibility as proving that "there is an infinite number", and that the success of "the hypothesis of infinites and the infinitely small" in geometry "increases the likelihood that they exist". In December of the same year (see Chap. 2), he finally rejects as absurd the interpretation of "worlds within worlds to infinity" as entailing worlds of descending degrees of infinity coexisting with each other, each being bounded by the next greater.[61] This circumstance takes on significance when Leibniz is confronted with Johann Bernoulli's position on the infinite and infinitely small in their exchange of letters on this issue in 1698 (text [32]).

For there he urges Bernoulli to see that "if we posit infinitely small real lines, it follows that we must also posit straight lines bounded at both ends, which are nonetheless to our ordinary lines as infinite to finite; assuming this, it follows that there is a point in space which cannot ever be reached from here in any assignable time by an equable motion",

---

straight line than which there is a greater is bounded", it follows necessarily "that there are bounded straight lines of infinite magnitude".

[59] The proof begins "*Quantitates autem infinite parvas esse fictitias [facile] ita ostendi potest*" and concludes with a clear "*Absurdum est autem rectam utriusque terminatur punctis G et H ipse magnitudine infinitam*" (the straight line GH being the hypothetical infinite bounded line terminated at G and H).

[60] —as claimed by Esquisabel and Raffo Quintana (2021) and Katz et al. (2021).

[61] Leibniz's argument is that to posit co-existing worlds, one bounding the other, would amount to positing a common space in which they all co-exist (by the very definition of what co-existence means)—an hypothesis which is ruled out by Leibniz.

adding "ôr, it would be possible for someone to live in such a way that they would die after a number of years that is not ever assignable, and yet they would die at some time—all of which I would not dare to admit, unless forced to do so by indubitable proofs" (text [**32a**]). But Bernoulli remains unmoved by this argument, replying that "I can easily conceive that there could exist in the least speck of powder a world in which everything would be in proportion to this large one, and on the other hand our world could be nothing other than a speck of powder in another infinitely larger one" (text [**32b**]), thus confronting Leibniz with the very same arguments for the reality of infinitesimals that he had countenanced in February 1676. He brushes aside Leibniz's rejoinder that the assumption in the calculus of straight lines bounded at both ends bearing ratios to infinite and infinitely small lines, does not prove that the latter exist in nature. For, Bernoulli asserts, our thought concerning the existence of the infinite or infinitely small is only relative: "because it is nothing in itself, and as neither large nor small, so neither infinite nor finite, and because finally there is no argument against the infinity of worlds which could not equally well be used by the inhabitants of another world to demonstrate that they alone exist" (text [**32d**]).[62]

Let us call this position the "relativity of the infinite". The idea is that a line that is infinitely small for us will appear finite to beings of a grade of infinity lower than us, and our finite lines will appear infinitely small to beings infinitely greater than us. The calculus is thus treated realistically, but relatively: there are no fixed infinitesimals, since comparability is relative: dx is incomparably small relative to x, but may have a finite ratio to dy. Bernoulli does not positively assert the infinity of worlds (of ascending and descending degrees of infinity), but cannot see that it is ruled out: he wishes only to assert its possibility (text [**32d**]). In response to this, Leibniz says, "If I were to concede such things as the infinite and the infinitely small that we are discussing to be possible, I would also believe them to exist." (text [**32e**]).

Returning now to the argument against infinitesimals in text [**28**], we cannot say for sure why Leibniz deleted it. Perhaps he simply saw that it was not a good argument, for reasons like those we have given above. But it is also possible that he was troubled by the relativity of the infinite. For in his Annotations on White's *Euclides Physicus* (text [**29**]) written at the end of the 1680s,[63] he argues that if there are infinite lines one greater than another standing in ratios, then, since "every straight line than which there is a greater is bounded", it follows necessarily "that there are bounded straight lines of infinite magnitude", from which "it

---

[62] Bernoulli had already presented Varignon with his metaphysical views about worlds embedded in worlds, the larger appearing infinite to the creature living in the smaller, in a letter from May 1698 (*Der Briefwechsel from Jean Bernoulli*, vol. 2, pp. 168–170). Varignon answered by approving this view, but Leibniz received just a copy of this passage without knowing to what Varignon was giving his assent. Hence some misunderstanding at the beginning of the exchange with Bernoulli, which the latter spells out in the text [**32d**].

[63] The significance of these annotations, and their connection with the argument in *Quicunque attentius ea considerabit*, was noted by Brad Bassler in his article on the latter (Bassler 2008, p. 151).

5.3 Further Thoughts on the Existence of Infinitesimals

seems to follow that there are infinitely small lines".[64] But, he remarks, "I am always brought up short by the difficulty that if an infinite number is once admitted, or an infinite body, there is not for me an essential mark for distinguishing the finite from the infinite, except with respect to us".

Earlier in the same annotations Leibniz had considered White's argument against infinite lines that "if from some point there are infinitely many spans, some part will be infinitely distant, but nothing can exist which is not in itself determinate". Leibniz refuses the first part of the reasoning (because an infinite length just means that one can take it as large as one wants),[65] but finds the link between the second premise and the conclusion very stimulating. He notes: "N.B. I like this; it seems to follow from it that those infinitely small lines cannot really exist, for there is no principle of determining the exact proportion. But this needs examining" (text [29]). Notice that Leibniz is pleased to find a new argument against infinitely small lines.

In the subsequent development, Leibniz reflects further on White's argument that "if there is an infinite multitude, a transition (*transitus*) needs to be made at some stage from the finite to the infinite".[66] Leibniz first recalls his initial doubt that once an infinite aggregate or infinite number is admitted, there will be no way of distinguishing the finite from the infinite "except with respect to us". This is just the repetition of the difficulty raised in *Quicunque attentius ea considerabit* (text [28]), but notice the following: here the problem does not come from the fact that an infinite number is a contradictory notion in and of itself, but that *if* one admits it, then one will not have a way to distinguish the finite from the infinite. This inspires him to add a new difficulty (*Difficultas nova*), which is of particular interest to us: "if at some time, that is, at a bounded time, it turns from being a finite line into an infinite one, it is necessary that at some time it ceases to be finite and starts to be infinite, which cannot be conceived. Therefore these things are imaginary: this to me seems demonstrative" (text [29]).

This is a new path for formulating a proof of the absurdity of infinitely small lines, which Leibniz finds demonstrative but which he does not develop. Fortunately enough, we can exhibit a text in which Leibniz finally elaborated this proof in full detail: the annotations he made on Froidmont's *Labyrinthus Continui* in the 1690s.[67]

---

[64] Cf. Leibniz's assertion in the scholium to Prop. 9 of the early draft of the *DQA*: "I call infinite that which is greater than anything intelligible by us, or greater than a quantity expressible by numbers, even though there is something else greater than it, which, since it extends farther when applied to the former, that to which it is applied will itself also be bounded." (A VII, 6, N. 20, 212).

[65] White writes "For if from some point there are infinitely many spans, some part will be infinitely distant" and Leibniz comments: "this does not validly follow, rather, there will always be a part more distant". (A VI, 4, 2092).

[66] The context is that of parts in a place (*loci partes*).

[67] This piece was first edited (although not identified at the time as notes on Froidmont) in (Leibniz 1996). A revised transcription and translation are given in Arthur and Ottaviani (forthcoming). The date of the text is not known, but it has to be from after late 1693 and, according to watermark evidence, was probably composed within a couple of years of that date.

The general context is a part of the notes in which Leibniz goes back to the possibility of interpreting physical endeavours [*conatus*] as infinitely small lines traversed in a moment. Leibniz first repeats his doubts about the fact that this would amount to accepting infinite bounded lines, an hypothesis which in turn invites the blurring of the distinction between finite and infinite. But he then expounds his new argument in the form of a proof: Suppose such an infinite but bounded line is traversed in a time that is also infinite but bounded. In traversing a point at the beginning of its motion "it will move for some time in such a way that it will have an infinite ratio to the space traversed in completing the whole" (text [**30**]). Thus, if dx is the initial "point" of motion covered in an unassignable time, x is some finite distance it covers in a finite time, and b_∞ is the infinite but bounded line traversed in an infinite but bounded time, these will be in the proportion $dx : x = x : b_\infty$. Now, since the infinite line is bounded, one can take its half, or any other assignable part of it. When it reaches such a halfway point, however, (or traverses any other assignable fraction of the line), the distance elapsed will be in an assignable ratio to the whole infinite line:

> Therefore at some time the ratio of what is traversed to the unassignable whole must necessarily begin to be assignable, but such a point or instant in which that transition comes about is impossible. Therefore such a motion, and so such a line, must also be held to be impossible, and there is no infinite yet bounded straight line. Whence it also follows that infinitely small straight lines are chimeras, although nevertheless useful, like imaginary roots... (text [**30**])

There is no ambiguity about the conclusion here: Leibniz considered such a reasoning as establishing the impossibility of infinite bounded lines, so that they should be ranked with infinitely small lines as chimerical mathematical entities. But the most important fact is to note that from 1676 to the reflections on Froidmont, we see Leibniz refusing to admit infinitesimal lines as non-contradictory objects, even when they are not taken as genuine magnitudes. On the contrary, he regularly attempted to prove that they involve a contradiction.[68]

---

[68] In this connection, notice that, contrary to what is claimed in (Katz et al. 2021), Leibniz has no problem equating the acceptance of infinitely large and small lines with the acceptance of an infinite number. The same thing is repeated in the *De Scientia Infiniti*: "And in general it can be said that an infinite number, an infinite line, a series composed from infinitely many terms, or an aggregate of an infinite multitude of things, is in metaphysical rigour not one thing, *since they always involve the greatest number, which is impossible*" (Gerhardt 1876, pp. 595–608. Our emphasis). Notice also that Leibniz then explains that in mathematics, one treats these entities "taken as one thing as an abbreviation of speech, since they have a foundation in reality"—a declaration which goes directly against any attempt to explain that there are "bad" (contradictory) and "good" (non contradictory) fictions in mathematics, and that only the latter have a *fundamentum in re* or else are "well founded fictions". In a passage just before the one we quoted, Leibniz has also put on the same footing the fiction of imaginary roots (one of the paradigms of well founded fictions) with... the fact of taking the Universe as a whole!

5.3 Further Thoughts on the Existence of Infinitesimals

Although these arguments involve motion and time, it is important to see that they rely on the contradictory nature of such behaviours and not on physical considerations such as those related to the principle of sufficient reason. In fact, Leibniz was happy to include motion and time in the very first notions of mathematics, as testified by the *Initia rerum mathematicarum metaphysica* (1715),[69] but already in many papers on *Analysis situs* dating back from the beginning of the stay in Hanover. In fact, we already saw that Leibniz had no problem stating that "the incremental or differential calculus takes place wherever there is some change" (A III, 4, 488), and it is this abstract notion of change or motion which is involved in the previous reasoning. At any rate, the argument itself is not based on any non-logical principle. It concludes that there must be a point at which a ratio is at the same time assignable and unassignable. And this is a contradiction, something which cannot be true whatever the world we consider is.

In his exchange with Bernoulli Leibniz does not enter into such considerations, but is content "to leave things in the middle" (A III, 8, 884). There he acknowledges that his arguments based on the mean proportional depend on the assumption of bounded infinite lines, and repeats that, as he had said in the *De Quadratura Arithmetica*, "it is possible to doubt" whether there are such lines, but that they can nevertheless be assumed as fictions, and this can be justified by the Lemmas on Incomparables without assuming their existence. One possible explanation for this relates to his reasons for being uneasy with the relativity of the infinite adopted by Bernoulli.[70] This is that to combat this he would have to appeal to metaphysical considerations, namely his distinction between the *unbounded* (as the source of quantity, but not in itself "subject to geometrical considerations"), and the *infinite but bounded* (which can be treated as if it is a quantity, consisting, fictionally, in an infinite aggregate of finite lines); whereas the former is an aspect of the Absolute, the latter is no part of really existing things.[71]

But another possible explanation for Leibniz's reluctance to go further with Bernoulli relates to the distinction we discussed earlier, according to which infinitesimals might be interpreted as magnitudes in some extended sense, not involving their having a proportion to the finite. Here it might be useful to analyse a last example regularly mentioned in support of the existence of nonstandard quantities in Leibniz: horn angles (also called "angles of contact", "angles of contingency").

---

[69] GM VII, 18–20.

[70] This may also explain why Leibniz cancels the scholium to Prop. XI of the *De Quadratura Arithmetica* (text [6]). Since he elaborates on this metaphysics of the infinite in much greater detail elsewhere, particularly in the *De Scientia Infiniti* of ca. 1698, it is clear that, if he represses the scholium because of its metaphysical content, this is not because he later repudiates that metaphysics.

[71] See, for example, the cancelled scholium to Prop. XI of the *DQA* (text [6]). This scholium is a reworking of what appeared in the earlier draft as the scholium to Prop. IX from April–June 1676, where the mean proportional argument does not appear. Leibniz explains this distinction in much more detail in his *De Scientia Infiniti*; for commentary on Leibniz's deep metaphysics of the infinite, see Arthur (2021).

As is well known, Euclid proved in prop. III, 16 of the *Elements* that such an angle of contact (between a circle and its tangent at a point) is "smaller" than any rectilinear one.[72] This prompted a lively debate amongst scholars to determine whether or not this type of angle could be ranked amongst quantities.[73] Leibniz thought a lot about this question, but unfortunately his foundational studies from the 1680s are still unpublished.[74] We cannot enter here into all the details of this discussion, but will content ourselves with addressing two main claims made by proponents of the idea that infinitesimals are for Leibniz a special sort of quantities. The first one is that since angles of contact can be compared with each other (by comparing the radii of the corresponding circles),[75] they are "quantities" (in a general sense of "quantity" which could entertain both assignable and unassignable ones). The second one is that horn angles can be compared (in a non-technical sense this time) with infinitesimals and provide us with a good model of the way infinitesimals relate to ordinary quantities. These two claims are based on confusions which, interestingly enough, Leibniz diagnosed very clearly and dismissed. We will just recall his arguments against them.

The first misunderstanding is already addressed in the manuscript *De Angulis curvarum*.[76] In this text, Leibniz expresses his first thought that the angle of contact is to the rectilinear angle as the line is to a surface. Just as a line can be compared to another line, angles of contact could be considered as quantities in relation to one other.[77] Yet, he notes, they would not be quantities in relation to rectilinear angles. There is no encompassing domain of "quantities" in which horn angles and rectilinear angles could be put together. In

---

[72] In fact, what Euclid proves is that the angle of contact is contained (in the sense of situated in) any rectilinear angle and from that, he concludes that it is "smaller".

[73] On these debates, see Jesseph (1999, p. 162 ff.) and Leibniz (2018, p. 182, n. 2).

[74] These, however, are easily accessible now that all the mathematical manuscripts from Leibniz have been digitalized. Some of these texts were already mentioned by Hess (1991). They are in the process of being transcribed by Sandra Bella in the framework of the project Philiumm (ERC Adg n° 101020985), see https://eman-archives.org/philiumm/angles-of-contact.

[75] This proportionality of angles of contact to the radii of the corresponding circles was argued for by Christopher Clavius. Leibniz developed this idea by fostering the notion of the osculating circle to a curve, the angle of contact being in that case the minimum angle between a circle and the tangent of the curve at a point, with the second and higher differentials of the curve at that point representing its curvature. See the manuscript "There can be infinite degrees of souls" (A VI 4, 1624-26/ LLC 299–303); this piece is now thought to date from the 1690s.

[76] LH 35, 1, 19, fol. 3–6 (1683).

[77] To endow angles of contact with a proper measure is not an easy task, though, and much of Leibniz' reflections in the 1680s is dedicated to this issue. His final conclusion is that it cannot be done in a general way, see *In Euclidis πρῶτα* (1712) where Leibniz explains that angles of contact can be endowed with quantity only in a loose way of speaking (*crassius loquitur*): "Indeed, it is not possible to assign a line that would be to another line as is the angle of contact *DCB* to the angle of contact *DCF*, which we already showed to be possible for rectilinear angles" (GM V, 191).

5.3 Further Thoughts on the Existence of Infinitesimals

the manuscript entitled *De Angulo Contactus et curvedine et de natura quantitatis*,[78] Leibniz insists on the fact that these questions cannot be treated properly if one does not elucidate the various meanings of quantity, which are confused in the controversies on the angle of contact. The fact that something can be called "smaller than" another thing is certainly not enough, and in a stricter (*arctior*) sense, only things endowed with parts could be called quantities. But this is still not enough since one could argue that an angle is part of another when it is situated in it (and in this sense the angle of contact could be called "part" of a rectilinear angle). One must therefore have this part-whole relationship behave in a properly quantitative way, i.e. according to exact ratios and proportions.[79]

Now, this leads us to the second misunderstanding. Leibniz comes to realize that, contrary to what happens with lines and surfaces, angles of contact and rectilinear angles cannot co-exist in the same continuous domain of variation, in the sense that there could exist a continuous transition from one to the other. In this sense, he was wrong to compare them to lines and surfaces. As he repeats later on numerous occasions, when one diminishes a rectilinear angle until one reaches the null angle, one never passes through the angles of contact.[80] This already rules out a reading of them as exemplifying the *status transitus* in the continuous diminution of rectilinear angles: there is no way one can rely on the existence of these geometrical entities in order to support an "extended continuum" in which non-Archimedean and Archimedean quantities could, so to speak, live together.

Interestingly enough, this is precisely the argument Leibniz finally put forward in order to dismiss the comparison between angles of contact and infinitely small quantities occurring in differential calculus. Although infinitely small quantities and horn angles are both examples of incomparable magnitudes (in the relational meaning that we explained), and can thus be put on the same footing in this respect, as Leibniz regularly does,[81] they cannot be put on the same footing when it comes to their relation to ordinary quantities. This is explained in the *In Euclidis πρῶτα* (1712), a text which was already published by Gerhardt at the end of the nineteenth century (GM V, 183–221). First Leibniz repeats that angles of contact are *not* quantities when related to the measure used for the estimation of rectilinear angles. When one attributes a quantity to them, it is based on a

---

[78] LH 35, 1, 19, fol. 7–8 (1683–1684).

[79] "Dico ergo ut dicatur aliqua res esse quantitas, seu habere quantitatem, non sufficere ut habeat partes sed eo quo diximus modo, sed etiam ut possit habere partem dimidiam, partem tertiam, etc. seu ut partes habeant rationem aliquam ad totum." Similarly, Peletier had argued that since no matter how far a rectilinear angle is divided, it will not make an angle less than an angle of contact, they could not be magnitudes of the same kind, and that therefore an angle of contact is not a magnitude in the proper sense, i.e. it does not obey the Archimedean axiom. See Jesseph (1999, p. 162).

[80] About the role of the *generatio uniformis*, which allows one to define what Leibniz will call "congenious" or "homogonous" entities, see, for example, the *De Situ et Distantia* (1685?), *Mathesis*, 147. See also, Leibniz (2018), pp. 197–209.

[81] See, for example, *De Legibus naturae et vera aestimatione virium motricium*, [1691], (GM VI, 214).

completely different principle.[82] Leibniz then expounds his idea that when one says that an angle of contact is smaller than a rectilinear one, this is only a loose way of speaking (*crassius loquitur*) and an "imperfect" kind of quantity (*recurrit ad quantitatis genus imperfectum*), in which one cannot determine exact ratios and proportions. Hence, the crucial conclusion: "And therefore, it is of a completely different genre, and in relation to a rectilinear angle, it [*scil*. the angle of contact] cannot even be considered as an infinitely small, which is always located between zero and an assignable."[83]

By contrast, we see here very clearly what characterizes infinitesimals in a Leibnizian setting: they can be arranged in a sequence representing a continuous transition amongst ordinary quantities. It is precisely because horn angles cannot be arranged in a sequence of rectilinear angles that one cannot compare them with infinitesimals. Meanwhile, although infinitesimals can be *conceived* as middle terms between points and lines, since they are all "congenious" with one another,[84] Leibniz takes himself to have demonstrated, in the argument he provides in his reflections on Froidmont's *Labyrinthus*, that taking them to *exist* as such middle terms would involve an impossible transition from an unassignable to an assignable ratio at some definite time.

---

[82]"quantitatem, inquam, non habet [angulus contactus] quae per mensuram anguli rectilinei aestimari possit. Si qua vero est ratio aestimandi angulos contactus comparandique inter se, oritur ex diverso plane principio et ad aliam plane mensuram refertur." (GM V, 191).

[83]"Ac proinde plane est diversi generis, et respectu anguli rectilinei ne quidem ut infinite parvus considerari potest, qui utique inter nullum et assignabilem collocatur" (GM V, 191).

[84]"Congenious" or "homogonous" is a technical term Leibniz introduced in this context in order to designate those entities which can be transformed continuously one into the other, although they do not need to be homogeneous (a typical example being a line and a point), cf. GM V, 385 and Leibniz (2018), p. 209.

Leibniz's Mature Justifications of the Calculus

6.1 Leibniz's Syncategorematic Conception of Infinitesimals

The rapid development of Differential Calculus in the 1690s, thanks to the joint efforts of the Bernoulli brothers, the Marquis de l'Hôpital and other members of the Malebranche circle, provoked a series of criticisms culminating in the great quarrel amongst proponents and opponents of the "new style" at the French *Académie des Sciences* at the turn of the century.[1] In this context, Leibniz expressed his views on the foundation of the calculus on several occasions. His first public declaration in the quarrel took the form of a letter to Pinsson in reaction to criticisms raised by Gouye on the new methods,[2] which was partially published in the issue of the *Mémoires de Trévoux* from November 1701 (text [37]).[3] In it, he declared that what L'Hôpital had published was enough to ground the certainty of the method and that, at any rate, it was not necessary to take the various forms of infinity used in this calculus "à la rigueur". He then repeated the same kind of comparisons as in the "Lemmas on Incomparables" (with a ball, the earth and the stars) and concluded accordingly that:

> instead of the infinite or infinitely small we take quantities as great or as small as is needed for the error to be less than the given error so that we differ from the style of Archimedes only in the expressions, which are more direct in our method, and more in keeping with the art of discovery. (text [37])

---

[1] See Blay (1986), Mancosu (1989), and Bella (2022).

[2] Gouye expressed his criticism as a reaction to a paper published by Johann Bernoulli in November 1700 in the *Acta Eruditorum*.

[3] "Memoire de Mr Leibnitz touchant son sentiment sur le calcul differentiel," *Memoires pour l'histoire des sciences et des beaux arts [Mémoires de Trévoux]*, Nov.-Dec. 1701, pp. 270–272.

This statement caused a lot of puzzlement amongst the members of the Académie (and also his correspondent Des Bosses), since it was not clear what Leibniz had in mind when saying that it was not necessary to take the infinite "à la rigueur". The fact that he took examples of finite entities could leave the impression that he thought that there was no need for the infinite at all and that he equated his infinitesimals with fixed finite quantities. This triggered an anxious letter from Varignon, the first of their correspondence,[4] to which Leibniz replied in a more precise way in a series of letters from 1702 (text [**38**]). Although all the texts from this late period are very well known (most of them were published either already at that time or not too long after Leibniz's death), we will briefly recall some of them so that the reader can see, firstly, that they are in perfect continuation of the material described in the preceding sections and secondly, how consistent they are in their denial of the existence of infinitely small quantities.

When writing to Grandi in 1713, Leibniz declares:

> Furthermore, my opinion, very often expounded, is that infinitely small as well as infinite quantities are indeed fictions, although useful for reasoning compendiously and at the same time safely. And it suffices that they are understood to be truly as small as is necessary in order for the error to be smaller than any given; from which it is shown that there is no error. (text [**42**])

Indeed, at the time, Leibniz had already expounded this conception on many occasions. Just the year before, as we saw, he published a paper in the 1712 *Acta Eruditorum* meant to clarify the "true meaning of the infinitesimal method (*de vero sensu Methodi infinitesimalis*), in which he wrote: "Moreover, just as I deny the reality of a ratio one of whose terms is a quantity less than nothing, I also deny that properly speaking there exists an infinite or infinitely small number, or an infinite or infinitely small line" (*Observatio*; text [**41**]). The argument he provides on this occasion is the one he devised during his Parisian stay against infinite quantities: "The *infinite*, whether *continuous* or *discrete*, is properly neither one, nor a whole, nor a quantity [*quantum*]", and the general strategy grounding its use is also the one he devised early on: "and if we take it as such by a certain analogy, it is only, so to speak, a way of speaking; when namely, there are more things than can be comprised by any number, yet by analogy we attribute to these things a number, which we call infinite." (text [**41**]; GM V, 389).[5]

In the *Nouveaux Essais* in 1704–1705, one finds the following claim:

---

[4]"M. L'Abbé Galloys (…) repand ici que vous avez déclaré n'entendre par *differentielle* ou *infiniment petit*, qu'une grandeur à la verité tres petite, mais cependant toujours fixe et déterminée" (A III 8, 798). The letter was sent in Nov. 1701, but reached Leibniz only in Feb. 1702.

[5]The fact that fictitious entities, typically imaginary roots of equations, are introduced by a form of analogy is already expressed in several texts from Leibniz's time in Paris: see A VII, 7, 560; A VI, 3, 462.

6.1 Leibniz's Syncategorematic Conception of Infinitesimals

> But we are mistaken in wanting to imagine an absolute space that is an infinite whole composed of parts; there is no such thing, it is a notion that implies a contradiction; and these infinite wholes, *and their opposites, the infinitely small*, have no place except in the calculation of geometers, just like the imaginary roots in algebra. (text [**34**]; our emphasis)

And in his only published philosophy book, the *Essais de Théodicée* from 1710:

> We embarrass ourselves in the same way with *number series* which go to infinity. We conceive a last term, an infinite number, *or an infinitely small one*; but this is all nothing but fictions. Every number is finite and assignable, every line is so likewise, and the infinite or infinitely small signify only magnitudes that one may take as big or as small as one wishes, to show that an error is less than that which has been assigned, that is to say, that there is no error: or else by the infinitely small is meant the state of vanishing or of beginning of a magnitude, conceived in imitation of magnitudes already formed. (text [**36**])

This accords with Leibniz's statements in his letter to Varignon of February 1702, published in the *Journal des sçavans* (text [**38a**]). In it, Leibniz answered the latter's perplexities by conceding that the comparison with fixed finite entities was just a way of rendering his reasoning about incomparability sensible to everybody ("*rendre le raisonnement sensible à tout le monde*"), but that even these ordinary incomparables should not be understood as fixed entities:

> one must consider that these common incomparables themselves, which are not at all fixed or determined, and which can be taken as small as one wishes in our geometrical reasonings, have the effect of rigorous infinitely small things, since whenever an opponent tries to contradict this statement, it follows from our calculus that the error will be less than any error he could assign, it being in our power to take this incomparably small as small enough for this purpose, inasmuch as one can always take a magnitude as small as one wishes. (text [**38a**])

This is the strategy based on the Principle of Unassignable Difference that we have explained in preceding chapters, and which is repeated over and over in all of these texts. Leibniz conjectures that this must be what Varignon had in mind when explaining that he considered the infinite in mathematics as being the same as the "inexhaustible" ("inépuisable"), adding: "and it is certainly in this that a rigorous proof of infinitesimal calculus consists [*et c'est sans doute en cela que consiste la demonstration rigoureuse du calcul infinitesimal*]" (text [**38a**]).

Just after this passage, Leibniz famously explains that one just has to take infinitesimals as fictions, which he calls in this letter "*notions ideales*".[6] The term itself will be used in a subsequent letter from April 1702, in which Leibniz explains that he did not change his views in order to reply to criticism (as suspected by Gouye after reading his response to Varignon, which he took to be of a different position from that expressed in the mémoire in the *Journal de Trevoux*), but that he had expressed them already before the quarrel:

---

[6]The first version, crossed out by Leibniz, gave "des fictions analytiques ou estres de raison".

> Moreover, I wrote some years ago to Mons. Bernoulli of Groningen that the infinite and infinitely small could be taken as fictions, similarly to imaginary roots, without that having to do a wrong to our calculus, these fictions being useful and founded in reality. (text [**38c**])

As we have seen in Chap. 4, §3, this passage is a reference to an exchange with Bernoulli in which Leibniz referred explicitly to the positions expressed in the *DQA*. So not only are Leibniz's positions in the 1700s consistent with the views we expounded in the preceding sections, they are situated *by him* in direct continuation of a strategy that he first developed in the *DQA*.

In the continuation of the letter of February 1702 to Varignon, Leibniz mentions fictions (which he replaced by "*notions ideales*" as we just saw) only in a rather negative way. This passage is also the one in which he explains his views on the syncategorematic infinite:

> Yet we must not imagine that the Science of the Infinite is debased by this explanation and reduced to fictions, for there always remains a syncategorematic infinite, as they say in the Schools, and it remains true, for example, that 2 is as much as $\frac{1}{1}+\frac{1}{2}+\frac{1}{4}+\frac{1}{8}+\frac{1}{16}+\frac{1}{32}$ etc., which is an infinite series in which all the fractions whose numerators are 1 and whose denominators are in a double geometric progression are comprised at once, although only ordinary numbers are used in it, and no infinitely small fraction, or one whose denominator is an infinite number, ever occurs in it. (text [**38a**])

The same reference to the views of the Schools is mentioned in the *Nouveaux Essais* in order to advocate for a positive conception of the infinite (as opposed to Gallois' understanding according to which Leibniz was simply denying any use of infinite by saying that one does not need the infinite in mathematics "à la rigueur"):

> Properly speaking it is true that there is an infinity of things, that is to say that there are always more of them than can be assigned. But there is no such thing as an infinite number or line or other infinite quantity, if we take them to be true wholes, as is easy to demonstrate. The Scholastics wanted to say this, or ought to have done so, in admitting a syncategorematic infinite, as they call it, and not the categorematic one. (text [**34**])

Notice that "syncategorematic" is not an observer's category, but the way Leibniz proposes to put a name on his particular view on the infinite. So the question is not whether or not one can interpret Leibniz's use of infinity in mathematics "syncategorematically", but what he had in mind by saying that his conception of the infinite was the same as that recognized by the Scholastics. This is not an easy task since there were various understandings of this term in scholastic philosophy.[7] Moreover, not only does Leibniz give no precise reference here, but he adds that if the Scholastics did not have precisely the view he has in mind, they "should have been doing so"! But several aspects are nonetheless pretty clear from what

---

[7] See Uckelman (2015) for discussion of medieval accounts of the syncategorematic understanding of the infinite, and for references; further discussion is given in Arthur (2018).

6.1 Leibniz's Syncategorematic Conception of Infinitesimals

Leibniz says: first, it is not just a way of talking about a potential infinite (by contrast to an actual one); second, a syncategorematic infinite for Leibniz is not a genuine whole; third, it is not the collection or set of its terms, yet it can be understood to include each of them; that is, it involves all of the terms provided the 'all' is understood distributively.

Let us dwell a little bit upon these three important features.

The term 'syncategorematic' has its origin in medieval grammar. While certain words (such as nouns) are *categorematic* terms, having meaning in isolation from other words, others (such as conjunctions, quantifiers and articles) are *syncategorematic*, deriving their meaning from the context in which they are used.[8] Other terms, like 'infinite', are ambiguous. Thus when Spinoza says that the divine attributes are infinite, if this is understood categorematically, he is saying that every attribute is itself infinite; understood syncategorematically, it can be taken to mean that there are infinitely many divine attributes. Likewise, one could claim, as William of Ockham did, that there are syncategorematically infinitely many parts in the continuum without this entailing either that the continuum is infinite in magnitude or that its parts are infinitely small. But Ockham proposed that these parts in the continuum are *actual* parts "protruding" into one another, contrary to the prevailing view, held by the Thomists, that there are only *potential* parts in the continuum. Now, although in his mature work Leibniz held that matter is actually divided into discrete, actual parts, he held that there are only potential parts in the continuum. So what was his understanding of the syncategorematic infinite?

According to one influential interpretation, the infinite occurring in Leibniz's mathematics is the Aristotelian potential infinite:[9] this, it is held, applies only to ideal entities, and is therefore distinct from the actual infinite that applies to the actual world. It denotes only the possibility of further progress in dividing, adding, and so forth. This infinite, it is claimed, is distinct from the actual infinite applicable to entities in the actual world, such as the monads that Leibniz takes to be the fundamental constituents of reality. Each of these infinites can be described as "syncategorematic", although in different senses. The former, which "concerns the abstract, ideal entities treated by mathematics", is "strictly speaking, only 'indefinite' or 'indeterminate' rather than genuinely infinite" (Antognazza 2015, p. 10). The latter, according to this view, applies to entities in the actual world, and is syncategorematic in the sense that it must be understood distributively: there are not actually infinitely many monads in the sense that there is an infinite number of them, so this kind of infinite is a purely qualitative one. The main text on which this interpretation rests is the "Supplementary Study" which Leibniz drafted for a letter to Des Bosses of 1 September 1706 (text [**35c**]):

---

[8] Leibniz is familiar with this meaning and uses it sometimes. See, for example, A VI, 4, 742 and GM VII, 54.

[9] This has been maintained particularly by Herbert Breger (see, for instance, Breger 1986, 1992, 2016). See also Bosinelli (1991) and Antognazza (2015).

> *There is a Syncategorematic infinite*, that is, a passive power having parts, namely the possibility of further progress in dividing, multiplying, subtracting, or adding. ...
>
> There is also an actual infinite in the sense of a distributive whole, not a collective one. Thus, something can be stated of all numbers, though not collectively. In this way it can be said that for every even number there is a corresponding odd number, and vice versa; but it is not therefore accurately said that there is an equal multitude of even and odd numbers.

The first point to note here is that in giving his example of the actual infinite understood distributively, Leibniz gives a *mathematical* example. He does the same in *Towards a Science of the Infinite* (text [**33**]), where he writes

> when I say that the infinite series of fractions 1, $\frac{1}{2}$, $\frac{1}{4}$, $\frac{1}{8}$ etc., is equal to 2, I mean that if each of these fractions is assumed and none besides, then neither more nor less is assumed than what is in 2. And in this sense it is understood that the whole infinite series is equal to 2, so that what in fact is called a collective whole, is understood to be a distributive one...

Such a series, like *any* "aggregate of an infinite multitude of things", "is in metaphysical rigour not one thing, since [such multitudes] always involve the greatest number, which is impossible". So, far from distinguishing different meanings of syncategorematic infinite depending on whether it applies to actual entities or to the ideal entities of mathematics, Leibniz explains how it applies to aggregates of all kinds, whether that is the multitude of even numbers, or "a series composed from infinitely many terms, or an aggregate of an infinite multitude of things" (text [**33**]), such as the substances making up the world. That matter is infinitely divided is explained mathematically: it means that there result more parts than there are unities in any given natural number.[10] Moreover, since each actual part presupposes a principle of activity (a monad within it), there are actually infinitely many monads in any part of matter—they are infinite in exactly the same sense as an infinite number is infinite. All such multitudes should be understood distributively, and not as constituting a collection (or infinite set, in a modern rendering). Thus the second paragraph of Leibniz's Supplementary Study appears to be a gloss on what it means to say that there are actually infinitely many entities of any kind. When we understand that claim syncategorematically, we are not referring to a collection, and so are necessarily using the term distributively.

Secondly, it is true that in the first statement of the Study Leibniz is recognizing a sense of the syncategorematic infinite as "a passive power" having merely potential parts. This is

---

[10] As Leibniz wrote in the late 1670s, "So, bodies are actually infinite, i.e. more bodies can be found than there are unities in any given number" (A VI 4, 1393/ LLC 235). Cf. also *Towards a Science of the Infinite*: "So even if it is conceded that there is no end of bodies, we do not thereby concede that the corporeal world or universe is really one substance whose immensity would be as great as God's; but only this, that whatever bodies are assumed, in addition to them still more bodies can be taken which really exist, yet one greatest body is no more constituted by them than a greatest number is constituted by unities" (§6).

6.1 Leibniz's Syncategorematic Conception of Infinitesimals

the conception of the infinite applicable to *continuous quantity*, whose parts (as we have seen) can be divided, added, etc. at will. But mathematics is not concerned only with continuous quantity; there is also *discrete quantity*. The contrast Leibniz always insists on is not that between a mathematical and a non-mathematical infinite, but that between the continuum as an ideal entity possessing indeterminate parts, and the actual as consisting in discrete unities or determinate parts. The parts of the mathematical continuum are those into which it *can* be divided, but as a continuum it is not actually divided. Leibniz gives a clear rendition of his mature position in his letter to Nicolas Remond in July 1714:

> In the ideal or the continuum, the whole is prior to the parts, as the arithmetical unity is prior to the fractions which divide it, and which can be assigned arbitrarily; its parts are merely potential. In the real, however, the simple is prior to the aggregates, parts are actual, and are prior to the whole. (GP III, 622).

That the latter alternative of the contrast applies to numbers as well as the actual parts of matter is made clear by Leibniz in various places, such as in the following:

> *Space*, or immobile place, is an ideal thing, like time, and concerns the possible as well as the actual. This is what makes it a *continuous quantity* (a magnitude in which there is no separation), being indifferent to all possible divisions, just as number is indifferent to all the fractions that one could make in it. But *matter*, which is real, is a *discrete quantity* (a magnitude that is already divided), as a whole number is in relation to the unities from which it results.[11]

Thus an arithmetical unity, like a continuous line, is divisible into whatever parts one wishes, but these are potential, not actual, parts. Things are otherwise with something that has determinate parts, such as the actual parts of matter determined by their differing motions, or the terms of an infinite series determined by the law of the series.

But an infinite aggregate of such determinate parts still does not add up to a genuine whole. No infinity of parts, whether they are actual or potential, can comprise a true whole. In Leibniz's example of the infinite series of fractions whose sum is 2, there is no question of adding the fractions one by one until the sum 2 is reached, since that would require there to be a last, infinitieth fraction, and therefore an actually infinitely small fraction and an infinite number of terms in the series. And yet there are infinitely many terms: no matter how many one takes, there are more; and taking the series as a distributive whole, each of the fractions—all of them finite—is included, and no others.

---

[11] "L'Espace *ou le lieu immobile est une chose ideale, comme le temps, et regarde le possible comme l'actuel. C'est ce qui fait qu'il est* Quantum continuum *(une grandeur où il n'y a point de separation) estant indifferent à toutes les Divisions possibles, comme le nombre l'est par rapport à toutes les fractions qu'on y pourra faire. Mais la* Matiere, *qui est* reelle, *est* Quantum Discretum *(une grandeur deja divisée) comme le Nombre Entier l'est par rapport aux Unités dont il resulte*" (Leibniz to Sophie, 24 November 1705, A I, 25, 330).

As we saw in Chap. 2, in 1676 Leibniz had already stressed the significance of such a conception of infinite series while he was working on quadrature, when he criticized the expressions that Brouncker and Wallis had found for circle quadrature as yielding only approximations, and not an exact value. In *Towards a Science of the Infinite* (text [**33**]) he gives a detailed exposition of this distinction between such an "indefinite finite series" and one that is properly infinite. The former he defines as "one produced arbitrarily far, yet always further producible—of the kind I introduced for general calculations", in contradistinction to "an infinite series, that is, a series beyond which none of the terms of a given nature can be taken" (A VII 3, 824–5). Leibniz gives numerical examples, the second of which is the decimal expansion of 1 as 0.99999 etc. This may be expressed as the infinite series 9/10 + 9/100 + 9/1000 + 9/10000 + 9/100000 etc. Now, if its terms "are understood to be continued to infinity, [they] make 1. But the same series taken indefinitely finitely would signify 99999/100000, or 999999/1000000, or 9999999/10000000, and so on, by which fact a fraction less than one is always designated." (text [**33**], §8). He explains:

> when we say that the infinite series 9/10 + 9/100 + 9/1000 + 9/10000 etc. equals 1, the sense is only that in an entire unity, as in a foot, there are in it 9/10 of a foot, and furthermore 99/100 of a foot, and 999/1000 of a foot, and no matter what other finite fraction of this kind, but nothing further. Nor, therefore, do we hold that there is some number of all fractions of this kind, which there is not; in the same way as we acknowledge any unity whatsoever or any created thing whatsoever taken one at a time, even though we deny that an infinite thing is compounded from them. (text [**33**], §9)

We are not dealing here with a distinction that makes no difference. As Leibniz explains, irrationals can be exactly expressed as roots of algebraic equations or by an infinite series, but *not by an indefinite finite series;* whereas transcendental numbers cannot be expressed as roots of algebraic equations, but can be expressed by infinite series (as in the Leibniz series for $\pi/4$), in contrast to the approximations offered by Wallis and Brouncker.[12]

This distinction of a syncategorematically understood actual infinite from the merely potential infinite was also extremely important to Leibniz in his metaphysics. The Thomists held that matter was merely potential until actualized by a form, so that it would be impossible for a corporeal substance with its own form to exist as an actual part of another body made actual by its own substantial form.[13] Leibniz, by contrast, sided with the

---

[12] This, we think, amply answers Monica Ugaglia's criticism "that Leibniz's notion of the syncategorematic actual infinite cannot be made mathematically sound: not because it is not sound, but because it is not mathematical at all", and that "difficulties with this [sc. Arthur's] interpretation derive from the fact that no genuinely mathematical characterization is proposed of Leibniz's syncategorematic actuality which allow to distinguish it from Aristotle's syncategorematic potentiality" (Ugaglia 2022, p. 268).

[13] See, for example, Des Bosses's characterisation of Thomas' doctrine in his letter of 12 February, 1706 (GP II, 296-99/ LDB 13-1).

6.1 Leibniz's Syncategorematic Conception of Infinitesimals

Plurality of Forms doctrine, according to which there could exist actual parts in a body, each actualized by its own form, whether or not that body had its own substantial form. Leibniz extrapolated this to infinity: every body is divided into actual parts that presuppose subsidiary forms, forms which constitute these parts as actually existing bodies in their turn, and so on down.[14]

This difference of interpretation emerges right at the beginning of Leibniz's correspondence with the Jesuit Des Bosses, who thought Leibniz would accept that "there is no actual infinity in nature", in agreement with the orthodox Aristotelian position that there is only a potential infinite.[15] Leibniz demurred, insisting that he did not doubt that there is an actual infinite in nature, for, because of the uniform divisibility of matter, the varying motions in it would ensure that "any point whatever is moved by a motion different from any other assignable point" (GP 300/ LDB 21). Des Bosses responded that what Leibniz had said elsewhere about not needing to take the infinite "à la rigueur" in his calculus, and not interpreting entities like the infinite degrees of impetus as being "actually found as such in nature", suggested to him that perhaps the infinite Leibniz was providing "could be confined to the syncategorematic, for what prevents us from transferring what you say about degrees of impetus to the multitude of substances?" (text [**35a**]; GP II, 302/ LDB 27). The suggestion appears to be that when Leibniz talked of the actual infinite in nature, perhaps all he had in mind was an indefiniteness of further addition or division, otherwise his claims about there being infinitely many substances would entail that actually infinitely many of them could be found in nature, which Des Bosses seems to suppose Leibniz could not have intended.

In response, Leibniz explains his interpretation of the syncategorematic infinite as applying not just to the mathematical entities used in physics, but as applying to any multitude at all, whether of numbers, lines, or substances:

> Arguments against an actual infinity assume that if this is admitted, there will be an infinite number; likewise, that all infinites are equal. But it must be recognized that an infinite aggregate is not one whole or endowed with magnitude, and is not consistent with number. And, accurately speaking, instead of 'infinite number' it should be said that there are more than can be expressed by any number, and in place of 'infinite straight line', that it is a line produced beyond any magnitude that can be assigned, so that there is always a longer and longer straight line. It is of the essence of number, of line, and of any whole whatsoever, to be bounded.

---

[14] For an account of Leibniz's metaphysics in relation to the Plurality of forms doctrine, see Arthur (2018, chs. 4 and 6).

[15] Ugaglia (2022) contests this as a correct interpretation of Aristotle, whom she believes has as much title to an actual infinite as Leibniz. But it is clearly Des Bosses's interpretation: that "there is no actual infinity in nature" is the third of five axioms "on which the Aristotelian philosophy depends", he tells Leibniz in his letter of 12 February 1706, axioms which "I believe agree well enough with your system, if you omit the third, about which I am not sure, for you seemed to me to defend the actual infinite somewhere. I believe, however, that your meaning can be satisfactorily explained through the potential infinite" (GP II, 297/ LDB 13-15).

Consequently, even if the world were infinite in magnitude, it would not be one whole... (text [**35b**]; GP II, 304-05/ LDB 31-33)

What is crucial in this syncategorematic interpretation of the infinite is that there are actually more things in an infinite multitude than can be expressed by any number, not merely potentially more. Yet neither a multitude of discrete parts, like an infinite series, nor an infinite magnitude, can be understood as a true whole. Leibniz continues:

It is therefore an abbreviation of speech when we say 'one' when there are more things than can be comprised in an assignable whole, and treat as a magnitude something that does not have its properties. For just as it cannot be said of an infinite number whether it is even or odd, so it cannot be said of an infinite straight line whether it is commensurable with a given straight line or otherwise; so that these are only improper ways of speaking of infinity, as though of one magnitude, based on some analogy, but which, if you examine them more carefully, cannot be sustained. (text [**35b**])

Turning next to Des Bosses's remarks about his calculus, Leibniz replies:

Speaking philosophically, I maintain that there are no more infinitely small magnitudes than there are infinitely large ones, that is, no more infinitesimals than infinituples. For I hold both to be fictions of the mind through an abbreviated way of speaking, suitable for calculation, as imaginary roots in algebra are too. (text [**35b**])

These expressions "cannot lead to error," Leibniz insists, since it suffices to substitute for the infinitely small something as small as one wishes, so that the error is less than that given, "whence it follows that there an be no error". This claim, which we saw him also make later in the *Théodicée*, is commensurate with the application of such notions in physics, where there are no infinitely small bodies or motions, but only arbitrarily small ones. *In rerum natura*, he argues, there are just finite parts and motions, some smaller than others to infinity. And "this is demanded by the nature of matter and motion, and by the whole frame of the universe, for physical, mathematical, and metaphysical reasons" (text [**35b**]). Thus it is that using the fictions of infinite and infinitely small in the calculus does not commit you to thinking that "these mathematical entities are actually found as such in nature", as he had written in the *Specimen Dynamicum*.

6.2 Leibniz's Mature Justifications of the Use of Infinitesimals

An important issue, often neglected by commentators, and which should be apparent from the previous section, is that Leibniz modulates his justifications according to the context in which such a justification is requested. This remark is in fact of great importance for assessing, by contrast, the texts usually mentioned in discussions of the justification of the differential calculus. Indeed most of these justifications, dating from around 1700, occur in

6.2 Leibniz's Mature Justifications of the Use of Infinitesimals

the context of the crisis occurring in the *Académie Royale* at the turn of the Century, after several attacks by mathematicians "of the old style" such as La Hire, Gallois or Rolle.[16] When he finally intervenes in the debate between Rolle and Varignon in 1701, Leibniz does not present a translation into the Archimedean style, it is true, and he seems to be content to allude to the possibility of such a translation. But there may be a very good reason for that, which is rarely mentioned. This is that for a Cartesian like Rolle, as was already the case for critics like Clüver, the reference to Archimedes was of little value, if any.[17] Recall that Descartes in fact considered the Archimedean Geometry to be a kind of Metaphysics (of Geometry), the value of which could only be heuristic.[18] Recall also that, although he was perfectly able to use indivisibilist methods, Descartes rejected them in geometry, tackling questions such as the quadrature of the cycloid only in his correspondence (AT II, 135–136). Recall finally that he thought of the *reductio* as the lowest form of reasoning in mathematics.[19] This may give an idea as to why a mathematician like Rolle could hardly be "convinced" by demonstrations "in the manner of the Ancients".

This will also help us in identifying a specific kind of justification introduced in these debates. Indeed, Leibniz emphasizes in this context a different argument in favour of infinitesimals. It does not amount to defending their use directly, but to objecting to the supporters of Cartesian techniques or the like that they use similar foundations in "ordinary" algebra. Accordingly, if this kind of foundation is not reliable in the Differential Calculus, it should not be reliable in their own practice either.

The underlying idea itself is not new and goes back to the Parisian period. It appears for the first time in the *Méthode de l'universalité* (1674), where Leibniz explains that infinitely small and infinitely large quantities are hidden in ordinary algebra as long as one wishes certain formulae to be general.[20] This forms the core of a text which Leibniz sent to Varignon for publication in the spring of 1702[21] and which was significantly entitled: "*Justification du calcul des infinitesimales par celuy de l'algebre ordinaire*" (text [**47**]).

---

[16] The expression is used by Varignon in a letter to Bernoulli, giving us a *terminus a quo* for these debates: (to Johann Bernoulli, 6 August 1697, *Der Briefwechsel von Johann Bernoulli*, Band 2, op. cit., p. 124). On these debates, see Bella (2022).

[17] Clüver was in fact criticizing Archimedes on a par with the new calculus, see Mancosu and Vailati (1990) and Mancosu (1996, p. 157).

[18] To Mersenne, 9 January 1639 (AT II, 490). Leibniz is familiar with this letter and recalls it strategically at the beginning of the *Cum prodiisset*, adding that Descartes did indeed rely on an Archimedean argument in his Metaphysics (Leibniz 1846, p. 42).

[19] Letter to Hardy (AT I, 490).

[20] See *Méthode de l'Universalité*, chap. VI: "Cavalieri, Mr. Fermat, Mr Wallis, and others suppose certain letters to be either infinitely small lines or equal to nothing. I have put the same thing into use, and have adjusted the letters which represent an infinite magnitude, or lines equal to rectangles, as are the asymptotes of the hyperbola" (A VII 7, 79).

[21] The *Justification* was conceived as a follow up to the exchange of letters from 1701–1702, which we studied in the preceding sections. It was sent to Pinsson via Varignon for publication in the *Journal des sçavans*, but the project did not succeed.

Fig. 6.1 From *Justification du calcul des infinitesimales par celuy de l'algebre ordinaire*

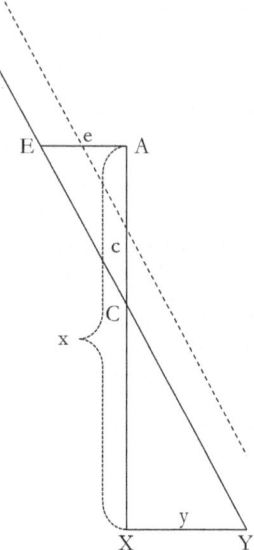

In it, Leibniz takes the example of two lines *AX* and *EY* meeting at *C*, draws the perpendiculars $EA = e$ and $XY = y$ and joins *A* to *X* (with $AX = x$ and $AC = c$). Hence he obtains two similar triangles *CAE* and *CXY*, or, in equation $\frac{x-c}{y} = \frac{c}{e}$ (see Fig. 6.1). Now, he considers what happens when *EY* moves toward *A*, assuming that the two triangles are not equal and that the angle in *C* is not 45 degrees. It is clear that the equation holds whatever the size of *CEA* might be. But what happens when *C* and *E* finally coincide with *A*? In this case, *c* and *e* vanish and $x - c = x$. If we take the vanishing of *c* and *e* to amount to "absolute zeros", this leads us to an absurd statement: $\frac{x}{y}$ will be equal to $\frac{c}{e} = \frac{0}{0}$ (which Leibniz takes to signify that they are equal or of ratio 1, hence contradicting the hypothesis). This means that *c* and *e* should not be considered as absolute nothings, but as relative ones, nothing in comparison with *x* and *y*. However, they need to keep a certain ratio, if we want the equation to be general, "and so they are treated as infinitesimals, exactly like the elements that our calculus of differences recognizes in the ordinates of curves that is to say, as momentary increments and decrements. Thus we find in the calculations of ordinary algebra traces of the transcendent calculus of differences, and those same singularities about which some scholars have scruples." (text [47])

We find here the same kind of continuity argument that already appeared in Prop. 8 of the *DQA*, when it came to introducing the degenerate case into the ordinary one. But the strategy is completely different: there is no question of translating this kind of reasoning into a proof "in the manner of the Ancients", but it is argued rather that its refusal would prevent some general equation from holding in ordinary algebra. The passage continues:

> ... about which some scholars have scruples. And even the algebraic calculus cannot do without them if it wants to retain its advantages, one of the most considerable of which is the

6.2 Leibniz's Mature Justifications of the Use of Infinitesimals

generality that is proper to it in order for it to comprehend all the cases, even that in which some given straight lines vanish. (text [**47**])

Leibniz then mentions that the same argument, when used in Physics, is what he termed the "Law of Continuity" in the *Nouvelles de la République des Lettres* from 1687. In this case, one relies on the introduction of the fiction of nascent and evanescent quantities, "taking equality for a particular case of inequality, and rest for a particular case of motion". It is crucial to note, however, that Leibniz immediately recalls that one can easily translate this procedure into an "Archimedean" procedure by relying on the *PUD*, that is,

> by supposing not that the difference between magnitudes that become equal is already nothing, but that it is in the act of vanishing, and the same with motion, which again is not absolutely nothing, but something that is on the point of being nothing. And anyone who is not satisfied with this can be made to see, in the style of Archimedes, that the error is not assignable and cannot be given by any construction. (text [**47**]).

The same kind of reasoning is repeated at the beginning of a text to which we shall return in greater detail later on, the *Defense du calcul* (text [**49**]). In it, Leibniz first announces he will provide "a quite palpable means of justifying our ways of calculus by means of the ordinary calculus of Algebra". He then takes up exactly the same example as in the *Justification*[22] (changing only the lettering) and announces: "As regards the first point, I will show that without noticing it enough, in the ordinary Calculus of Algebra one has long been practising the method contested in relation to infinitesimals, when one applies a general calculus to some particular cases, where some magnitudes vanish".

Following this thread, a new strategy for justifying the differential calculus against the attacks of algebraists such as Michel Rolle appears clearly: instead of trying to convince them that the apparent use of infinitesimals was harmless, one could posit the Law of Continuity as a postulate common to ordinary algebra and Differential Calculus, and then show that the rules of the calculus could be derived upon this supposition.[23] This is exactly what the famous *Cum prodiisset* proposed (text [**50**]).[24]

---

[22] This parallel has led some scholars to read the *Defense* as a version of the *Justification* (See Pasini 1988, p. 698). This has had the unfortunate consequence of hiding another new strategy announced in the *Defense* (and absent from the *Justification*) in which Leibniz proposes an interpretation devoid of infinitesimals. We'll give our hypothesis about the proper status of this text below.

[23] In one draft Leibniz thought about the possibility of positing the Law of Continuity as an axiom (LH 35, 8, 29 fol. 1).

[24] Let us note, for now, that when doing so, Leibniz will be very explicit about the general context in which he takes the Law of Continuity to hold in mathematics mentioning the examples dating back from the Parisian stay: parallel lines seen as meeting at infinity or ellipses transforming into parabolas when one of the focus goes at infinity.

Some commentators have interpreted this new strategy as forcing a specific interpretation in terms of the existence of infinitesimals.[25] This is a typical confusion between the issues of use and existence. Leibniz, for his part, was very clear about the neutrality of this new strategy as regards the question of existence, as was the case for his previous strategies also. We just saw that he mentions in the *Justification* that everything he said about "vanishing quantities" can be translated into a proof in the style of Archimedes by using the *PUD*. But it is especially in the *Defense* that he comments about the two possible interpretations of the "foundations" ("*fondement*") he just presented:

> the foundation of all of this can be explained by taking z and x, i.e. ba, hA, *in the very act of vanishing* and falling on A. It is like in a nascent motion, since an instant of motion is different from an instant of rest, and in this element of time, there is an element of nascent progress, which is more than nothing. But even if the instant of this act of vanishing or emerging (*naissance*) is in metaphysical rigor only a fiction (in order not to sink into the labyrinth *de compositione continui*), it suffices that no error could arise from it, and that these fictions could always hold in place of truths in the calculus, in a like manner to imaginary roots. *Since by rejecting all the infinitely smalls, and employing in their place only magnitudes as small as one wishes, one will always show that the error would be less than any given error. That is to say that there is none.* (text [**49**])

Hence we cannot agree with the reading of this text as presenting a "semantic" strategy, which would amount to the acceptance of a model in which infinitesimals are introduced as genuine entities. What Leibniz says here, as he does elsewhere, is that infinitesimals can be introduced as "fictions" and that this use is innocuous, not because we have a way of establishing the constistency of the extended theory (by what means?), but because we can "reject all the infinitely smalls" and take in their place only ordinary (variable) quantities. By describing the strategy in terms of semantics, one is just forcing this necessity into a picture in which Leibniz thought and said it was *absent*. But there is more, since we saw in

---

[25]Thus in his 2015 paper, D. Jesseph distinguishes two strategies that might reconcile Leibniz['s] requirements for rigorous demonstration", a "syntactic" or "proof-theoretic" one, and a "semantic" or "model-theoretic" approach (Jesseph 2015, pp. 196–197). The former is roughly the idea that "a symbol like 'dx' is simply a placeholder for a much more elaborate line of reasoning that makes reference only to finite differences of finite quantities", while the latter is the idea that "use of infinitesimals would never lead from truth to falsehood", which would require "something like the proof of a principle that adding infinitesimals to the standard geometry yields a model-theoretic conservative extension of standard geometry" (197). On that view the infinitesimal would be "something like a Hilbertian ideal element" (202). Similarly, Katz and Sherry identify two different approaches, one relying on the Method of Exhaustion, the second embracing infinitesimals as fictions in the sense of "modern formalist positions such as Hilbert's and Robinson's" (Katz and Sherry 2012, p. 1553). They refer this dichotomy to Bos (Katz and Sherry 2012, p. 1551; Katz and Sherry 2013, p. 575), although Bos (1974-5) seems to have been more cautious on this issue. As we have shown in Sect. 5.2, the semantic strategy is, however, not consistent with Leibniz's characterization of what quantities, magnitudes and numbers are.

6.2 Leibniz's Mature Justifications of the Use of Infinitesimals

Sect. 5.2 that the semantic strategy amounts to introducing a conception of "quantity" or "magnitude" that is incompatible with Leibniz's.

Proponents of the view that Leibniz held infinitesimals to be ideal elements in an extended continuum have drawn support from the fact that in describing his infinitesimals as incomparables, Leibniz makes reference to proposition 5.4 of Euclid's *Elements* (5.5 in the editions of Euclid available to Leibniz). Thus in explaining his conception of incomparables to Nieuwentijt in his famous letter of 1695, Leibniz writes:

> I believe those things to be equal, not only whose difference is absolutely zero, but also those whose difference is incomparably small; and even though this difference should not be said to be nothing at all, there is however no quantity comparable with them whose difference it is. ... For only those homogeneous quantities are comparable, I hold with Euclid Book 5 Definition 5, one of which can be made greater than the other when multiplied by a finite number. (text [**45**])

Commenting on a similar passage in his letter to L'Hôpital,[26] Katz and Sherry claim that in these passages "Leibniz describes such entities [infinitesimals] as 'incomparable quantities' and defines them in terms of the violation of what today is called the Archimedean property." (Katz and Sherry 2012, p. 1555).[27] But Leibniz has not thereby defined infinitesimals as non-Archimedean elements of an extended continuum. What he has defined is what it is for one quantity (say, dx) to be incomparable *in relation to* another (say, x), as we have explained in Chap. 5 above. Leibniz makes no assertion of the absolute existence of infinitesimals here. Rather, as he explains to Guido Grandi, infinitesimals are "relative nothings",[28] in that dx is nothing compared to x, although not compared to dy (if y is a function of x). And of course the same goes for ddx relative to dx, as opposed to relative to ddy, etc. And the justification for holding that dx is nothing relative to x is provided by the *Lemmata incomparabilium* and fully in accord with Archimedean principles, as the continuation of the above-quoted passage makes clear:

---

[26] "I have referred him in the meantime to my Lemmas on Incomparables inserted in the *Acta* of Leipzig, February 1689, and I count as equal quantities whose difference is incomparable to them. I call those magnitudes incomparable of which one multiplied by any finite number whatsoever cannot exceed the other, in the same manner that Euclid takes it in the fifth definition of the fifth book." (Leibniz to L'Hôpital, GM II, 288).

[27] Similarly, Karin and Mikhail Katz comment "Leibniz repeatedly asserted that his infinitesimals, when compared to other quantities, violate the Archimedean property, viz., Euclid's Elements, V.4... This appears directly to contradict Ishiguro's claim that Leibniz was working with an Archimedean continuum." (Katz and Katz, 3-4).

[28] Cf. Leibniz to Grandi, September 1713: "Meanwhile we conceive infinitely small [quantities] not as simply and absolutely nothings, but as *relative nothings* (as is well known), that is, as indeed vanishing into nothing, yet retaining the character of that which is vanishing." (text [**42**]). (More on this below.)

And those things that do not differ by such a quantity I hold to be equal, as Archimedes also assumed, as have all others after him. And this is the very thing which is said to be a difference smaller than any given. And, by a process that is indeed Archimedean, the matter can always be confirmed by an inference *ad absurdum*. But since the direct method is easier to understand and more useful for discovery, once this way of reducing is known it suffices that afterwards the method is applied in which incomparably small things are neglected, which certainly also brings with it its demonstration according to the lemmas communicated by me in February 1689. (text [**45**])

This is in complete accord with the use Leibniz makes of infinitesimals (and points at infinity) in his Law of Continuity.[29] Here it does not have to be accepted that they are non-contradictory in order for them to be used successfully, since their use as "relative nothings" is justifiable by appeal to the Lemmas on Incomparables, and ultimately, the *PUD*. What this means is that nothing can be inferred about the existence of infinitesimals from their appearance in the surface language. As far as mathematics is concerned, they may be used (under the conditions stated) *as if* they exist. If one wants to infer existence, one cannot just rely on the nominal definition of "incomparables" (as not respecting the definition of Archimedean quantities), but one has to establish the possibility of such objects, i.e. that they are non-contradictory. But, as we have shown in Sect. 5.2, Leibniz always claimed that infinite entities, be they infinitely large or infinitely small, could not be considered as genuine quantities without violating a constitutive property of quantities given by the part-whole axiom. Hence, they cannot be introduced into the system without contradiction (even as pure geometrical entities, measured indirectly, as we have seen in Sect. 5.3). To show that one can build a consistent model including such infinite entities is not a way of "vindicating" Leibniz's point of view. It is a way of showing that, from a modern point of view (rejecting the part-whole axiom for infinite sets), he was wrong in assuming that these infinite entities were contradictory.

A stronger argument arises when supporters of the non-standard view object that in fact all of this is pure wishful thinking on the part of Leibniz and that the kind of justification of the differential algorithm which is sketched in texts like the *Cum prodiisset* under the postulation of the Law of Continuity, *cannot* be translated into a justification compatible with the syncategorematic view. In this sense, one "vindicates" Leibniz's approach by showing that, even if he thought of infinitesimals as non-existing, he had to commit himself to an "extended continuum" to support the kind of justification he provided through the Law of continuity. This is the objection that we will address in the last part of our study.

---

[29] Leibniz makes this point to Grandi in the same letter (text [**42**]).

6.3 The Justifications of the Differential Algorithm

So far, we have dealt mainly with the justifications of the *use* of infinitesimals in general and we have not yet tackled the crucial issue of the justification of the differential algorithm in and of itself (i.e. the justification that its rules are sound and trustworthy). As we hope we have shown in the preceding developments, one benefit of proceeding in this way is precisely to disentangle issues that are too often confused (typically that of use and that of existence). More importantly, it clarifies what remains to be done. As we have indicated, Leibniz finally elaborated two strategies for the defence of the use of infinitesimals, which are presented in turn in the texts he sent in 1701–1702: the first one is a translation of the discourse using the fictions (of infinitely small or infinitely large quantities) into proofs in the manner of the Ancients (the prototype of this strategy being given by prop. 8 of the *DQA* to which he refers his interlocutors);[30] the second one relies on the Law of Continuity taken as a postulate, which should be accepted by both parties. In both cases, we have emphasized, Leibniz claims that the question of existence is not settled and that one can either believe in real infinitesimals or not. But he is also very clear about the fact that he does not himself believe in the existence of infinitesimals, and that in every justification one can provide an interpretation in which they do not enter constitutively.

Interestingly enough, these various aspects are explicitly described at the beginning of the *Defense du Calcul des differences* and it provides a nice place to begin. But before doing so, let us say a few words about the status of this manuscript. In the margin of the *Cum prodiisset*, Leibniz has written the following commentary (not translated by Child):

> All this must be edited very carefully so it could be published, omitting what is harsher in contradicting others. It is to be joined by my method for the law of continuity shown by the tracing of lines, and also by the tract [*schediasma*] I had sent the Parisians in order to show that in the common example the ratio between nothings is feigned to be something. (text [**50**])

It is highly plausible that the tract sent to the Parisians mentioned here is the *Justification*—not only because the descriptions of the contents match, but also because this is indeed the only *schediasma* which was sent by Leibniz to the Parisians (the other texts being in fact letters of which extracts were made in order to be published). Now the *Defense* is precisely a rewriting of the *Justification* after an introduction which announces very nicely the content of the *Cum prodiisset* We therefore surmise that it was conceived as the planned edition of the *Cum prodiisset* with the addition of the *Justification*. Be that as it may, the most interesting aspect of the *Defense* is the way in which it spells out the various issues at

---

[30]That this transfers to the use of the differential algorithm is immediate since we just have to express the relationship in the symbolism of this calculus: this is what Leibniz does in his letter to Bodenhausen from 1690 and in the *Compendium* of the *Quadratura* written at the same period, see Sect. 4.3.

stake. For convenience, we will indicate these various stakes and Leibniz' strategies for responding to them with numbers in the text:

> I learn that talented people are opposed to the Calculus of Differences, (**1**) because it seems that in it one necessarily proceeds by infinitesimals, or by infinitely small magnitudes, and (**2**) because they believe that in it one makes elisions at pleasure. One can always show them (**1a**) that everything that is concluded by this calculus can be proved by a *reductio ad absurdum* in the style of Archimedes, and by using the Lemmas of Incomparables proposed in the Leipzig *Acta*; and (**2a**) that it is always easy to recognize what one can neglect with impunity without any error arising from it, so that the elisions are made according to certain rules, and not as we see fit (…). But (**3+4**) without here making use of these Archimedean-style demonstrations, which are extremely long and not really suitable for enlightening the mind, I want to propose here two things: (**3a**) a very palpable way of justifying our method of calculation by means of ordinary algebraic calculus; and then (**4a**) an interpretation of the Calculus of Differences where instead of infinitesimals one understands only very small magnitudes, and this does not stop one from reaching a conclusion (text [**49**]).

What is very clear from this introductory remark is that there were various criticisms addressed to the differential calculus, and that Leibniz did not respond to them with only one answer. In particular, he details here two different strategies: (**1+2**) is based on a combination of the Archimedean style of proof and the *lemmata incomparabilium*; (**3+4**) is presented as a different (and quicker) way and relies on the fact that one can justify the algorithm "by means of ordinary algebraic calculus"; in addition, Leibniz proposes an interpretation of this justification in which, as in the first strategy, one does not rely on genuine "infinitesimals".

This constitutes a beautiful summary of what has been shown in the previous development: Leibniz was consistent in asserting that one could *always* eliminate the reference to infinitesimals that his adversaries claimed to be necessary to his procedures. Moreover, he claimed that the rules of his algorithm could be justified on that very basis. The recourse to the Archimedean procedure (plus the *Lemmata*) was the standard procedure to provide such a justification. But Leibniz also developed an alternative strategy in which the foundations of the calculus were justified by the fact that "ordinary algebra" proceeded on the very same foundations. Now, one could doubt whether this latter strategy was consistent with the first one in the sense that it was expressible without recourse to infinitesimals, and this is what the last sentence expresses very clearly: one can produce an interpretation of the second strategy where "*in place of infinitesimals*, one understands only very small magnitudes and one no less reaches a conclusion" (our emphasis). This should be sufficient to achieve our demonstration: not only did Leibniz claim that one could justify the calculus without recourse to infinite quantities, but he provided precisely such a justification *even in the case where the basic postulate was the Law of continuity*.

Now, as we have seen, what remained to be shown is that one can provide justifications of the rules of the algorithm presented in 1684 following these two strategies. Interestingly enough, this is precisely what we find in the known texts from the same period. But there is more, as we will now demonstrate: these justifications were both compatible with the

6.3 The Justifications of the Differential Algorithm

syncategorematic view (which, we insist, is a thesis about existence). The first strategy is typically the one expounded in the famous letter to Wallis from March 1699 and relies on the *Lemmata incomparabilium*. The second strategy is the one expounded in the *Cum prodiisset* and relies on the *Lex continuitatis*. Let us take a detailed look at each of these texts in turn.

As we have also seen, Leibniz's justification strategy is consistent with the fact that "what is concluded by means of the infinite and infinitely small can be evinced by a *reductio ad absurdum* by my method of incomparables".[31] It is of first importance to notice that in all of these passages, Leibniz does not mention only *one* of these two foundations (*reductio* or method of incomparables). Indeed, the kind of justification presented in the *DQA* (*reductio*) does not transfer immediately to the Differential Algorithm. On the other hand, as Leibniz regularly emphasizes, the *Lemmata incomparabilium* are neutral, in and of themselves, as regards the question of the complete eliminability of infinitesimals. It is therefore the connection of *both* principles which supports, according to Leibniz, a justification of the differential algorithm in which infinitesimals can be "evinced". This strategy is clearly presented to Wallis in 1699. The whole exchange between the two authors, beginning in 1695, is very interesting.[32] We cannot enter into all of its details, but would like to emphasize that, once again, there are good reasons why Leibniz did not present the strategy in terms of "nascent" quantities to Wallis. Indeed, Wallis agreed about this model, but thought that one could infer from it that these nascent quantities were, at the end of the day, absolute zeros:

> For my quantity *a* is the same as your *dx*, except that my *a* is nothing and your *dx* infinitely small. Then when those things are neglected which I warn should be neglected for abbreviating the calculation, that which remains is your minute triangle, which according to you is infinitely small, but according to me is nothing ôr evanescent. (Wallis to Leibniz, 30 July 1697; GM IV 37; cf. Jesseph 1998, p. 25)

The method Wallis is here alluding to is his own method of tangents, which is an improvement on Fermat's method that he thinks to be fully equivalent to Leibniz's. As some readers of Fermat did at the time, he introduced the increments into the calculation and then, in the last step where the limit case was reached, discarded them as absolute zeros.

According to Wallis, what the "characteristic triangle" stood for was a triangle whose shape was preserved in the process, but not the magnitude (since it became a triangle of magnitude "zero") (GM IV 50). In the limit case, what remained henceforth was a shape without magnitude, the kind of "pure geometrical entities" we studied already in Sect. 5.3. Leibniz noted that this idea was obscure. More profoundly, he made Wallis notice that

---

[31] Letter to Johann Bernoulli, 29 July 1698 (text [**32e**]). The same statement may be found in the letter to Bodenhausen A III 5, 149, or at the beginning of the *Defense du calcul (quoted above)*. See also, in addition to the letter to Wallis of 1699, the *Responsio* GM V, 322.

[32] For a general presentation, see Jesseph (1998).

taking the *dx* to stand for an absolute nothing would prevent the possibility of taking its difference and the difference of its difference, etc.

> In my view, the form of the Characteristic Triangle on a curve can be correctly explained by the degree of declivity, but for the calculation it is useful to feign [*fingere*] infinitely small quantities, ôr as Nicolaus Mercator called them, infinitieths [*infinitesimas*]: which are of such a kind that, since a ratio between them is sought that is certainly assignable, it is already illicit to hold them to be nothings. Meanwhile they are rejected whenever they are added to incomparably greater ones, according to the Lemmas on Incomparables once proposed by me in the *Acta* of Leipzig; which foundation is also used by the Marquis de l'Hospital. (text [**46**])

Quoting this passage, D. Jesseph concludes that Leibniz is here lining up on the side of the Marquis de l'Hôpital and hence as a defender of the reality of the infinitesimals.[33] But, strangely enough, he fails to quote the rest of the passage in which Leibniz departs from l'Hôpital by explaining how the infinitesimals could be evinced by a combination of the *Lemmata* and of a proof "in the Archimedean style":

> ...there remains $xdy + ydx + dxdy$. But this $dxdy$ should be rejected, as it is incomparably smaller than $xdy + ydx$, and this gives $d(xy) = xdy + ydx$, so that, if someone wished to translate the calculation into the Archimedean style, it is always evident that, when the thing is done using assignables, the error that could result from this would always be smaller than any given. (text [**46**])

Now, in order to make the whole procedure fully clear, let us write out this tedious Archimedean proof.

What the *Lemmata incomparabilium* tell us is where to look for the "difference" which has to be made "smaller than any given error" (this is the issue numbered 2 in our account of the *Defense du calcul*). It is a crucial issue since the whole algorithm is dealing with "differences", and a main reason why the transfer of reasoning in the style of the *DQA* is not immediate is that we need to identify which of these "differences" will be considered as the "error" to render equal to zero.[34] In the case under study, considering that we are dealing only with assignable quantities, the *Lemmata* indicate to us that the "error" will be located in the $dxdy$, which is incomparably smaller than dx and dy. Proceeding in the manner of the Ancients as exemplified in *DQA* prop. 7, we then just have to suppose that the equality does not hold and that $d(xy)$ differs from $xdy + ydx$ by a given quantity Z. Now, since we can take dx and dy as small as we wish, let's take them smaller than $\frac{1}{2}\sqrt{Z}$. We hence have $d(xy) = xdy + ydx + dxdy < xdy + ydx + \frac{1}{4} Z$. And since $d(xy)$ was supposed to be equal to $xdy + ydx + Z$, we obtain that $Z < \frac{1}{4}Z$. *Quod est absurdum*. The same reasoning applies, evidently, to all the rules of the algorithm and is just a matter of boring routine.

---

[33] This argument plays a crucial role in Jesseph's idea that Leibniz was fluctuating in his acceptance of infinitesimals as eliminable or not (Jesseph 2008, p. 215).

[34] We saw in our study of Prop. 8 that this is not a trivial matter.

6.3 The Justifications of the Differential Algorithm

Let us now turn to the second strategy, that which relies on the Law of Continuity and is presented in the *Cum prodiisset*. Here what Leibniz has to do is simply to generalize the kind of example that he expounded in the *Justification* (and which may have been chosen for that precise reason). The core of the strategy is henceforth to rely on proportionalities, typically the one given by the ratio between the "characteristic triangle" of sides dx, dy and the triangle made by the ordinate and the subtangent.[35] This calls for a different approach than in the letter to Wallis since we then need to reach an equation in which the proportion dy/dx is made apparent. This is particularly salient in the first example which Leibniz proposes to tackle, the finding of the tangent to the parabola of equation $xx = ay$ (or $y = xx : a$). By introducing the increments dx and dy, one obtains: $y + dy = xx + 2xdx + dxdx, : a$.[36] But Leibniz does not remain content with subtracting y and $xx:a$ from both sides. He also divides both sides by dx to obtain:

$$dy : dx = 2x + dx, : a$$

which is the general rule expressing the ratio of the difference in ordinates to the difference in abscissas. That is, if the chord $_1Y_2Y$ is produced until it meets the axis in T, then the ratio of the ordinate $_1X_1Y$ to T_1X, the intercepted part of the axis between the point of intersection and the ordinate, will be as $2x + dx$ to a. (text [50], see Fig. 6.2)

Up to this point, we have been dealing only with finite quantities and this is where the Law of Continuity will enter the scene to conclude the reasoning:

> Now, since by our postulate it is permissible to include under one reasoning also the case where the ordinate $_2X_2Y$, having been moved closer and closer to the fixed ordinate $_1X_1Y$ until it finally coincides with it, it is clear that in this case dx will be equal to zero or should be omitted, and so it is clear that, since in this case T_1Y is the tangent, $_1X_1Y$ to T_1X is as $2x$ to a. (text [50])

This first basic example was not mentioned in Henk Bos's seminal study (Bos 1974-5) and this had the unfortunate consequence of hiding the global structure of the reasoning. Indeed, after taking another example (the curve of equation $x^3 = aay$) and commenting on the fact that the differential calculus, *like ordinary calculus*, "applies equally to the case where the difference is something and to where it is zero", Leibniz continues:

> But if we want to retain dx and dy in the calculation *in such a way that they denote non-vanishing quantities even in the ultimate case*, let $(d)x$ be assumed to be any assignable straight line whatever; and let the straight line which is to $(d)x$ as y or $_1X_1Y$ is to $_1XT$ be called $(d)y$, so dy and dx will always be assignable to one another in the ratio D_2Y to D_1Y, which latter vanish in the ultimate case. (text [50])

---

[35] See Bos (1974-5) and Arthur (2013) for details.

[36] Leibniz uses the comma to play a role similar to our parentheses. Here it indicates that the whole expression $xx + 2xdx + dxdx$ has to be divided by a, and not just the last term.

This second method differs from the first one in the sense that it *does not rely on vanishing quantities* (**4a**). Moreover, it has the particularity of taking *(d)x* as constant.[37]

Here again, the presentation by Bos, as thorough and interesting as it is, had the unfortunate consequence of hiding some particularities of the reasoning by neglecting some developments that were apparently of no interest. Typically, he did not detail the boring justification of the rule for subtraction given by Leibniz. But this justification, as obvious as it may be, is nonetheless telling, since it arrived at the following equation:

$$(d)y-(d)z, : (d)x = (d)v : (d)x. \text{And so } (d)y-(d)z = (d)v, \text{ as was proposed}\ldots$$

In this equation, there are only assignable quantities, a situation which should be intriguing for those who claim that what derives from the law of continuity *commits us* to the existence of real unassignable quantities. Moreover, Leibniz then comments:

> Although we may be content with the assignable quantities *(d)y, (d)v, (d)z,* and *(d)x,* since in this way we can perceive the whole fruit of our calculus, namely a construction using assignable quantities, still it is clear from this that we may, at least by feigning, substitute for them the unassignables *dx, dy* by way of fiction even in the case where they vanish, since *dy:dx* can always be reduced to *(d)y: (d)x*, a ratio between assignable or undoubtedly real quantities. (text [**50**])

It is pretty clear here that Leibniz contrasts the proof relying uniquely on assignable quantities, with the one in which one can introduce the fiction of vanishing quantities. The justification of this second discourse is that all the vanishing quantities stand for "ratios between assignable or undoubtedly real quantities".

The situation is a little more intricate with the example of the derivation of the rule for the product *d(xv)*, the example chosen by Bos. Here Leibniz reached the following situation:

$$a(d)y/(d)x = x(d)v/(d)x + v + dv$$

> so that the only remaining term which can vanish is *dv*, and, in the case of vanishing differences, since $dv = 0$, we obtain *a(d)y = x(d)v + v(d)x*. (text [**50**])

It may seem that in this case the last equation involves irreducibly the existence of an unassignable quantity. But this is not the case. As in the strategy developed with Wallis, the whole interest of this procedure is to isolate the part of the equation which will constitute the difference to be rendered as small as one wishes. Moreover, the diagram (see Fig. 6.2 below) makes it clear to what it corresponds (the quantities other than *dv*, taken into quotients, correspond to finite ratios between ordinates and subtangents to the given

---

[37] Hence it does not seem possible to claim that "The assignable quantity *(d)x* passes via infinitesimal *dx* on its way to absolute 0" (Katz and Sherry 2013, p. 581).

6.3 The Justifications of the Differential Algorithm

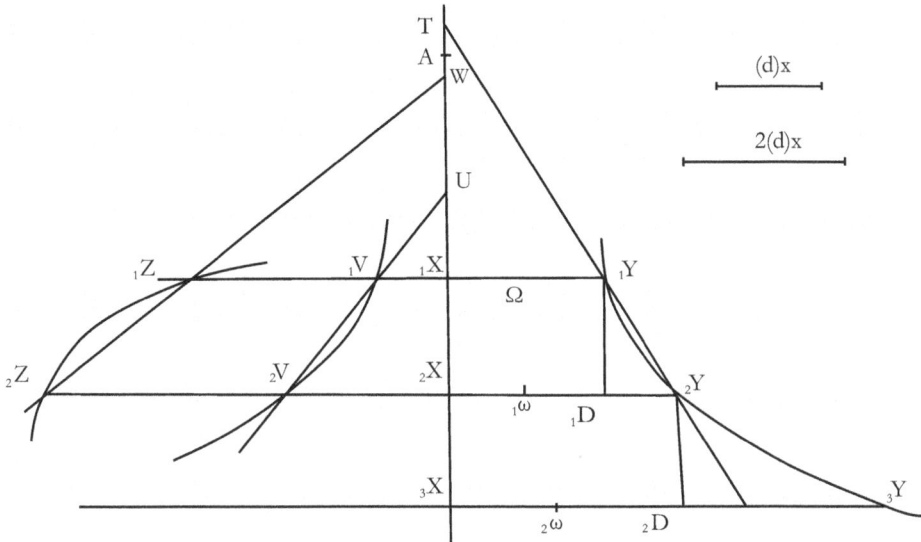

Fig. 6.2 From the *Cum Prodiisset*

curves). We henceforth are facing a typical "syncategorematic" interpretation in which all the "nascent" and "evanescent" quantities have been replaced by finite quantities which can be taken as small as one wishes.

Leibniz stops this derivation here, but he might have easily continued by providing another boring proof "in the Archimedean style": one just has to suppose that $(d)y/(d)x$ differs from $x\,(d)v/(d)x + v$ by a given error Z. Then by taking $d(x)$, $d(y)$, $d(v)$ as small as one wishes, for example, smaller than ¼Z, and since all the other terms are finite quantities whatever the size of $(d)x$, $(d)y$ may be, we easily reach the contradiction of having Z smaller than itself.

Two last comments: firstly, as can be seen in both examples (Wallis and *Cum prodiisset*), the rendering of the proof of the procedure into an "Archimedean" argument is completely trivial and this may explain why Leibniz never felt compelled to give it in detail. We have made it fully explicit only in order to counter the claim according to which it was only wishful thinking on Leibniz's part, and that he would have been completely unable to provide such a proof. Secondly, Leibniz, having realized that the form of his first order differential equations did not depend on the choice of independent variable, as explained in Chap. 5, wrongly supposed this to apply also to second and higher order differential equations. However, in his application of his calculus, especially in the analysis of motion, the calculations depend on taking time as the independent variable (so that dt is constant), and this saves the consistency of his calculations, if not the derivations he gives in *Cum prodiisset*.[38]

---

[38] See Arthur (2013), ("Leibniz's Syncategorematic Infinitesimals") for details.

Conclusion 7

It has long been maintained that the differential calculus was accepted largely because of its success, and with little attention on the part of its progenitors to foundational issues. This line of criticism began early. Since Leibniz's careful expositions had remained unpublished, scholars had little to go on for his views on foundations beyond his "Lemmas on Incomparables" of 1689 and his published letter to Varignon of 1702. Thus Berkeley's disparagement of infinitesimals as "ghosts of departed quantities" went largely unchallenged, and D'Alembert was apparently left to exclaim: "*Allez en avant, et la foi vous viendra!*" We hope we have shown that, on the contrary, Leibniz was intimately concerned with foundational issues regarding infinitesimals and the infinite even prior to conceiving the calculus, and paid close attention to the question of its justification once he had formulated it. While we cannot rule out that details of this picture may be challenged when more of his unpublished work in mathematics is transcribed and edited, the consistency of the excerpts we have provided here from his writings, both published and private, makes that less likely. On this basis, we believe we have established certain key features of Leibniz's views on the use of infinitesimals and infinities in mathematics, and the foundations he provided for his differential and integral calculus.

First, Leibniz treated the infinite and infinitely small as fictions beginning with his work on infinite series in Paris in the mid-1670s, and not as a result of later criticisms. Moreover, once this appeal to fictions is set in its historical context, we can see that the characterization of mathematical entities like negative numbers and irrational roots as fictions was far from being a startling innovation, but in fact quite commonplace among mathematicians of the seventeenth century. This consideration in itself did not determine the ontic status of such entities, which depended on the context in which they were used. For instance, negative numbers could be construed as debts in the context of accounting, whereas imaginary roots of equations, even if they had no direct geometric interpretation, could still sometimes be used in a calculation delivering results that were geometrically determinate. Using such fictions, moreover, one could treat cases in the same way whether or not the problem posed is a possible one. Thus, we maintain, in calling infinitesimals fictions, Leibniz was not assigning them a special kind of existence, but rather asserting that mathematicians could use them in computations without entering into the question of whether they exist. Leibniz's innovations, however, consisted not just in analogizing infinitesimals to imaginary roots, but in providing an account of how their use could not lead to error, since, as he repeatedly urged, one can always substitute for them expressions in which only finite quantities occur. Thus he consistently maintained that mathematicians could and should proceed in their calculations *as if* infinitesimals exist as infinitely small common measures of continuous magnitudes, and *as if* magnitudes were infinite wholes, notwithstanding his own rejection of the existence of either.

Second, Leibniz availed himself of a way of justifying the use of infinitesimals that he (like Newton) could already have found assumed by Pascal and Barrow, according to which quantities (or their ratios) whose differences become smaller than any assignable quantity are declared equal. This principle, which we call the Principle of Unassignable Difference (*PUD*), was never explicitly named by Leibniz, who ascribed it to Archimedes. But it assumed a foundational role not only in his justification of the use of infinitesimals in quadratures, but also in providing a rigorous foundation for the rules of differentiation in the calculus.[1]

By 1674 Leibniz had asserted that an infinite whole like the area under the Apollonian Hyperbola is a fiction, because a part of the area can be demonstrated to be equal to the whole. Given his insistence that the part-whole axiom is essential to the notion of quantity, this shows that the fiction in this case is an impossible one. Having established this, Leibniz was able to give a justification of using the fiction of infinite quantities as if they are wholes by showing how the same results could be obtained without this assumption. In the case of converging infinite series, this was done by calculating the general term of the series in question from the general terms of the series that are manipulated in the computation, and then showing how finite series with this general term have sums that converge on the value calculated as the number of terms is increased. An infinitely small "last term" is then to be understood as a variable term that is smaller in proportion as more terms are taken in the series. Accordingly, Leibniz defines the sum of converging series in terms of what we call a sequence of partial sums, whose difference from the limiting value can be made smaller than any assignable value by taking further terms. The "sum" of a diverging series is understood as a variable that becomes greater without limit as more terms are taken. In this

---

[1] It is also the first lemma that Newton provides for his (geometrical) Method of First and Ultimate Ratios, although he had earlier assumed the *PUD* in his work in analysis as well. See, for example "Those ultimate ratios … are not actually ratios of ultimate quantities, but limits … which they can approach so closely that their difference is less than any given quantity…." (Newton 1999, p. 442); see Pourciau (2001) and Arthur (2008) for discussion.

way, one can use infinities and infinitesimals without being committed to them as existing entities.

Third, Leibniz's approach to the calculus can be seen as emerging naturally from these foundations. Thus in his treatise on quadrature from 1676, *De quadratura arithmetica*, Leibniz appeals to the *PUD* to justify what amounts to a Riemannian integration, where the difference between the area under a curve and the corresponding step-space figure of rectangles of various widths is shown to be smaller than any area that can be assigned, and therefore null. That demonstration is already a step beyond the traditional Method of Exhaustion, in that it appeals to only one proof, not the traditional two (corresponding to the circumscribed and inscribed figures assumed by Archimedes); and it is superior to indivisibilist methods in that it does not assume indivisibles or ratios between infinite aggregates of them, but rather differences that are divisible and arbitrarily small, yet still subject to the Archimedean axiom. This proof from first principles shows that one does not have to appeal to the existence of infinities and infinitesimals in order to justify quadratures. But, as Leibniz goes on to show by a *reductio* argument, a "direct" proof relying on a continuity argument together with the fiction of the infinitely small can also be shown to be equivalent to a *reductio* proof. This establishes the legitimacy of a direct proof using infinitesimals and infinite wholes as fictions, just as in the case of summing infinite series. Moreover, this *reductio* proof establishes that an element of area lying beyond the sequence in question and completing the quadrature can be included as if it is the last term in the sequence, even if it is of a different shape. This is thus an early form of a strategy Leibniz will later use to justify using infinitesimals by reference to his Law of Continuity.

Fourth, Leibniz's justifications of the rules of differentiation through his Lemmas on Incomparables can also be seen as a natural outgrowth of these early positions. Thus if y equals xv, then its difference dy will equal $(x + dx)(v + dv) - y = xdv + vdx + dxdv$. But $dxdv$ is incomparably smaller than the other terms, so $dy = xdv + vdx$. This is justified as follows. Suppose the equality does not hold, and dy differs from $xdv + vdx$ by some least assignable error Z. Now, since dx and dv are variable quantities satisfying the Archimedean axiom, there will exist values for these variables as small as we wish, such that $dxdv$ is smaller than Z. By r*eductio ad absurdum*, the error is therefore less than any assignable. So there is no assignable error: the difference is smaller than any assignable in the sense that any assignable error leads to contradiction. This is what it means for $dxdv$ to be said to be incomparably smaller than the other terms, xdv and vdx. Since a similar *reductio* proof can be given in any such case, one can simply assume that terms that are incomparably smaller than other terms in a resulting equation can be set equal to 0—this is what Leibniz calls *the direct method*: "But since the direct method is easier to understand and more useful for discovery, once this way of reducing is known it suffices that afterwards the method is applied in which incomparably small things are neglected" (GM V, 322).

The notion of incomparability is a relational notion. If y, s, and x are functions of a curve, for instance its ordinate, arc length, and abscissa respectively, then neither ds nor dx is incomparably small in itself, but only in relation to s, x, or y—as, indeed ddx, dyy, $dxdy$, or other higher powers of differentiation, are incomparably smaller than them. What this entails is that the rules of the calculus can be applied whether one takes infinitesimals to be existing things or not; in the latter case, the rules are justifiable by the *reductio* reasoning just sketched, where an assumed error is rendered smaller than itself, and therefore proven to be null.

This relational nature of comparability was not well understood by Leibniz's contemporaries. Wallis thought that this kind of reasoning showed that infinitesimals were in fact "nothings", whereas according to Leibniz's analysis in terms of incomparables, ds (for instance) does have an assignable ratio to dx, but not to quantities of a different order of infinity. Nieuwentijt, on the other hand, held that infinitesimals were non-zero entities whose squares were zero, and Rolle and Berkeley that discarding them at the end of a calculation which began by assuming they are non-zero is simply contradictory, both of which assertions could be shown to be untenable. Modern attempts to justify the use of infinitesimals have also often assumed that Leibniz's "unassignables" must be interpreted as actual infinitesimals (elements in a non-Archimedean continuum). Like Johann Bernoulli, they have assumed that the successful use of infinitesimals entails their existence. But, as we have argued, Leibniz held that the use of infinitesimals in mathematical reasoning was neutral with respect to their existence. His insistence that the question of use is independent of the issue of existence is a constant refrain in his work, from 1676 till the end, and not a sop to appease his supporters.

Fifth, Leibniz's justification of his rules in terms of incomparability also explains something that puzzled his supporters and detractors alike: the fact that in his publication of his rules for differentiation in 1684, he defined his differences in terms of "an arbitrarily assumed straight line", and in later publications gave examples of incomparables that were finite, not infinitesimal, such as taking the Earth as a point, or the diameter of the Earth as incomparably small with respect to the heavens. This is because Leibniz understood his rules for differentiation to be applicable to any incomparable differences, whether these were finite or infinitely small. In the former case, the differences could be taken so small as to be considered not comparable, as for instance the size of the Earth is not considered in comparison with the distance of the fixed stars. In the latter case, differences treated as infinitely small could also be understood as finite differences that become arbitrarily small, so that the resulting error is smaller than any that can be assigned, and therefore null. So the Method of Incomparables may be used whether or not one accepts the existence of infinitesimals.

Sixth, despite the fact that Leibniz regarded his calculus as justifiable without assuming the existence of infinitesimals, this is not an indication that he thought the issue of existence unimportant. Indeed, he had insisted from the beginning of his career that the concepts of the infinite and infinitely small were replete with metaphysical significance. But by 1676 he had become convinced that neither infinitely small things nor infinite wholes (in the sense of infinite collections) in fact exist. These are facets of a remarkably subtle philosophy of the infinite. On the one hand, Leibniz had rejected infinite number and infinite wholes as contradictory already in 1672, but he still believed that there were infinitely small things in the continuum, ones "infinitely smaller than any given sensible thing" (text [3]), and which had to be defined in terms of lines proportional to (infinitesimal) endeavours at the same instant. On the other hand, however, his reasons for believing in the existence of infinitesimals so defined fell away one by one during his last months in Paris, even after his devising of his infinitesimal calculus. He realized that matter could be regarded as actually infinitely divided without issuing in minima if it were divided in such a way that each part were further divided into further (finite) parts without end. This corresponded with his justification of quadratures in terms of polygons whose sides could always be further "inflected to such a degree that even assuming an infinitely small difference, the error would be smaller", with the resulting error rendered "not infinitely small, but nothing at all" (text [**8**]). In the same way, angles, considered as spaces comprised between two intersecting lines smaller than any assignable, could be regarded as fictitious, rather than serving as instances of actual infinitesimals, as he had previously thought.

Leibniz also held that if infinitesimals exist as common measures, perhaps ones proportional to irrational roots, then a rational quadrature of the circle (and of the hyperbolic and logarithmic curves) would be possible. But he became convinced that such a rational quadrature was impossible, and this became another reason for rejecting actual infinitesimals. A final clinching argument derived from his establishing of a "mean proportional". This consisted in the consideration that the proposition that a finite line stands in an infinite ratio to an infinitesimal line (that is, treating the latter as a quantity violating the Archimedean axiom) stands or falls with the proposition that a bounded infinite line would stand in an infinite ratio to a finite line. Accordingly, the finite line x is the mean proportional between an infinitesimal line dx and an infinite but bounded one: $dx : x = x : b_\infty$. But, as we have seen, a bounded infinite line cannot be interpreted as an existing whole without contradiction, notwithstanding the wonderful fruitfulness of this fiction, and that of infinitesimals, in the calculus. It seems, however, that Leibniz harboured doubts about the efficacy of this argument in the face of those who (like Galileo or Johann Bernoulli) would not accept that infinite number is contradictory. Later, though, inspired by White's reflections on the infinite, Leibniz formulated an argument in terms of the impossibility of a transition from finite to infinite, which he finally set out as a clinching demonstration of the chimerical nature of infinite wholes and infinitesimals in the 1690s.

Thus Leibniz's settled conception of the infinite and infinitely small were as follows. The actual infinite, insofar as it pertains to created things, must be understood syncategorematically. That is, to say there are infinitely many things of a certain kind is to say that there are more of them than can be expressed by any number; to say that a magnitude is infinite is to say that it can be produced beyond any magnitude that can be assigned. But there is no infinite number, nor any infinite magnitude, insofar as these are taken to be genuine wholes. The only true or absolute infinite is God, whose attributes are without bounds or limits, in keeping with his perfection. Created things, on the other hand, are bounded, in keeping with their lack of perfection; similarly, in geometry all magnitudes are conceived by Leibniz as bounded. Consequently, the infinite occurring in mathematics—for example, the area under the Apollonian hyperbola—must be bounded, even if its bound is fictional, in order for it to be treated mathematically. Thus Leibniz contrasts unbounded lines, which are not amenable to quantity, with infinite but bounded lines, which are fictions. The latter may be treated as distributive wholes, whose fictional parts can be correlated 1-1 with those of other infinite aggregates, also conceived distributively—such as the multitude of natural numbers and that of even numbers; but any such infinite aggregate, treated as a whole in the collective sense, is an impossible fiction, since it contravenes the part-whole axiom.

Correspondingly, infinitely small quantities cannot be interpreted as quantities in the strict sense, since they are incomparable with any finite quantity. Insofar as they occur in the calculations of geometers, however, they are fictions; they may stand in a finite ratio with other fictional quantities of the same order of infinity, even while being incomparably smaller than the whole of which they are fictionally parts. And just as there are no infinite numbers, so there are no infinitesimal numbers, since they are incomparable with 1. During the early 1670s Leibniz had believed that the continuum contains actual infinitesimals defined in terms of their proportionality to endeavours (beginnings of motion). At that time, he held that they exist in the same way as angles of contact, which can be greater or less than one another even while they are smaller than any rectilinear angle. But he later came to reject this analogy, distinguishing quantities in this broad sense (*lato sensu*) from quantities in the strict sense (*stricto sensu*), which must be able to stand in proportions to each other. Thus there is no way of obtaining an angle of contact through the continuous diminution of a rectilinear angle; they are not magnitudes in the same sense, not "congenious" with one another. Congenious things, like a point and a line, or a line and a plane, can be generated from one another through a continuous process of transformation. And while an infinitely small line is intermediate between a point and a line and congenious with them, the assumption that it actually exists leads to the paradoxical situation where there would have to be a transition from an unassignable to an assignable ratio at a certain time. Therefore, Leibniz concludes, infinitesimals, like infinite but bounded lines, are not possible. Although we can conceive infinitely small lines as states of vanishing or beginning of lines, in imitation of magnitudes already formed, they are not in fact existents, but "chimeras".

These considerations inform three different strategies that Leibniz developed in order to defend the calculus from criticism in the early 1700s. The first consists in demonstrations by *reductio* in the style of Archimedes, coupled with his Lemmas on Incomparables proposed in the *Acta eruditorum*. Such demonstrations show precisely what terms can be neglected in a calculation without any error arising from this, so that elisions are made according to certain rules, and not arbitrarily. Secondly, recognizing that Archimedean-style demonstrations are unwieldy and not accepted by everyone, Leibniz provided a different strategy, appealing to the Law of Continuity that is assumed in the ordinary algebraic calculus. This strategy licenses the inclusion, as fictions, of limiting cases in a converging sequence of terms that excludes them: a point at infinity (to include parallel lines as a fictional limit of lines making a finite angle to each other), rest (included as a fictional last term in motions of ever-decreasing velocity), and so forth. This, Leibniz asserts, is exactly the same strategy as that of including zero as the fictional last term of a

7 Conclusion

convergent decreasing infinite series, or an infinitesimal magnitude as a fictional smallest magnitude in a sequence of ever-decreasing finite magnitudes. In the same way, when providing a "universal" equation for conic sections in ordinary algebra, one includes, from projective considerations, the limiting cases of focuses at infinity and infinitely short distances. So if this strategy is rejected in his differential calculus, then by the same right it should be rejected in ordinary algebra as well.[2] Finally, as a third strategy, he shows that even in these cases where one appeals to the fiction of infinitely small quantities, one can give a justification where these infinitely small quantities are understood syncategorematically: that is, not as non-Archimedean quantities, but as standing for quantities that may be made as small as is needed for the error to be proven null.

---

[2] As expressed to Grandi: "For as imaginary roots are necessary for upholding equations that contain possible cases and impossible ones on a par, so extraordinary quantities are necessary for general rules which include intermediate values together with extreme ones, e.g. as parallelism, although an extreme of convergence, is comprehended under convergence." (text [**42**]).

Part II

A Selection of Translations of Key Texts

Texts for Chapter 2, The Question of the Existence of Infinitesimals (1669–1676)

8.1 (for 2.1) Early Views on the Continuum [1669–1672]

Text 1. From Notes on Galileo's *Discorsi* [Fall 1672][1]

P. 24 [EN 77]: One indivisible added to another makes nothing unless an infinity of indivisibles are added together. In the infinite there is neither greater nor smaller, he says on p. 25 [EN 78], which seems to me contrary to what Galileo himself says. He thinks that one infinity is not only not greater than another infinity, but not greater than a finite quantity [EN 80]. And the demonstration is worth noting: Among numbers there are infinite roots, infinite squares, infinite cubes. Moreover, there are as many roots as numbers. And there are as many squares as roots. Therefore there are as many squares as numbers, that is to say, there are as many square numbers as there are numbers in the universe. Which is impossible. Hence it follows either that in the infinite the whole is not greater than the part, which is the opinion of Galileo and Grégoire de Saint-Vincent, and which I cannot accept; or that infinity itself is nothing, ôr that it is not One and not a whole. Or we will say: distinguishing among infinites, that the most infinite, ôr all the numbers, is something implying a contradiction, for if it is a whole it can be understood as made up of all the numbers continuing into infinity, which will be far greater than all the numbers ôr the greatest number. Or we will say that one ought not to say anything about the infinite as a whole, except where there is a demonstration.

---

[1] A VI, 3, 168/ LLC 7-9.

Text 2. From *Accessio ad Arithmeticam Infinitorum* [ca. December 1672][2]

It is established that the Science of Minimum and Maximum, ôr of the Indivisible and the Infinite, is one of the most important pieces of evidence by which the human mind convinces itself of incorporeality. For those who are led by the senses would persuade themselves that there cannot be *given* a line so short that there are in it not only infinitely many points, but also infinitely many lines (and therefore actually infinitely many parts separated from one another), having a finite ratio to a *given one*; unless demonstrations forced them. This would be also to enter wonderfully into the sum of infinites continually decreasing, either to prescribe limits to increasing or decreasing to infinity yet within a finite space; or to generate finite figures by drawing them in each other, and to demonstrate their proportions.

...

Since, then, in this infinite number there are as many even numbers as even numbers and odd numbers taken together, that is, as many as numbers taken simply, it follows that in this infinite number the axiom that the whole is greater than the part fails (just as Father Grégoire de Saint-Vincent contends that it fails for the angle of contact); but it is impossible for that axiom to fail, ôr what is the same thing, that axiom never fails and only fails for nought or *Nothing*. Therefore an infinite number is impossible, not one thing, not a whole, but *Nothing*. And in fact in 0 or zero there is not only this property of the infinite observed by Galileo in unity, but also all others, for the square and the cube of 0 is 0, and the double or triple of 0 is 0, and $0 + 0 = 0$, the whole equal to the part. Lest I seem to have digressed too far from the topic, the same thing is confirmed by the collection into a sum of series progressing to infinity. For in summing the fractions of the Geometric progression, it is said that the sum of the following series is the first fraction of the preceding series, and $1/3 + 1/9 + 1/27$ etc. $= 1/2$, likewise $1/2 + 1/4 + 1/8$ etc. $= 1/1$. Therefore $1/1 + 1/1 + 1/1$ etc. $= 0/0$. Now $1 + 1 + 1$ etc. constitutes an infinite number. The same thing clearly happens in the immediately preceding Table of Fractions of the replicated arithmetic progression, where it is obvious that $1/1 + 1/2 + 1/3 + 1/4 + 1/5$ etc. $= 1/0 = 0$. And $1/1 + 1/1 + 1/1$ etc. $= 0/0 = 0$.

The same situation should warn us that if this is to be done seriously, if philosophy is to be perfected, no proposition should be accepted unless it is either established by immediate observation through sense perception, or is demonstrated by a clear and distinct imagination, that is, by an idea, or by a definition, which is the signification of the idea (: with the exception of the definitions themselves, which—as has often been insisted by Galileo, the restorer of philosophy, in his writings—are arbitrary, and not to be disputed for their falsity, but only for their inaptness and obscurity:).

---

[2] A II, 1, 342-43, 349-352.

8.1 (for 2.1) Early Views on the Continuum [1669–1672]

Text 3. From *On Minimum and Maximum; on Bodies and Minds* [Nov. 1672-Jan 73][3]

There is no minimum ôr indivisible in space and body.
For if there is an indivisible in space or body, there will also be one in the line *ab* (Fig. 8.1). If there is one in the line *ab*, there will be indivisibles in it everywhere. Moreover, every indivisible point can be understood as the indivisible boundary of a line. So let us understand infinitely many lines parallel to each other, and perpendicular to *ab*, to be drawn from *ab* to *cd*. Now no point can be assigned in the transverse line or diagonal *ad* which does not fall on one of the infinitely many parallel lines extending perpendicularly from *ab*. For, if this is possible, let there exist some such point *g*: then a straight line *gh* may certainly be understood to be drawn from it perpendicular to *ab*. But this line *gh* must necessarily be one of all the parallels extending perpendicularly from *ab*. Therefore the point *g* falls—i.e. any assignable point will fall—on one of these lines. Moreover, the same point cannot fall on several parallel lines, nor can one parallel fall on several points. Therefore the line *ad* will have as many indivisible points as there are parallel lines extending from *ab*, i.e. as many as there are indivisible points in the line *ab*. Therefore there are as many indivisible points in *ad* as in *ab*. Let us assume in *ad* a line *ai* equal to *ab*. Now since there are as many points in *ai* as in *ab* (since they are equal), and as many in *ab* as in *ad*, as has been shown, there will be as many indivisible points in *ai* as in *ad*. Therefore there will be no points in the difference between *ai* and *ad*, namely in *id*, which is absurd.

There is no minimum ôr indivisible in time and motion.
For let us suppose a space *ad* is traversed with a uniform motion in a time *ab*. Then in half the time *ae* half the space *af* will be completed, and in a thousandth of the time, a thousandth of the space, etc. Therefore in an indivisible of time, an indivisible of space will be traversed, since time and space are divided proportionally. For let us suppose that in a minimum of time the amount of space traversed is not a minimum: then in a time however

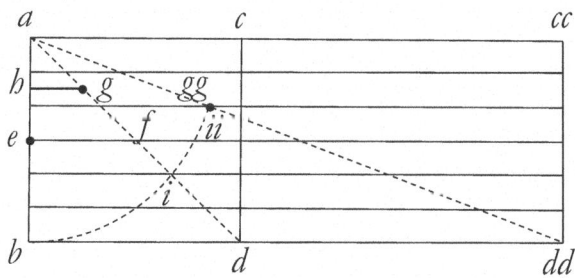

Fig. 8.1 The "Diagonal Paradox"

---

[3] A VI, 3, 97-99/ LLC 9-15.

small, provided it is not a minimum, infinite divisible spaces would be traversed, and in some perceptible time, an infinite space would be traversed. For the ratio of an indivisible—if such a thing is understood to exist—to the divisible, or the ratio of the minimum in the continuum to whatever is not a minimum, is that of the finite to the infinite.

...

Scholium
We therefore hold that two things are excluded from the realm of intelligibles: minimum, and maximum; the indivisible, and what is entirely *one* and *everything*; what lacks parts, and what cannot be part of another.

There are in the continuum infinitely small things, that is, things infinitely smaller than any given sensible thing.
First I show this for the case of space as follows. Let there be a line *ab*, to be traversed by some motion. Since some beginning of motion is intelligible in that line, so also will be a beginning of the line traversed by this beginning of motion. Let this beginning of the line be *ac*. But it is evident that *dc* can be cut off from it without cutting off the beginning. And if *ad* is believed to be the beginning, from it again *ed* can be cut off without cutting off the beginning, and so on to infinity. For even if my hand is unable and my mind unwilling to pursue the division to infinity, it can nevertheless in general be understood at once that everything that can be cut off without cutting off the beginning does not involve the beginning. And since parts can be cut off to infinity (for the continuum, as others have demonstrated, is divisible to infinity), it follows that the beginning of the line, i.e. that which is traversed in the beginning of the motion, is infinitely small.

The same thing may be understood from angles, and from contacts, which, if the bodies are understood to be perfectly finished (which no one will deny to be possible by the nature of things), will necessarily occur at a point.

> A *point* is of a length, breadth and depth infinitely smaller than any that can be sensed; a *line* is of a breadth and depth infinitely smaller than any that can be sensed; a *surface* is of a depth infinitely smaller than any that can be sensed.

This follows from the premisses, since it has been shown that there are no indivisibles, yet that there are infinitely small things. But I do not wish to define a point as a line of length smaller than any given length, since a centre should not be conceived as a line, but as a figure smaller than any given figure, as, for example, the centre of a circle should be conceived as a circle smaller than any given circle, the portions of which are angles.

There is one point smaller than another.

Thus the vertex of a larger angle is greater than the vertex of a smaller angle. And the angle is the quantity of a point, namely of its centre. For it is evident that even if the lines are perpetually diminished, the angle still remains intact. Therefore, for an angle only lines infinitely smaller than any given sensible lines are required, and the space intercepted by them is in the same way infinitely smaller than any sensible space, and is consequently a point. Likewise, since one body moves faster than another, it is necessary that the very beginnings of the motions by which a line infinitely smaller than any given sensible line is traversed, be unequal—i.e. that these lines be unequal. But these lines are points.

One point can be infinitely smaller than another.

For an angle of contact is a point, and a rectilinear angle is a point, and yet any rectilinear angle, however small, is greater than any angle of contact whatsoever.

Text 4. *A Noteworthy Observation about the Infinite* [Feb. 1676?][4]

Here is a noteworthy observation about the infinite. Since there is one infinity greater than another, will there be something more eternal than something else? For instance, a thing can exist before any time imaginable, and yet [not][5] from eternity, because its time [in existence] will not be absolutely infinite, but infinite only in relation to ours. Therefore there was a time when it did not exist, but that time is infinitely remote from now. This is just like an infinitely small line in relation to a point.

8.2 (for 2.2) Infinite Series and the Infinitely Small in Mathematics [1672–1676]

Text 5. From *De serierum summis et de quadraturis pars octava* [October 1674][6]

Let there be a hyperbola *GBE* with centre *A*, vertex *B*, and radius, so to speak, *AC* or *CB* = a. *CD* or *CH* or *CF* = y. *AD* = $a - y$. *AF* or *AH* = $a + y$. *DE* = $\frac{a^2}{a-y}$. *HL* = $\frac{a^2}{a+y}$. Therefore, supposing $a = 1$, *DE* = $1 + y + y^2 + y^3 + y^4$ etc., and the sum of all the *DE*, applied to *AC*, will give $1 + \frac{1}{2} + \frac{1}{3} + \frac{1}{4} + \frac{1}{5}$ etc. = the space produced to infinity, *ACBEM* (Fig. 8.2).

But *HL* = $\frac{1}{1+y}$ = $1 - y + y^2 - y^3 + y^4$ etc. and the sum of all the *HL* applied to *CF* will be:

---

[4] A VI, 3, 481/LLC 45.

[5] The addition of this 'not' seems necessary in order to make sense of the example.

[6] A VII, 3, 468.

Fig. 8.2 Quadrature of the Apollonian Hyperbola

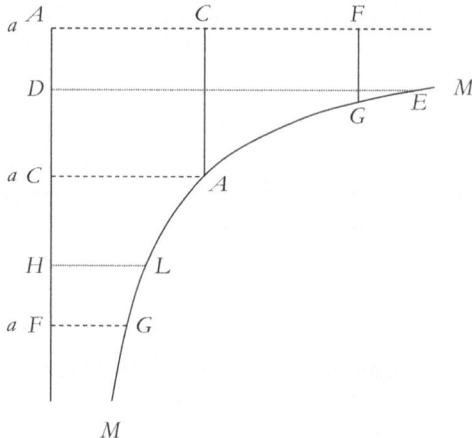

$$1 - \tfrac{1}{2} + \tfrac{1}{3} - \tfrac{1}{4} + \tfrac{1}{5} - \tfrac{1}{6} \; etc. = \text{the space } CFGLB.$$

$$\Downarrow \quad \Downarrow \quad \Downarrow$$

$$\tfrac{1}{2} \quad \tfrac{1}{12} \quad \tfrac{1}{30}$$

Let the space *CFGLB* be subtracted from the space *ACBEM*, giving:

$\tfrac{1}{1}\left(-\tfrac{1}{1}\right) + \tfrac{1}{2}\left(+\tfrac{1}{2}\right) + \tfrac{1}{3}\left(-\tfrac{1}{3}\right) + \tfrac{1}{4}\left(+\tfrac{1}{4}\right)$ etc. or $\tfrac{2}{2}\tfrac{2}{4}\tfrac{2}{6}$ etc. $= 1\tfrac{1}{2}\tfrac{1}{3}$ etc. Which is rather amazing, and shows that the sum of the series 1, $1/2$, $1/3$, etc. is infinite, and consequently that the area of the space *ACGBM* remains the same even when the finite space *CBGF* is taken away from it, that is to say, nothing noticeable is taken away.

By this argument it is concluded that the infinite is not a whole, but only a fiction, since otherwise the part would be equal to the whole.

Text 6. The Cancelled Scholium to Prop. 11 of the *DQA* [June 1676][7]

The contemplation of spaces infinite in length but finite in magnitude is quite remarkable. No such thing was known to the ancients that I know of, and to them it seemed wonderful enough that there are certain straight lines that are asymptotes, which approach closer and closer to a curve, without ever reaching it. First, I believe, Torricelli measured an acute hyperbolic solid as infinite in length, and reduced it to a certain finite cylinder: in the plane, Father Grégoire de Saint-Vincent squared the infinite space enclosed between two hyperbolas in a certain ratio, and a gentleman very famous on his own merit, Christian Huygens, reduced an infinite cissoidal space to a circle. And the eminent geometer, John Wallis, shows how there are innumerable hyperboloids whose area can be found despite their being of infinite length, and how these can be distinguished from those which do not

---

[7] A VII, 6, 548-49.

have this property by a probable method, which, although it uses induction, is nonetheless most ingenious (see under Prop. 22, where we have demonstrated it). Proposition 7, by the doubling of a certain segment or sector, however small, of any given curve, gives us a way of exhibiting, in infinitely many ways, equal figures infinite in length—and we do not know if this can also be done by any other methods.

Besides, our frankness does not allow us to pretend that these things are as surprising as appear to people at first sight. Rev. Father Pardies of the Society of Jesus, known to the learned for his elegant writings, and deserving of a longer life, attributed so much to meditations of this kind that he believed they were sufficiently effective to provide an argument for proving *the immateriality of the soul*, as he asserted in the preface to his compendium of geometry.[8] It seems to me that the very nature of mind per se, and its operations, especially that by which it acts back on itself, suffice for it to be distinguished from body, ôr from a thing endowed with only the two [attributes], extension and mass. Although I do not deny that certain particular operations are more remarkable than others, or at least appear so, these are of more value for persuading, I would have said, than for proving. Indeed, what pertains to this action of the mind by which we measure an infinite space, does not involve anything extraordinary, since it attempts it by a kind of fiction, and proceeds with no trouble by supposing a certain *line that is indeed bounded, yet infinite*, which gives it no more difficulty than if we were to measure a space finite in length.

It would be more remarkable if someone could take the absolutely unbounded space between two curves and reduce it to a finite space. But no such thing is known to me, nor, I believe, to anyone else (but see Prop. 14, Coroll.). Indeed, since this locution concerning lines that are infinite and yet not unbounded will seem a paradox to some, it should be advised, therefore, that just as there is a very great difference between the indivisible and the infinitely small, so there is a big difference between the infinite and the unbounded. The Geometry of Indivisibles is fallacious unless it is explained using the infinitely small; for neither points nor truly indivisible things may be safely used, but instead one must use lines that are indeed infinitely small, yet still lines, and therefore divisible. In the same way, plainly, an unbounded quantity differs from an infinite one. For the magnitude of an unbounded line is in no way subject to geometrical considerations, any more than the magnitude of a point. For just as points, even though infinite in number, are added or subtracted in vain from a bounded line, so a bounded line, no matter how many times it is repeated, can neither make nor exhaust an unbounded line. But it is otherwise with a bounded yet infinite line, which can be understood as constituted by some multiplicity of finite lines even if this multiplicity exceeds every number. And just as an infinite bounded line is composed out of finite ones, so a finite line is composed out of infinitely small yet divisible ones. Hence it cannot be said that a bounded line is the mean proportional between a point ôr minimum line and an unbounded ôr maximum line. But it can be said that a finite

---

[8] I. Pardies, *Elemens de geometrie*, 1671, préface, [a vii] v⁰. Compare with Leibniz's Scholium to Prop. 9 in the earlier draft of *De quadratura arithmetica*:" (A VII, 6, 212-13).

line is the mean proportional, not in a certain way, but truly and exactly, between some infinitely small line and some infinite one; and it is true that a rectangle made up of infinitely many infinitely small ones can be equal to a certain finite square; and this in fact occurs in the conic hyperbola. For if the curve 4Dδλ is hyperbolic, having a centre μ and some infinitely small abscissa μ(μ), there will be an ordinate (μ)λ that is of course infinitely long, that is, greater than any designatable straight line, and an infinite rectangle μ(μ)λ comprised of infinitely many infinitely small ones will, by the nature of the squared hyperbola, be equal to some constant finite square. And so I call unbounded that in which no ultimate point can be supposed, at least on one side. On the other hand, I call a quantity infinite, whether bounded or unbounded, whenever we conceive it to be greater than any quantity assignable by us, or expressible in numbers. Now, whether the nature of things tolerates quantities of this kind is for the metaphysician to discuss; for the geometer it suffices to demonstrate what follows from supposing them.

8.3 (for 2.3) Leibniz's Rejection of the Existence of the Infinitely Small in 1676

Text 7. From *On the Secrets of the Sublime* [February 1676][9]

It seems to follow from a solid in a liquid that a perfectly fluid matter is nothing but a multiplicity of infinitely many points, i.e. bodies smaller than any that can be assigned, or, that there must be such a thing as an interspersed vacuum—a metaphysical one, which is not contrary to a physical plenum. A metaphysical vacuum is an empty place however small, only true and real. A physical plenum is consistent with an unassignable metaphysical vacuum. Perhaps it follows from this that matter is divided into perfect points, that is, into all parts into which it can be divided. Nothing absurd follows from this. For it will follow that a perfect fluid is not a continuum, but discrete, i.e. is a multiplicity of points. But it does not therefore follow from this that a continuum is composed of points, since liquid matter will not be a true continuum, even if space is a true continuum; from which it is again clear how great a difference there is between space and matter. Matter alone is explicable by a multiplicity without continuity. And matter does in fact seem to be a discrete entity; for even if it is assumed solid, still, insofar as it is matter, when its cement—motion, for example, or something else—ceases to exist, it will be reduced to a state of liquidity, i.e. of divisibility, from which it follows that it is composed of points. This I prove as follows: every perfect liquid is composed of points, since it can be dissolved into points, which I prove by the motion of a solid in it. Therefore matter is a discrete entity, not a continuous one; it is only contiguous, and is united by motion or by a mind of some sort.

...

---

[9] A VI, 3, 473-75, 477/ LLC 47-53.

If it is true that any part of matter, however small, contains an infinity of creatures, i.e. is a world, it follows also that matter is actually divided into an infinity of points. But this is true, provided it is possible, for it increases the multiplicity of existents and the harmony of things, or admiration of divine wisdom. Hence it follows further that any part of matter is commensurable with any other, which again is an admirable effect of the harmony of things. It must be seen whether this truly follows. In that connection, I should examine the line of reasoning I have used elsewhere, according to which it seems to follow that a circle, if it exists, has a ratio to the diameter as one number to another. It must be seen whether this inference is a good one. It must be seen, on the other hand, whether there does not follow in a liquid a subdivision that is now greater, now less, in accordance with the various motions of a solid in it; accordingly it must be rigorously examined whether there follows a perfect division of a liquid into metaphysical points, or only one into mathematical points. For mathematical points could be called Cavalierian indivisibles, even if they are not metaphysical points, ôr minima. But if a liquid can be shown to be divided to a greater or lesser degree, it will follow that a liquid is not resolved into indivisibles. Yet one could defend the composition of a liquid out of perfect points, even if it is never absolutely resolved into them, on the grounds that it is capable of all resolutions, and will cease to exist when its cement—namely mind and motion—ceases to exist.

...

The whole labyrinth of the composition of the continuum (see Froidmont's book) must be unravelled as rigorously as possible, and the angle of contact must also be treated, for that dispute is of concern not to geometers, but to metaphysicians. We must try to see if it can be demonstrated that there is something infinitely small, yet not indivisible. If such a thing exists, there follow some wonderful consequences concerning the infinite: namely, if we feign creatures of another world that is infinitely small, we will be infinite in comparison with them. Whence it is clear in turn that we could be feigned as being infinitely small in comparison with another world that is of infinite magnitude, and yet bounded. Whence it is clear that the infinite is—as, of course, we commonly assume—something other than the unbounded. The latter should more properly be called the Immensum. This is a wonder, too: that someone who has lived for infinitely many years can have begun to live, and that someone who lives for a number of years that is greater than any finite number can at some time die. From which it will follow that there is an infinite number. Another way of establishing that there is necessarily an infinite number is from the fact that a liquid is actually divided into parts infinite in number: if this is impossible, it will follow that a liquid is impossible. Since we see the hypothesis of infinites and the infinitely small is splendidly consistent and successful in geometry, this also increases the likelihood that they really exist.

...

The number of finite numbers cannot be infinite. Whence it follows that the finite squares, taken one after another in order from the number one, cannot be infinite in number. Whence it seems to follow that an infinite number is impossible. It seems that all that needs to be proved is that the number of finite numbers cannot be infinite. If numbers can be assumed as continually increasing by one, the number of such finite numbers cannot be infinite, since in that case the number of numbers is equal to the greatest number, which is supposed to be finite. It must be responded that there is no such thing as the greatest number. But even if they increase otherwise than by ones, nevertheless, provided only that they always increase by finite differences, it is necessary that the number of all numbers always has a finite ratio to the last number; indeed, furthermore, the last number will always be greater than the number of all numbers. Whence it follows that the number of numbers is not infinite.[10] Neither, therefore, is the number of unities. Therefore there is no such thing as infinite number, that is, it is not <possible>.

Text 8. On the Infinitely Small [26 March 1676][11]

We need to see exactly whether it can be demonstrated in quadratures that a difference is , however, not infinitely small, but nothing at all. And this will be shown if it is agreed that a polygon can always be bent inwards to such a degree that even when the difference is assumed infinitely small, the error will be smaller. Granting this, it follows not only that the error is not infinitely small, but that it is nothing at all—since, of course, none can be assumed.

I call that quantity *undesignatable* whose magnitude cannot be expressed by any signs with characters detectable by the senses. For every designatable magnitude whatsoever will always be writable in a sufficiently small book with the aid of abbreviations and representations. It was in order to show this that Archimedes undertook his investigation in *The Sand-Reckoner*.

Text 9. From *On Motion and Matter* [1–10 April 1676][12]

I do not see how a perfect circle can be described unless one is already presupposed. Indeed, it seems that if two bodies with contrary motions, DA and CB, run into a body AB, a certain circular motion will arise from this (Fig. 8.3). Likewise it seems to be proved that curved lines would arise from motion in a plenum (even if one admits an interspersed vacuum). But on the other hand there is the great difficulty that endeavours are along tangents, so that motions will be too. For I have demonstrated elsewhere very recently that endeavours are true motions, not infinitely small ones. From this it will follow that there is no really curvilinear motion in things which endeavour along tangents. Therefore one of two things is true: either there are things which do not endeavour along tangents; or there is

---

[10] *Leibniz's note*: "No, N.B., this only proves that such a series is unbounded."
[11] A VI, 3, 434–5/ LLC 65.
[12] A VI, 3, 492, 494 /LLC 75-77, 81.

Fig. 8.3 Producing Circular Motion

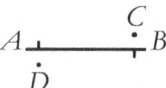

no curvilinear motion. As long as there is no method for directly establishing the quadratures of curved lines, there will be a strong suspicion that none exist. For otherwise something could be established about them directly. The squaring of Hippocrates' lunule must be examined; it supposes circles to be as the squares of their diameters, which latter, if it can be established absolutely, allows one to conclude that some quadrature can be established directly.[13] And this difficulty will cease.

We must see whether it is possible for a new endeavour to be impressed at any moment whatever. It could, if there were such a thing as a perfect fluid. But if this is conceded, time will be actually divided into instants, which is not possible. So there will be no uniformly accelerated motion anywhere, and so the parabola will not be describable in this way. And so it is quite credible that circles and parabolas and other things of that kind, are all fictitious entities. We must see whether there is not an argument which concludes that some way of producing a parabola is impossible: by this you would show that the parabola itself is impossible. For supposing a point moves in a parabolic line, it will by all means be true of it that at any instant it is moving with a uniform motion in one direction, and with a uniformly accelerated motion in another, which is impossible. But if the parabola is therefore impossible, it will be quite understandable for the same fortune to befall the circle. We should learn from this how often we are deceived when we think we perceive things clearly and distinctly, for what geometer will not regard the circle as perfectly understood in itself? To this we should add the reasonings of Hobbes.[14]

...

But [if motion is supposed continuous], it is also inexplicable how it is possible for one motion to be faster than another, and a great many difficulties will occur. One proves that there is one motion faster than another by the motion of a radius about some centre, in which radius various points are designated; now we may easily reply that there is no circular motion. Assuming there are no curved lines, what is said about them will be the properties of polygons. And a particular circle will not be an entity, but [should be] taken in general, that is, what is demonstrated about it will have to be understood of any polygon inscribed in it, or of one having a greater number of sides than we have employed.

---

[13] Hippocrates' lunule is a figure bounded by the arcs of two circles, the smaller of which has as its diameter a chord spanning a right angle on the larger circle. Hippocrates of Chios proved in his *Elements* (now lost, but preceding Euclid's *Elements*) that the area of this figure is equal to the right-angled triangle whose hypotenuse is that chord. As Leibniz notes, this shows that at least some quadratures of curved lines can be calculated directly.

[14] *Leibniz's note:* "—who treated these matters, and made a collection of the difficulties, but did not select them convincingly enough to determine what was to be established about the remaining ones."

Text 10. From *Infinite Numbers* [ca. 10 April 1676][15]

Now let there be two figures, one rectilinear *QRST*, the other a mixed curvilinear one *QTVQ*, of the same height *QT*. Let us suppose that the curvilinear one is a polygon *QTυβγω*, that is, a gradiform figure divided into an infinity of infinitely small squares, such as *αυβγ*. In the same way, let the rectilinear figure be understood to be divided into squares equal to these, such as *XQZ*. In this way, the height *QT* is understood to be divided into an infinity of parts, and however many ordinates the rectilinear figure has, e.g. *ZY*, so does the curvilinear one, e.g. *Zω* (Fig. 8.4).[16] Now let us assume any arbitrary curvilinear ordinate, such as *Zω*, to be in rational proportion to the corresponding rectilinear ordinate, *ZY*, which will be possible if the equation of the curve allows it; then the number of infinitely many squares of one will also be commensurable with the number of squares of the other, and if one is exhausted by a repetition of squares, then so will the other be. And since in this way the individual ordinates taken one by one have a common figure, so the whole figures will also have a common measure, namely the assumed square. If all these ordinates are extended in a straight line, i.e. if the squares are juxtaposed in a straight line, the number of them in the rectilinear figure will be to the, number in the curvilinear one as one commensurable infinite line to another, and so as we showed above, as finite number to finite number. But a curvilinear figure of the above kind is congruent with the circle, as I have shown elsewhere. Therefore the circle is to the square as finite number to finite number, which is absurd.

And now I see the reason for the error. It must be denied that two commensurable finite lines are as finite number to finite number. It is possible for them to be as infinite number to infinite.[17] Two infinite numbers which are not as two finite numbers can be commensurable, namely if their greatest common measure is a finite number—as if both are prime. Meanwhile, this much is certain here, that these two figures, the circle and the square, are

Fig. 8.4 Attempt at a Quadrature with Rational Common Measure

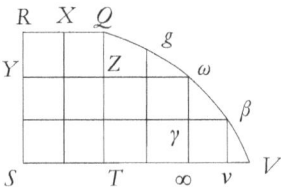

---

[15] A VI, 3, 497–99, 502–04/LLC 85–91, 97–101.

[16] Thus *QT*, the abscissa common to the two figures, is drawn vertically, and the ordinates *ZY* and *Zω* are drawn horizontally on either side of it. This is the inverse of the modern convention, where abscissae are drawn horizontally and ordinates vertically.

[17] Leibniz had begun by assuming that that if two infinite numbers that obey the Archimedean axiom, and are thus "as one designatable line to another", are commensurable, then "they will be as finite number to finite number", before seeing that this assumption leads to an absurdity. He then writes "error" over each occurrence of "finite", and begins anew.

commensurable, i.e. have a common measure, whether (i) finite and ordinary (in which case they would be as finite number to finite number, which I believe to be completely irreconcilable with approximations); or (ii) infinitely small, which I believe to be necessary.

Hence now at last there seems to be a way open for a marvellous demonstration that it is impossible for there to be a quadrature of the circle of the kind we are seeking: namely, one which would express the relation by an algebraic [*equabile*] equation. And in order for this to be done, it must be shown that the diameter and the side do not have even an infinitely small common measure, even in the kind of line which is as an irrational root, whether quadratic or cubic—as, for example, the side of a double cube, or of some higher power. Hence here we have a splendid use for demonstrations about incommensurables using lines, for they can also be carried over to the infinitely small, which those of arithmetic cannot. Supposing this, it follows that the magnitude of a circle cannot be expressed by an equation of any degree. By the same argument it is proved that not even a portion of a circle can be squared by this means; and it is the same with the logarithm and the hyperbola.[18]

The circle—as a polygon greater than any assignable, as if that were possible—and other things of that kind, are fictive entities. So when something is said about the circle we understand it to be true of any polygon such that there is some polygon in which the error is less than any assigned amount a, and another polygon in which the error is less than any other definite assigned amount b. However, there will not be a polygon in which this error is less than all assignable amounts a and b at once, even if it can be said that polygons somehow approach such an entity in order. And so if certain polygons are able to increase according to some law, and something is true of them the more they increase, our mind feigns[19] some ultimate polygon; and whatever it sees becoming more and more so in the individual polygons, it declares to be perfectly so in this ultimate one. And even though this ultimate polygon does not exist in the nature of things, one can still give an expression for it, for the sake of abbreviating propositions.

For the rest, it must be seen whether there are not still other things that are infinitely small, such as angles. Here, for instance, is an angle; is it not in a point? For the length of the sides is irrelevant to it, and it remains even if they are shortened forever. Therefore there is a quantity in a point, for it is the quantity of an angle. First, it should be replied that there is no angle in a point by itself unless lines are added. Now if these lines are infinitely small, but are lines, the difficulty will remain, for I will cut them back in the same way. Moreover, an angle is not the quantity of a point. For we have supposed a point to be that whose part is nothing, an extremum; for we have already shown that there is nothing else unassignable.

---

[18] *Marginal note by Leibniz:* "This reasoning, which seems to prove that a circle is not squarable, should not be relied on as long as it has not been proved that the diagonal cannot—at least, by subtracting an infinitely small quantity—be rendered commensurable to the side, assuming an infinitely small measure. And the same holds for the other roots."

[19] As noted in the Note on Texts and Translations, we translate "*fingere*" as "feign" (and not "imagine", as is often proposed), in order to maintain the link between the verb (*fingere*), the noun (*fictio*) and the adjective (*fictitiia*), which are used by Leibniz in this context.

Therefore the quantity of the angle will be nothing but the quantity of the proportion of the sine, which is the same however far you might produce it, so that the angle itself, it seems, is a fictitious entity. If we take it for some thing existing in the point itself, that is if we suppose that the angle is in the point, i.e. that it subsists when any assignable line has been cut back, the quantity of the angle is given. The angle is the same when the sides are produced. If these things are assumed, an angle will be that which is in lines smaller than any that intersect, i.e. the space comprised between two intersecting lines that are smaller than any assignable. But such an entity is fictitious, since lines of this kind are fictitious.

Even though these entities are fictitious, geometry nevertheless exhibits real truths which can be expressed by other means, and without them. But these fictitious entities are excellent abbreviations for expressions, and for this reason extremely useful. For entities of this kind, i.e. polygons whose sides do not appear distinctly, are made apparent to us by the imagination, whence there arises in us afterwards the suspicion of an entity having no sides. But what if that image does not represent any polygons at all? Then the image presented to the mind is a perfect circle. Here there is a surprising and subtle difficulty. For even if the image is false, the entity in it is nevertheless true; and so it follows that in the mind there is a perfect circle, or rather, there is a real image. Therefore everything else will also exist in the mind: and in it everything that I denied to be possible will now be possible. Instead, what must be said is that in the mind there is a thought of uniformity, yet no image of a perfect circle: instead the uniformity is applied by us to this image afterwards, a uniformity we forget we have applied after sensing the irregularities. Were we then conscious at some time that we had sensed them? for consciousness is necessary for forgetting. But this is not the case. Therefore it must be said, rather, that when we sense a circle or polygon, we never sense uniformity in it, but neither do we even sense a non-uniformity, that is to say, we do not remember having sensed anything non-uniform in it, since the inequality did not immediately strike us. And because of this memory we now ascribe the name of uniformity to it. It must be seen whether we might not be conscious for very small intervals of time of many things we do not remember, or about which we are unable to speak or write, which we cannot express in characters on account of their extremely small size, since they have little relation to such things. But they are not on this account any less sensed by us consciously. Rather, we forget about these things, just as we forget about the things we dream about.

[...]

The reason why the unbounded, i.e. that which is greater than anything finite, is something, and the infinitely small is not, is that in the continuum the maximum is something, and the minimum is not; the most perfect is something, the least is not, God is something, nothing is not. In the continuum, the whole is prior to its parts; the absolute is prior to the limited; and so the unbounded is prior to that which has a bound, since a bound is a kind of addition. There is no greatest number, and no minimum line.

We must still investigate whether and to what extent the following is true, namely that the square is to the circle as 1 to $1 - \frac{1}{3} + \frac{1}{5} - \frac{1}{7} + \frac{1}{9} - \frac{1}{11}$ etc. For when we say 'etc. to infinity', the last number is not really understood to be the greatest number, for there isn't one, but it is still infinite. But since the series is not bounded, how can this be the case? For something must be added, even if an infinite number is assumed, therefore it must be said that this is not rigorously true. And since the circle is nothing, this series will of course also be nothing.

Meanwhile there remains this difficulty. Diagonal to square is a certain ratio, since the diagonal is a line, a real quantity, and so too is the side. If this is to be expounded by means of numbers, there will also be a need for infinite numbers—indeed, for all numbers in general. But to say all numbers is to say nothing; and for this reason that ratio also means nothing, unless it is something as close as desired. Still the ratio of these two lines is not thereby eliminated, even if no measure is assigned. Unless (there being no measure) you also say of the ratio what you said of the angle, that in itself it is nothing but the very agreement of the divisions; an agreement that always remains, as did the sine above. Indeed it seems that the ratio always subsists, since it is through this ratio that two figures are similar. [...]

Whenever it is said that a certain infinite series of numbers has a sum, I am of the opinion that all that is being said is that any finite series with the same rule has a sum, and that the error always diminishes as the series increases, so that it becomes as small as we would like. For numbers do not *in themselves* go absolutely to infinity, since then there would be a greatest number. But they do go to infinity when applied to a certain space or unbounded line divided into parts. Now here there is a new difficulty. Is the last number of a series of this kind the last one that would be ascribed to the divisions of an unbounded line? It is not, otherwise there would also be a last number in the unbounded series. Yet there does seem to be, because the number of terms of the series will be the last number. Suppose to the point of division we ascribe a number always greater by unity than the preceding one, then of course the number of terms will be the last number of the series. But in fact there is no last number of the series, since it is unbounded; especially if the series is unbounded at both ends. Therefore we conclude finally that there is no infinite multiplicity, from which it will follow that there is not an infinity of things, either. Or it must be said that an infinity of things is not one whole, i.e. that there is no aggregate of them. If an infinity of things could not exist, the world would be necessarily finite in time and place, but for the world to be finite in time does not seem possible. Indeed, it would then follow also that at some time things will come to an end, and everything will be reduced to nothing, for otherwise [the number of] all future things would be infinite. Thus if you say that in an unbounded [series] there exists no last finite number that can be written in, although there can exist an infinite one: I reply, not even this can exist, if there is no last number. The only other thing I would consider replying to this reasoning is that the number of terms is not always the last number of the series. That is, it is clear that even if finite numbers are increased to infinity, they never—unless eternity is finite, i.e. never—reach infinity. This consideration is extremely subtle.

Text 11. From *Pacidius to Philalethes* [October 1676, OS][20]

(*Ch.*: Perhaps leaps through infinitely small spaces are not absurd, and nor are little rests for infinitely small times inserted between these leaps. For assuming the spaces of the momentaneous leaps to be proportional to the times of the rests, they will correspond together in the way in which we explained the leaps and rests through ordinary times and lines above.

Pa.: I would indeed admit these infinitely small spaces and times in geometry, for the sake of discovery, even if they are imaginary. But I am deliberating whether they can be admitted in nature. For there seem to arise from them infinite straight lines bounded at both ends, as I will show elsewhere; which is absurd. Moreover, since some infinitely small things can also be assumed smaller than others to infinity, again no reason can be provided why some should be assumed rather than others; but nothing happens without a reason.)

Text 12. *Geometry with the Metaphysics of the Continuum Omitted* [1679][21]

The whole of Geometry can be treated and demonstrated with the metaphysics of the continuum omitted, that is, whether lines are composed of points, motion from little rests, or not. For instead of a point we may take a body sufficiently small that even if it is repeated a million times, it will still make a sufficiently small body. Instead of a line [we may take] a motion of this body, if not a continuous one, at least a motion through leaps of such a kind that the intervals between two neighbouring places are also sufficiently small, and so that the leaps are equal to one another in equal times. Intervals, that is, I understand as certain lines that are drawn from one to another, not caring whether the lines are straight. And it is sufficient that as many leaps as possible be taken, all of them congruent, that is to say, equal and similar, to each other.

A straight line is made if different bodies *abc* and *def* are at first congruent, then *def* is moved over *abc* so that it can be held to be congruent to it, that is to say, recedes from it so little that the recession is also a sufficiently small amount. Now again another [body] *ghi* at first congruent to *def* is moved away from it in the same way, so that it departs from *abc* and so forth, and the space or body thus formed may be held to be a straight line to which the former definition is accommodated, because, assuming only two of the points congruent, all the parts would be congruent with each other.

By the motion of a straight line of this kind a circle can also be described, and everything else in Geometry is achieved with equal certainty, for it will always be demonstrated rigorously that the error is as small as we wish at the beginning of the construction of each problem.

---

[20] A VI, 3, 564-65/ LLC 207.

[21] LH 35, 1, 1, fol. 12; *Mathesis* (2021, 68).

It comes down to the same thing whether someone composes a body only from points, or on the contrary holds points to be bodies.

Transcription of the Latin text:

Geometria omissis metaphysicis de continuo [1679][22]

Tota Geometria tractari potest et demonstrari omissis Metaphysicis de continuo, id est sive ponamus lineas ex punctis, motus ex quietulis componi sive non. Nam pro puncto sumamus corpus satis parvum, ut millies millieses repetitum, nihilominus adhuc faciat corpus satis parvum; pro linea motus hujus corporis, si non continuum, saltem per subsultationes ejusmodi, ut intervalla inter duo loca proxima etiam satis parva sint et aequalia inter se iisdem temporibus. Intervalla intelligo id est ut lineae quaedam ducantur satis parvae ab uno ad aliud nihil curando an sint recta. Et satis est ut quam plurimae sumantur subsultationes omnes sibi invicem seu congruae seu aequales et similes.

Recta fiet, si corpora alia ut *abc*, *def* primum congruant, inde *def* moveatur super *abc*, ut pro congruente haberi possit seu parum admodum ab eo recedat, ut recessus etiam sit quantum satis parvus. Jam rursus alia *ghi* ipsi *def* primum congruens, eodem modo ab ipsa moveatur ut ipsa abiit ab *abc* et ita porro et poterit spatium vel corpus ita conflatum pro linea recta haberi, cui definitio illa accommodabitur, quod partes omnes invicem congruae sint positis tantum duobus punctis congruis.

Hujusmodi lineae rectae motu etiam circulus describetur, caeteraque omnia in Geometria resolventur, pari certitudine nam semper demonstrabitur rigorose errorem esse tam parvum quam initio constructionis cujusque problematis volumus.

Eadem res redit, si quis corpus ex punctis tantum componat, sive contra puncta pro corporibus habeat.

---

[22]LH 35, 1, 1, fol. 12.

Texts for Chapter 3, *Mathematical Fictions* 9

9.1 (for 3.2) Mathematical Fictions in Leibniz

Text 13. From *Little Tracts of a Mathematical Collection, Tract 3, Part 2: Quadrature of a logarithmic figure* [Late Spring 1673][1]

> Quadrature of a logarithmic figure, or, what is the same thing, of things in geometrical proportion.

If the ordinates of a certain figure are in geometrical proportion to the abscissae of the same [figure], the complements will be in logarithmic proportion.[2]

A logarithmic figure is necessarily quadrilinear, that is, it consists of three straight lines, two of which are parallel to the ordinate of the height, and of the curve. Since a ratio cannot be beginning with a point, a mean proportional is not intelligible except as a certain line that is infinitely smaller than a straight line, and infinitely smaller than a point, which kind of thing is imaginary. The ratio of the neighbouring ordinates is indeed infinitely small, and may be called a "ratiuncule", this certain Ratiuncule being always the same. And therefore the ratio of the maximum ordinate to the minimum, is equal to the ratiuncule multiplied infinitely many times into itself, or multiplied into itself as many times as the height is divided into points.

Even if the ordinates are in geometrical proportion they do not differ except by infinitely small straight lines, and yet these infinitely small lines are certainly quantities, even though

---

[1] A VII, 4, 298–99.
[2] Here "ordinates" renders what is more literally translated as "applicates to the height" and "abscissa" renders "applicates to the base".

their ratio to things that can be sensed is infinite. The sum of all the differences is the difference between the maximum and the minimum ordinates, as can easily be shown: for if there is a series of continuously increasing or decreasing quantities, infinite or finite, the sum of all the differences is always equal to the difference between the maximum and the minimum.

Now in a series continually decreasing in geometrical proportion, it is evident that this ratio is always between the terms and their differences, that is, it is the ratio of the first term to the difference from the second, which is the difference of the second from the third, so that it is the ratio of the sum of the terms to the sum of the differences, which is the ratio of the first term to the first difference. Now the ratio of the first term to the first difference here is that of a line to a point. Now if the line is divided by a point, the line remains, just as if something is divided by unity, it remains what is was. Moreover, a point to a line is as a unity to infinity. Being divided by a point, if this is being divided by unity itself, namely an equal part of the base, it does not matter what the difference is, for some line will be produced, but one only differing by an unassignable part from the product of a division by the unity expounded. And this is manifestly the case here. But for an accurate demonstration, it is evident that this maximum ordinate, insofar as it enters into the composition of the figure, is a rectangle whose length is the maximum ordinate, and whose width is unity.

9.2 (for 3.3) Fictions and the Question of Impossibility

Text 14. From *Arithmetic Quadrature of the Circle and the Hyperbola* [3 May 1676][3]

Moreover, assuming an infinite number (for a maximum number cannot at all be assumed), one will be able to undertake the sum not only of all the z, but also all the z^2, the z^3 etc., namely of the numbers in arithmetical progression having equal second, third, and fourth differences, etc. For it will always be possible to obtain the sum of such numbers from some last given number, in this case an infinite one, so that an expression will be obtained for x by the aid of powers both of z and of an infinite number; but the infinite number cannot coincide with the number of infinitieths[4] of z, since this varies, and differs according to the value of z assumed, unless you think the infinitieths are also in proportion. And hence an analytic method can already be obtained by converting expressions of this kind that use infinite number into other expressions in which there is no infinite number. For those things having an infinite number in this way do not even have parts that are treatable. If indeed we could feign an infinite number constant, the part would always be numerically the same as

---

[3] A VII, 3, 751.
[4] Leibniz's term '[*pars*] *infinitesima*' is the same term that he and Johann Bernoulli will later use in their dispute about the actuality of an infinitieth term of an infinite series. See text [**32**].

9.2 (for 3.3) Fictions and the Question of Impossibility   187

an infinitieth of z; if, on the other hand, these infinitieth parts are different in proportion as z differs and increase arithmetically, then the infinite number will be equivalent to z, and for its square we could write z^2, and so forth. If this is successful we will have a beautiful method for converting expressions of the kind $y^2 + y^4 + y^6 + y^8$ etc. into other expressions. From which a great many new theorems will follow.

Text 15. From *New Elements of a Universal Mathematics* [Summer 1683?][5]

Sometimes also the operation which must be made in actuality is impossible, either for the time being or not possible at all, even if it could be exhibited at least by a construction in our characters, or is already exhibited in nature. Thus it is impossible to subtract when there is nothing there, and yet this is represented in nature, for example when someone owes more than he has in credit. In the same way it is impossible for one prime integer to be actually divided by another; hence there arises a fraction, which represents the division to be made—with the thing which is designated by this number divided into parts more suitable for exhibiting this division. In the same way there arise incommensurable quantities, i.e. surd roots, where [root] extraction cannot take place. And some extractions are such that these surd roots do not even manifest themselves in nature. Then they are called imaginary, and the problem is then impossible; for instance, when analysis shows that the point sought must be exhibited by the intersection of a certain circle and a certain straight line, where it can be the case that that circle can in no way reach that straight line, and then the intersection will be imaginary; it is, however, useful for it to be considered exactly as if it were real, since imaginary roots can be computed together with real ones, so that an integer is obtained as the root of some equation; and many other uses of this could be determined. As a matter of fact it is sometimes discovered that the values of real quantities must necessarily be expressed by the intervening of imaginary quantities, and from this formulas arise that are no less useful for all works of analysis than common formulas. And these quantities I call *impossible in appearance*; since they are in fact real, and I relate the precepts by which this could be recognized.

There is, however, a great difference between imaginary quantities, ôr ones that are accidentally impossible, and those that are absolutely impossible, which involve a contradiction, as when it is found that they are needed to solve a problem—for instance, in order to prove that 3 equals 4, which is absurd. Indeed, imaginary ôr accidentally impossible things, that is to say, those which cannot be exhibited because of a lack of the sufficient arrangement needed for intersection, can be compared with infinite and infinitely small quantities, which arise in the same way. For let there be an indefinite straight line AB passing through the point A, and let CA be perpendicular to it. And a straight line drawn from the point C must cut AB at some point B, such as $_1B$, or $_2B$, or $_3B$, etc. It is clear that

---

[5] A VI, 4, 520–21.

Fig. 9.1 From *Elementa nova matheseos universalis* (A VI, 4, 521)

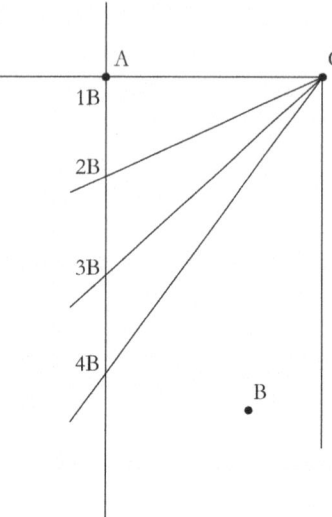

just as the straight line *CB* makes an angle [∠*BCA*] to the straight line *AB* that is closer to a right angle, so *AB* is smaller, to the extent that in the case of a right angle, $_1B$ falls on *A*, that is to say, $_1B$ is infinitely small ôr null. Conversely, the angle or inclination of the straight line *CB* to the straight line *AC* will become closer to being parallel to it the greater *AB* is, so that A_4B is greater than A_3B. And when the straight line *CB* is perfectly parallel to *AB*, the common point *B* is imaginary ôr null, that is infinitely distant from here, and the straight line *AB* is infinite (Fig. 9.1).

And yet imaginaries of this kind have an excellent use, not only in conic sections but also everywhere else, for the discovery of universal constructions; which is so true that it often happens that calculation necessarily leads to them, at which point those who still lack experience in these things are wonderfully troubled by them and think they have stumbled upon some absurdity. Those who understand, however, are well aware that this apparent impossibility only means that instead of the straight line making the angle sought, it is drawn parallel; and this parallelism is that sought angle or quasi-angle.

Texts for Chapter 4, *De Quadratura Arithmetica* (*DQA*)

10.1 (for 4.2) The *DQA* and the "Direct Method"

Text 16. Selections from the Scholia of *De Quadratura Arithmetica* [June-September 1676][1]

Scholium to Prop. 6[2]
I would willingly have refrained from giving this proposition, since nothing is more alien to my mind than these sorts of meticulous details, which are more ostentatious than fruitful. For they use up time with a kind of ceremony, take more labour than ingenuity, and envelop the origin of discovery in darkest night, while to me the origin of discoveries seems much more important than the discoveries themselves. Since, however, I do not deny that it is of interest to Geometry that these methods and principles of discovery, as well as certain very important theorems, be obtained by strict demonstration, I thought something should be granted to received opinions.

Scholium to Prop. 7[3]
Two powerful results will be noticed as arising from this, one concerning demonstration, the other concerning the proposition itself. The *demonstration* has the peculiarity that the problem is solved not through inscribed and circumscribed figures together, but only through inscribed ones. For my part I confess that up till now I have noticed no way by which even a single quadrature can be perfectly demonstrated without an inference *ad*

---

[1] A VII, 6, N. 51.
[2] A VII, 6, 533.
[3] A VII, 6, 537–538.

absurdum. Indeed, I have reasons for doubting that this would be possible in the nature of things without assuming fictitious quantities, namely infinite and infinitely small ones; but of all inferences *ad absurdum* I believe none to be simpler and more natural, and more proper for a direct demonstration, than that which not only simply shows that the difference between two quantities is nothing, so that they are then equal (whereas otherwise it is usually proved by a double *reductio* that one is neither greater nor smaller than the other), but which also uses only one middle term, namely either inscribed or circumscribed, rather than both together; and so brings it about that we have clearer comprehensions of these matters.

As regards this proposition itself, I hold that it is one of the most general and useful in geometry, inasmuch as it is so general that it applies to all curves, even in the case where they are drawn at will without a specific law; and given any figure it exhibits infinitely many others, the dimension of each of which depends on the former and vice versa. But it can also be reckoned among the most fruitful theorems in geometry; for from it one can immediately demonstrate the quadratures of all parabolas and hyperbolas to infinity—that is, of the figures whose ordinates or their powers are in a multiplicative or submultiplicative direct or reciprocal ratio to the abscissas or powers of the abscissas; and not to mention that we have transformed infinitely many other absolute or hypothetical quadratures—certainly the circle and by its aid any conic having a centre—into a rational figure, and from this we derive an Arithmetical Quadrature of the whole circle and any portion of it, and a true and perfect analytic expression for the arc from a given tangent, all of which it is the business of this treatise to demonstrate (Fig. 10.1).

Moreover, leading geometers,[4] in order to begin to treat conics in a universal way, understand by the name of ordinates to the curves not only parallel straight lines, like $_1C_1B$, $_2C_2B$, $_3C_3B$, as is customary, but also the straight lines A_1C, A_2C, A_3C, which all converge to a common point A (which indeed is already done correctly, since one may assume these very parallels to be convergent without error, inasmuch as their meeting point or common centre may be feigned to be infinitely distant, just like the other focus or vertex of a parabola). Hence now by the aid of this theorem of ours it happily transpires that each of these new ordinates, namely the convergent ones, can be used to perform quadratures, and that the figures can be resolved not only through parallel ordinates in the parallelograms $_1C_1B_2B$ or $_2C_2B_3B$, etc., as is customarily done by Cavalieri and others after him, but can also be resolved through the convergent ordinates in the triangles A_1C_2C or A_2C_3C in infinitely many ways, according as the point A is variously assumed. By this a whole field of new discoveries is opened up, whose foundations we give here, from which, for all I know, no fewer and no less important things can be inferred than those given here.

---

[4] The first draft (N. 20) mentions Desargues and Pascal by name here.

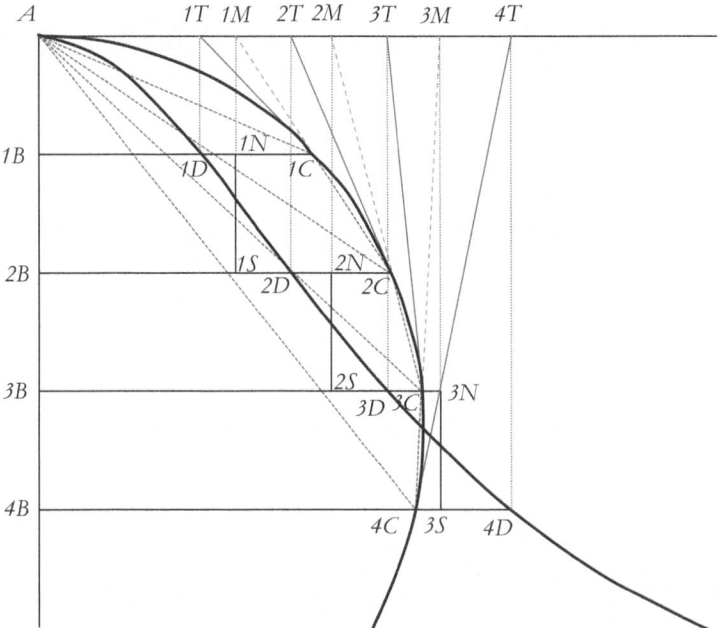

Fig. 10.1 The "transmutation" theorem

Scholium to Proposition 8:[5]
I have expounded these things in minute detail so that learned gentlemen might recognize that what had seemed suspect to them in this business can be strictly demonstrated, by which means Geometers can hereafter wholly refrain from these minutiae when a similar reasoning occurs.

Scholium to Proposition 22:[6]

I have gladly pursued this contemplation since it exhibits an instance of caution concerning reasoning about the infinite and the method of indivisibles, and shows that it is not always possible to jump from the parts of perpetually finite abscissas with a certain property to a property of the whole infinite space. As here in the conic hyperbola someone could reason as follows: zone $_1C_1B_2B_2C_1C$ is equal to the conjugate zone $_1C_1G_2G_2C_1C$, and zone $_0C_0B_1B_1C_0C$ is equal to the conjugate zone $_0C_0G_1G_1C_0C$ (supposing them always to be finite), as established by prop. 18, and thus any horizontal quadrilineal space is always equal to the transverse ôr perpendicular space. Now all the horizontal quadrilineal spaces to infinity up to A fill the infinite space $_2C_2BAM$ etc.$_1C_2C$, and all the perpendicular or

---

[5] A VII, 6, 544.
[6] A VII, 6, 583–584.

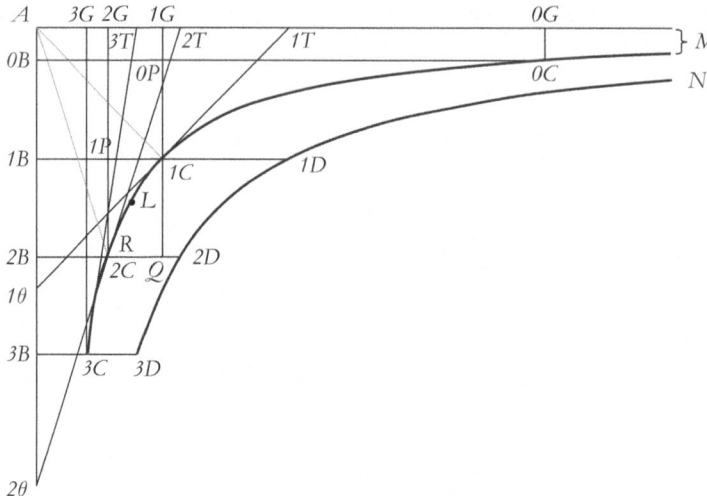

Fig. 10.2 The construction of an asymptote to the *Quadratrix*

conjugate spaces to infinity fill the infinite space $_2C_2GM$ etc. $_1C_2C$, therefore this infinite space will be equal to that, the part to the whole. And in other hyperboloids an absurdity will always be inferred by a similar argument. Now, for example, in the next Hyperboloid after the conic, or Antiparabola, if the ordinates BC are as the inverse squares of the abscissas AB, then by prop. 18 $_1C_1B_2B_2C_1C$ will be a half of $_1C_1G_2G_2C_1C$, and in the same way $_0C_0B_1B_1C_0C$ a half of $_0C_0G_1G_1C_0C$—supposing them always to be finite—and thus any quadrilineum will always be half the corresponding conjugate. Therefore the infinite space $_2C_2BAM$ etc. $_1C_2C$ completed from all the preceding quadrilinea taken to infinity will be a half of the space $_2C_2GM$ etc. $_1C_2C$ completed from all the conjugates, the whole half of the part. In the same way, it will be inferred in other cases that the whole is equal to its third or quarter part. By an elegant argument one sees how slippery reasoning about infinities is if it is not guided by the thread of a demonstration (Fig. 10.2).

Scholium to Proposition 23:[7]

What we have said up to this point about infinities and infinitely small quantities will seem obscure to certain people, as does everything new—although we have said nothing that cannot be easily understood by each of them after a little reflection: indeed, whoever has understood them will recognize their fecundity. It does not matter whether there are such quantities in nature, for it suffices that they be introduced by a fiction, since they provide abbreviations of speaking and thinking, and thereby of discovery as well as of demonstration, so that it is not always necessary to use inscribed or circumscribed figures and to infer *ad absurdum*, and to show that the error is smaller than any assignable. Nevertheless, it is

---

[7] A VII, 6, 585–586.

evident that the latter can easily be done by means of what we have said in *Props. 6, 7 & 8*. Indeed, if it is even possible to exhibit direct demonstrations of these things, I dare assert, it is not possible for these to be given except by admitting fictitious quantities, infinitely small or infinite ones. See *Scholium 7* above.

If then someone in the future were to question the use of these quantities, they would show themself to be either ignorant or ungrateful: ignorant if they do not understand how much light this throws on the whole method of indivisibles and the matter of quadratures; and ungrateful if they perceive the utility, but pretend not to. For here there is no danger of any mistake, as there is in Cavalierian geometry; nor are we forced for safety's sake, as was Cavalieri, to restrict the method to parallel ordinates, and to require that the intervals between two neighbouring ordinates always be equal, blocking ways of progressing to us, as he had done for himself. Rather, by the freest mental discourse we can treat curves no less boldly and safely than straight lines. If someone asks for the fruit of this method, this little book will be a whole specimen of it; but a specific example of its safety will be given by what we say in the Scholium to proposition 22, where Cavalieri's method, taken crudely, is unreliable. From this it could be understood that it could not be obtained in any other way except insofar as it can be resolved into this method.

If I had the time, I could show the difference by many outstanding examples. But I prefer that readers would learn it through their own experience rather than through my words, yet they can feel how much the field of discovery has been opened up when they correctly perceive this one thing, that every curvilinear figure is nothing but a polygon having infinitely many sides of infinitely small magnitude. If Cavalieri, and even Descartes himself, had sufficiently considered this, they would have produced or hoped to have produced, great things.

Scholium to Proposition 32:[8]

Here, finally, is the true quadrature of the circle in numbers, and I do not know whether a simpler one can be given than this, and one that more affects the mind. Up till now approximations have been produced. But no one that I know of has seen the true value, nor understood that it could be comprised by an exact equation, which we give here; although infinite, it is well enough recognized, since, consisting in a very simple progression, it penetrates the mind in one blow. In my view, posterity should not be prejudged without a certain demonstration. There are, however, distinguished men who despair of a better one than this seems to be; others judge that, if by this means a full geometric quadrature can be hoped for, a way into it seems open from here, especially when there can be obtained absolutely the sum of other series similar to this one, as I will say below in prop. 42.

---

[8] A VII, 6, 601.

10.2 (for 4.3) The Posterity of the *DQA*

Text 17. Extracts from the *Compendium of the Arithmetical Quadrature* [1690][9]

Prop. 6. The points ($_1C$, $_2C$, $_3C$, $_4C$, $_5C$ etc.) in the curves can be assumed so close that the rectilinear step-space ($_1N$ $_1B$ $_5B$ $_4P$ $_4N$ $_3P$ $_3N$ $_2P$ $_2N$ $_1P$ $_1N$) differs from the space comprised between the curve and straight lines ($_1D$ $_1B$ $_5B$ $_5D$ $_4D$ $_3D$ $_2D$ $_1D$) by a quantity smaller than a given quantity.

Here [I explain] in passing what the *reversion points* in a curve are, namely those in which the ordinate and tangent coincide. The *point of contrary curvature* [*inflexion*] is where the curve turns from being concave to convex, or vice versa, so that the whole curve is not bent in the same direction. A curve either has or does not have a reversion point according as it is related to different axes, but points of inflection are absolute.

Prop. 7. If from any point (*C*) on the curve to one side (*AB*) of a right angle supposed in the same plane (which side I will call the *directrix*) one draws ordinates (*CB*) normal to it, and draws the tangents (*CT*) to the other one (which I call the *codirectrix*), and from the points where they meet (*T*) let there be drawn the perpendiculars (*TD*) from the tangents to the ordinates (produced, if need be), and let a new curve ($_1D$ $_2D$ etc.) pass through the intersections of the perpendiculars with the ordinates, then the Zone ($_1D$ $_1B$ $_4B$ $_4D$ $_3D$ $_2D$ $_1D$) ôr space comprised between the axis, the two extreme ordinates, and the new curve, will be twice that of the sector of space ($_1CA$ $_4C_3C_2C_1C$) comprised between the first curve and the straight lines joining its extrema with the centre of the proposed right angle.

Here I explain in passing that the *abscissa* is the interval of the ordinate from a right angle, the *resecta* is the interval of the tangent from the same angle, with each one taken on the side of the angle, that to which the abscissa runs in the directrix, and that to which the resecta runs in the co-directrix. The new curve is the *resected figure*, since its ordinates are equal to the resectas.

Demonstration: Suppose the zone of the resected figure is not double the sector of the previous figure; and that the difference from double the sector of the single zone is *Z*. Let polygons be circumscribed to a sector by means of tangents and chords, and the normals to the ordinates drawn from the intersections of the tangents to the codirectrix will give a gradifrom rectilinear space, and let this be continued (*prop.*[6]) until the polygon differs from the sector, and the step-space from the zone, by less than a given quantity, and so by less than $^1/_4Z$. But the step-space is twice the polygon (by *prop. 1*). Let us call the zone ôr quadrilineal space *Q*, the step-space ôr twice the polygon *P*, the sector ôr trilnieal space *T*. Now the difference between *Q* and *P* is less than $^1/_4Z$, and the difference between *P* and twice *T* is less than $^2/_4Z$; therefore because of the schema of quantities Q P T, whose

---

[9]GM V, 100–101.

differences are less than $^1/_4Z\ ^2/_4Z$, by *prop.* 5 the difference between Q and T will be less than $^3/_4Z$, and so less than Z, and so less than any given quantity whatever, so that the difference is null.

Text 18. From the Appendix to a Letter for Bodenhausen [28 October, 1690][10]

In order to proffer a plausible specimen of our geometrical discoveries, framed for those who are accustomed solely to the methods of the ancients, it will be convenient to present a general theorem, on the basis of which I have also constructed an Arithmetical Quadrature of the Circle. From it there also follow the dimensions of all Paraboloids and Hyperboloids, and by means of it I have also squared a certain segment of your Florentine Cycloid (for Florence, if I am not mistaken, is its homeland, if eternal things can have a homeland), without supposing the dimensions of the circle, and I can show much else.

That theorem goes as follows (Fig. 10.3):

If from any point on any curve straight lines are drawn onto two indefinite co-directrices, the ordinate to one, and the tangent to the other; and on the ordinates (produced, if need be) one then takes parts from the axis that are equal to the corresponding resecta[11] cut off by the

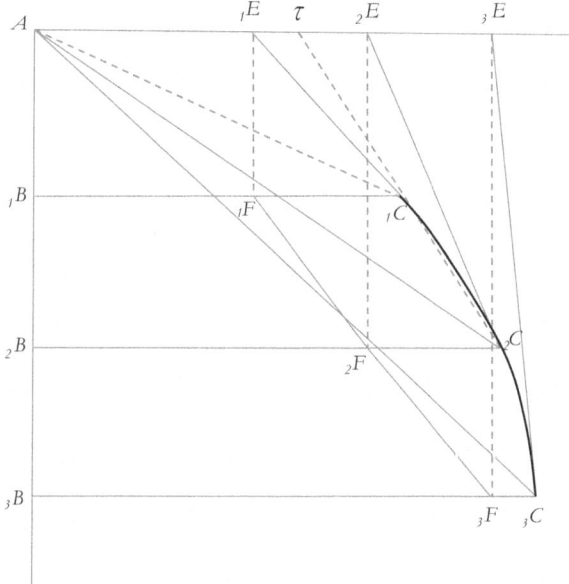

Fig. 10.3 Area of zone equal to twice that of the corresponding sector

---

[10] A III, 4, 634–635, 637–638.

[11] What is meant by the term "resecta" is made clear in text [20] below.

tangents from the other indefinite [lines], these parts are now the ordinates to a new curve passing through their endpoints, which we will call the curve of the resecta; there will be a zone of the new curve equal to twice the corresponding sector of the former curve, comprised by the straight lines drawn from the intersection of the co-directrices to the endpoints of the arc taken in the former curve.

This can easily be demonstrated through inscribed and circumscribed figures by the Archimedean Method, by adding *the following lemma*: (if you join the straight line $_1C_2C$): that in every triangle A_1C_2C through the three angles of which there pass three parallel straight lines A_1E_2E, $_1B_1F_1C$, and $_2B_2F_2C$: the rectangle included under $_1B_2B$ and AT, that is to say, under the interval of two straight lines and a portion of a third between the angle A through which it passes and the opposite side of the intercept $_1C_2C$ produced, is equal to twice the triangle A_1C_2C.

For from this it is not difficult for the proposition to be demonstrated by the Archimedean Method. Let as assume a number of points in the proposed curve, namely $_1C$, $_2C$, $_3C$, $_4C$, etc., and the same number of points corresponding to them in the curve of the resecta, $_1F$, $_2F$, $_3F$, $_4F$, etc. ... [*There follows a proof that the difference between the sum of the triangles and the sector is smaller than a given difference, and that the difference between the sum of the corresponding rectangles and the zone will be smaller than a given difference.*] ... And so the difference between half the zone and the sector will be smaller than the given difference, and so the sector is equal to half the zone. Q.E.D.

I admit, though, that I do not need a theorem of this kind, for what can be inferred from these things can already be understood in my calculus; I gladly make use of these things, however, because they reconcile the calculus to the imagination in a certain way. Of course, I am pleased to record how this theorem may be derived from my calculus (Fig. 10.4).

Fig. 10.4 Trilineum equal to twice the segment

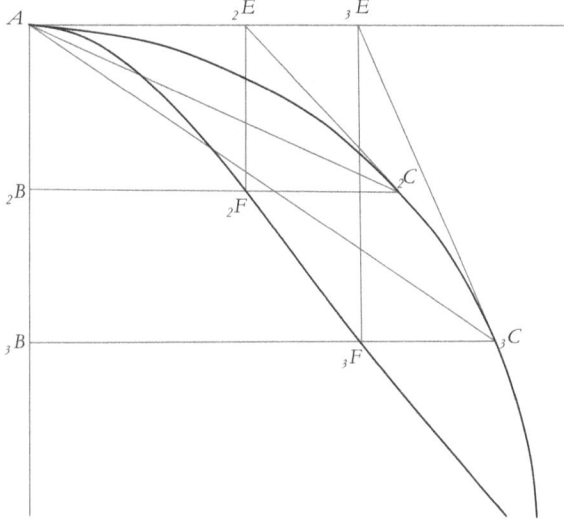

For the sake of abbreviation, we will show this now in the case where the beginning of the curve $_1C$ falls on the point A, in which case the sector will vanish into a segment, and the zone into a trilineum, so that the trilineum $A_3B_3F_2FA$ is equal to twice the segment A_3C_2CA. Now let AB ôr EF be x, BC y. Let AE ôr BF be z. We will have $FC = y - z$. And because the tangent EC is dx to dy, it is as EF to FC, ôr (1) $dx : dy = x : (y-z)$. Therefore (2) $ydx - zdx = xdy$, or (3) $2ydx - zdx = xdy + ydx$. Now, $\int xdy + \int ydx$ or (4) $\int (xdy + ydx) = xy$ (as follows from our calculus, since (5) $d,xy = xdy + ydx$). Therefore, from equ. 3 by the aid of eq. 4, we have (6) $2\int ydx - \int zdx = xy$, ôr (7) $\int ydx - \frac{1}{2} xy = \frac{1}{2} \int zdx$. But $\frac{1}{2} \int zdx$ is nothing but half the trilineum $A_3B_3F_2FA$, and $\frac{1}{2} xy$ is the triangle A_3B_3C, and $\int ydx$ is the trilineum $A_3B_3C_2CA$; whence $\int ydx - \frac{1}{2} xy$ is the segment A_3C_2CA; and therefore from equ. 7 we have (8) half the trilineum $A_3B_3F_2FA =$ segment A_3C_2CA, which is what was proposed to be demonstrated.

Text 19. An Arithmetical Quadrature common to conic sections that have a centre, and from which are deduced a canonical trigonometry for precision in numbers to any degree, and freed from the need for tables, with a special use for the nautical rhumb line, and appropriate for the planar projection of the sphere [April 1691][12]

Already in the year 1675 I had composed a Short Treatise on the Arithmetical Quadrature, read by my friends at that time, but which, with subject matter growing under my hands, I did not have the time to polish up for publication. Afterwards other occupations supervened, especially since now it did not seem worth the trouble to expound at length in the common fashion what our new analysis exhibits in a few lines. Meanwhile certain distinguished mathematicians, to whom the truth of our primary propositions had become known when they were published in these *Acta*, by virtue of their humanity candidly remembered our various discoveries.

And certainly, among other propositions contained in our unpublished and unseen Short Treatise, including both the latter series and several complex ones, there is one sufficiently memorable on account of its generality: to give an arithmetical quadrature of a sector comprised by a conic curve beginning from its vertex and rays drawn from its centre. ...

Text 20. From a Letter to Otto Mencke for Johann Christoph Sturm [November-December 1695][13]

Concerning my Quadrature, I am pleased to remove the difficulty that the distinguished gentleman has encountered. But since he seems to desire a demonstration, I attach here

---

[12] *Acta eruditorum*, April 1691; GM V, 128–132.

[13] A II, 3, 102.

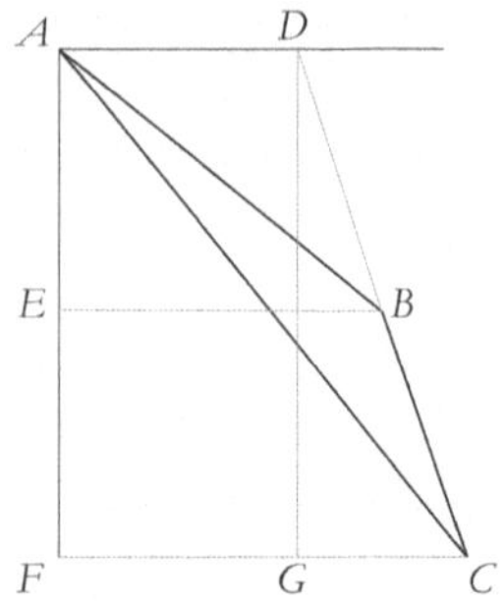

Fig. 10.5 Theorem for computing areas of triangles composing a polygon

Fig. 10.6 Approximating the curve by a polygon

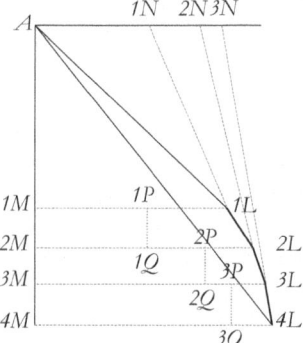

some foundations from which he will easily be able to resolve it, for there is no time to explain the whole thing now at length. I have, indeed, several ways of demonstrating the same thing, but this one seems most elegant. First I suppose this lemma: let the vertex A of any triangle ABC be in a certain right angle DAE, and let the base opposite, BC, be produced, then the straight line from one side of the right angle from the vertex A will be the resecta AD; and let the same base be projected through rays BE, CF parallel to the same AD, the side of the right angle, onto the other side AE; then the rectangle EG comprised by the resecta AD or FG, multiplied by the projection EF (: that is, assuming FG equal and parallel to AD :) will be twice the triangle ABC (Fig. 10.5).

Hence now from several triangles having a vertex A in common let us compose a polygon $A_1L_2L_3L_4LA$, which is a kind of sector whose arc or quasi-arc is $_1L\,_2L\,_3L\,_4L,$, and whose legs are A_1L, A_4L; and on the other hand, let the sides $_1L\,_2L\,_3L\,_4L$, projected onto one side of the right angle give $_1M\,_2M\,_3M\,_4M$; but the same sides will make the resecta AN when produced onto the other side of the right angle. Next, MP or MQ are taken equal to these resecta from the point M onto the projections LM (produced if need be); namely, taking $_1M_1P$ and $_2M_1Q$ equal to A_1N in the lines $_1M_1L$ and $_2M_2L$, and $_2M_2P$ and $_3M_3Q$ equal to A_2N in the lines $_2M_2P$ and $_3M_3Q$, and so on. It is clear that the ratcheted rectangle $_1M_1P_1Q_2P_2Q_3P_3Q_4M$ equals twice the quasi-sector $A_1L_2L_3L_4LA$ (Fig. 10.6).

Fig. 10.7 Theorem for computing quadratures of conics

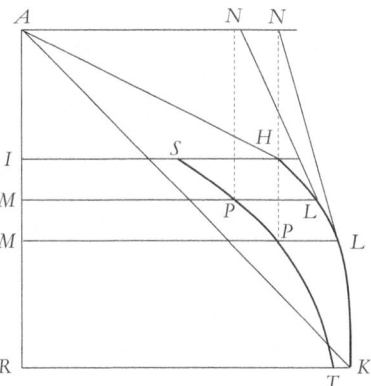

And so with the polygon finally disappearing into the curve this produces the following very general theorem: if from any point *L* of an arc of the curve *HLK*, the tangent is drawn to one side of the right angle, *AN*, and the ordinate *LM* is drawn to the other side, *AM*, and in this ordinate (produced if need be) *MP* is taken equal to *AN*, and every point such as *P* is understood to fall on a new curve *SPT*, then the sum of these *MP* applied to the axis *AM* in order, that is, the figure *SIRTPS* included under the new curve *SPT* and applied to it ordinately between the outermost ordinates *IS*, *RT*, will be equal to twice the sector *HAKLH*. But if the point *H* falls on *A*, the sector will disappear into a segment and yet they will have the same place, with the point *I* also falling onto *A* (Fig. 10.7).

From this theorem there follows almost at once the quadrature of all the paraboloids and hyperboloids which Fermat, Wallis and others have squared through sums of numbers by employing a kind of induction. For if *LK* be a parabola, a hyperbola, a paraboloid or a hyperboloid, then *SPT* makes a curve of the same nature: for instance, if *HLK* is an Apollonian parabola, then *SPT* will also be an Apollonian parabola, and so in the remaining cases, where the quadrature of the figure sought, reduced to itself, gives a kind of equation through which it is found absolutely, with only the Apollonian hyperbola frustrating our hope (as it should), while the unknown in the equation is eliminated. And from this theorem I also found the absolute quadrature—hitherto unknown—of that segment of a cycloid which is cut off if you draw from the vertex two straight lines to a point on the curve, on which a parallel to the base drawn through the centre of the generating circle meets the curve. But with these things dispatched we ascend to the circle (Fig. 10.8).

Let there be a semicircle *ARKLA*, and let the tangent *LN* at a point *L* of its arc meet the tangent of the arc at vertex *A*, parallel to the base, at *N*; and at *LM*, the sine of the arc *AL* (produced if need be), let a straight line *MP* be assumed, equal to *AN* (which is half the tangent of the arc and equal to *LN*); and in this way is described the curve *AP*, which never has an ultimate point, since the base *R* etc. is asymptotic to it; the trilineum *AMPA* will be twice the segment *AL*, comprised under the straight line *AL* and the arc *AL*. And the infinitely long space *AP* etc. *R*. will be equal to the circle. Now let the radius be 1 and let *AM* be *x*, and *MP* be *y*, and we will have the equation $x = \frac{2yy}{1+yy} = 2yy - 2y^4 + 2y^6 - 2y^8$ etc.

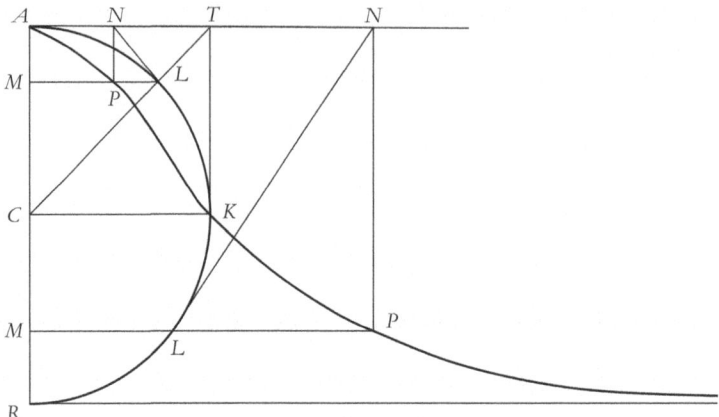

Fig. 10.8 The Arithmetic Quadrature of the Circle

Therefore the sum of all such terms, i.e. of all x, or NP (which constitutes the area of the figure APN and so also gives as a consequence the area of the figure AMP) will be twice (the sum of yy – the sum of y^4 + the sum of y^6 etc.), that is, twice $\frac{1}{3} y^3 - \frac{1}{5} y^5 + \frac{1}{7} y^7$ etc., by the presupposed quadratures of paraboloids. Hence, now changing the nomenclature, it will be found in general that if the tangent AT is y, and the radius 1, then the arc AL will be $= \frac{1}{1} y - \frac{1}{3} y^3 + \frac{1}{5} y^5 - \frac{1}{7} y^7$ etc. Thus with the vanishing of the degrees of the arc (in which case the tangent is the radius ôr unity) y is as its powers in unity, giving, $\frac{1}{1} - \frac{1}{3} + \frac{1}{5} - \frac{1}{7}$ etc. So the circle is to the square of its diameter as this series is to unity.

[*Leibniz adds that when he was in Paris, on the basis of the foregoing arguments Jean Prestet*, "author of the *Elements of a Universal Mathematics*,[14] had persuaded himself that the cycloid is an arc of this circle—which is impossible, as I easily showed him. ... But if the quadratrix were a parabola, a section of the angle in the given ratio would be given by an equation of a certain degree, which is impossible, since the equation would be higher [in degree] in proportion as the arc or angle would have to be cut into several parts."][15]

---

[14] Leibniz is referring to Prestet (1675).
[15] A II, 3, 104.

Texts for Chapter 5, Infinitesimals and Existence After 1676

11.1 (for 5.1) Publication and the Lemmas on Incomparables

Text 21. From *A New Method for Finding Maxima et Minima, As Well As Tangents, Unhindered by Either Fractions or Irrational Quantities, and a Singular Kind of Calculus for Finding Them* [1684][1]

Let there be an axis *AX* (Fig. 11.1), and several curves, such as *VV*, *WW*, *YY*, *ZZ*, and let their ordinates normal to the axis, *VX*, *WX*, *YX*, *ZX*, be called respectively *v*, *w*, *y*, *x*, and let the abscissa from the axis, *AX*, be called *x*. Let there be tangents *VB*, *WC*, *YD*, *ZE*, meeting the axis at the points *B*, *C*, *D*, *E* respectively. Now let an arbitrarily assumed straight line be called *dx*, and let the straight lines which are to *dx* as *v* (or *w*, or *y*, or *z*) is to *XB* (or *XC*, or *XD*, or *XE*) be called *dv* (or *dw*, or *dy*, or *dz*) ôr the differences of *v* (or of *w*, or *y*, or *z*). With these things supposed, the rules of the calculus are as follows.

Let *a* be a given constant quantity, then *da* will be equal to 0, and *d(ax)* will be equal to *adx*. If *y* is equal to *v* (that is to say, if any ordinate of the curve *YY* is equal to any ordinate of the corresponding curve *VV*) then *dy* will equal *dv*. Now *Addition and Subtraction*: if $z - y + w + x$ equal *v*, then $d(z - y + w + x)$ ôr *dv* will equal $dz - dy + dw + dx$. *Multiplication*: *d(xv)* equals $xdv + vdx$, ôr, supposing *y* equals *xv*, then *dy* will equal $xdv + vdx$. For either a formula, such as *xv*, or a letter abbreviating it, such as *y*, may be used at one's discretion. It should be noted that both *x* and *dx* are treated in the same way in this calculus, as are *y* and *dy*, or any other indeterminate letter with its differential. It should also be noted that one cannot always revert from a differential equation without some caution, on which more elsewhere. Next *Division*: if $d\frac{x}{y}$, if we assume $z = \frac{x}{y}$, then *dz* equals $\frac{\pm vdy \mp ydv}{yy}$.

---

[1] *Acta eruditorum*, October 1684, 467–473; GM V, 220–226.

Fig. 11.1 Diagram from the *Nova Methodus*

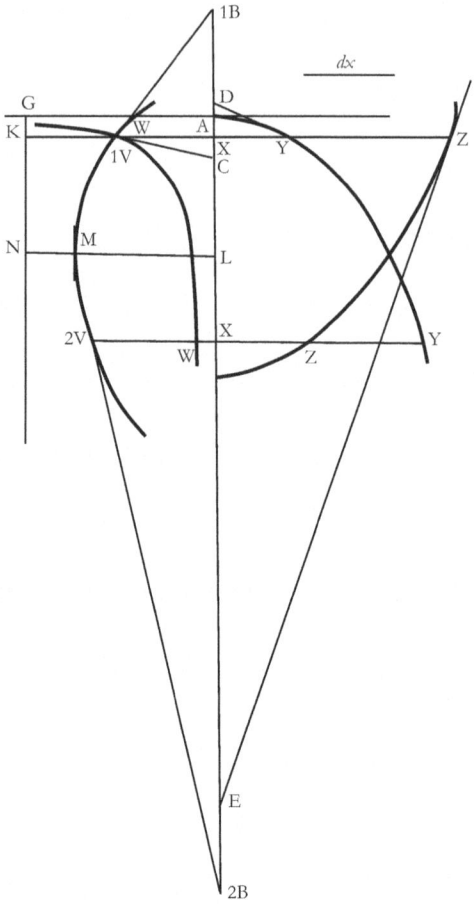

As regards the *sign*, it should properly be noted that when one substitutes a letter in the calculation simply by its differential, the same sign is used for it, and for +*z* we write +*dz*, for –*z* we write –*dz*, as appears in the rule for addition and subtraction proposed a little earlier; but when it comes to the explication of values, ôr when the relation of *z* to *x* is considered, then it must be seen whether the value of *dz* is a positive quantity, or less than zero ôr negative: and the latter happens when the tangent *ZE* is drawn from the point *Z* not towards *A*, but in the opposite direction, down from *X*, that is, when its ordinates *z* decrease with increasing *x*. And according as these ordinates *v* increase, or decrease, *dv* will either be a positive, or a negative quantity, and in the former case the tangent $_1V_1B$ is drawn towards *A*, in the latter $_2V_2B$ is drawn in the opposite direction. But neither happens in the intermediate case around *M*, at which moment these *v* neither increase nor decrease, but are stationary, so that *dv* equals 0, in which case it does not matter whether the quantity is positive or negative, since +0 equals –0: and at this place, *v*, namely the ordinate *LM*, is a *Maximum* (or if the convexity of the axis is inverted, a *Minimum*) and the tangent of the

11.1 (for 5.1) Publication and the Lemmas on Incomparables

curve at M is drawn neither above X towards A, to approach the axis there, nor below X in the opposite direction, but is parallel to the axis. If dv is infinite with respect to dx, then the tangent is at a right angle to the axis, ôr is itself an ordinate. If dv and dx are equal, the tangent makes an angle that is half a right angle to the axis. If with increasing ordinates v, their increments ôr differences dv also increase (that is, if when the dv are supposed positive, the differences of the differences ddv are also positive, and if when the dv are supposed negative, they are negative) the curve presents a *convexity* to the axis (in the contrary case, a *concavity*):[2] on the other hand, wherever there is a maximum or minimum increment, or wherever the increments turn from decreasing to increasing, or vice versa, there is a *point of inflexion*, and concavity and convexity are interchanged, provided the ordinates do not turn from increasing to decreasing, or conversely, for then the concavity or convexity will stay the same; but in order for the increments to continue to increase or decrease, it cannot happen that the ordinates in fact turn from decreasing to increasing, and conversely. So a point of inflection occurs when neither v nor dv are 0, yet ddv is 0. From this it also follows that the problem of the point of inflexion does not have two equal roots, like the problem of the maximum, but three. And all these cases depend on the correct use of signs.

Sometimes, however, it is necessary to employ *ambiguous signs*, as was done above in *Division*, that is, before it is established how they must be expressed. And, indeed, if with x increasing, the $\frac{v}{y}$, increase (decrease) there should be ambiguous signs in $d\frac{v}{y}$, ôr in $\frac{\pm v dy \mp y dv}{yy}$, to be expressed so that this fraction becomes a positive (negative) quantity. But $\mp$ signifies the contrary of $\pm$, so that if the latter is $+$, the former is $-$, and vice versa. Further ambiguities can also occur in the same calculation, which I distinguish by parentheses,. For example, if we had $\frac{v}{y} + \frac{y}{z} + \frac{x}{v} = w$, this would give $\frac{\pm v dy \mp y dv}{yy} + \frac{(\pm) y dz (\mp) z dy}{zz} + \frac{(\pm) x dv (\mp) v dx}{vv} = dw$, otherwise ambiguities arising from different sources would be conflated. Here it is to be noted that an ambiguous sign multiplied by itself yields $+$, multiplied by its contrary yields $-$, and multiplied by another ambiguous sign makes a new ambiguity dependent on both of them.

Powers: $dx^a = a \cdot x^{a-1} dx$; for example, $dx^3 = 3x^2 dx$. $d\frac{1}{x^a} = \frac{adx}{x^{a+1}}$, e. g. if $w = \frac{1}{x^3}$, we will have $dw = -\frac{3dx}{x^4}$.

Roots: $d, \sqrt[b]{x^a} = \frac{a}{b} dx \sqrt[b]{x^{a-b}}$, —(hence $d\sqrt[2]{y} = \frac{dy}{2\sqrt[2]{y}}$, for in this case [$a$] is 1, and b is 2; therefore $\frac{a}{b} dx \sqrt[b]{x^{a-b}}$ is $\frac{1}{2} \sqrt[2]{y^{-1}}$; now, y^{-1} is the same as $\frac{1}{y}$, from the nature of the exponents of geometric progressions, and $\sqrt[2]{\frac{1}{y}}$ is $\frac{1}{\sqrt[2]{y}}$)—, and $d\frac{1}{\sqrt[b]{x^a}} = \frac{-adx}{b\sqrt[b]{x^{a+b}}}$. Now, the rule for integral powers would have sufficed for determining those of fractions and roots, for the power

---

[2] In the original text in the *Acta* Leibniz mistakenly said this the wrong way round, writing *concavitatem* for *convexitatem*, and vice versa. On this Nieuwentijt later took him to task, and Leibniz corrected the error in his response to him.

becomes a fraction when the exponent is negative, and it changes into a root when the exponent is a fraction: but I have preferred to deduce those consequences myself, rather than leaving them to be deduced by others, since they are completely general and frequently occurring, and in a matter complicated in itself, it makes them easier to consult.

From a knowledge of this kind of algorithm of a calculus, so to speak, which I call the differential calculus, all other differential equations can be found through a common calculus, and maxima and minima as well as tangents can be obtained, so that there is no need to eliminate fractions or irrationals or other restrictions, whereas this had to be done according to published methods available hitherto. The demonstration of all these things will be easy for those experienced in these matters, when they consider this one thing that has not before now been sufficiently appreciated, that dx, dy, dv, dw, dz, can be obtained as proportional to the differences, ôr momentaneous increments or decrements, of the x, y, v, w, z (respectively). Thus it happens that when any equation is proposed, its differential equation can be written. This is done by simply substituting for any *term* of the equation (that is, any part which contributes to the constitution of the equation solely by addition or subtraction) the differential quantity of that term, while for any other quantity (which is not itself a term, but contributes to the forming of a term) by using its differential quantity, not taken simply, but as it occurs in forming the differential quantity of the term itself, according to the algorithm prescribed above. The methods hitherto available, on the other hand, have no such transition, for most of them employ a straight line such as DX, or something else of this kind, but not the straight line dy, which is a fourth proportional to DX, DY, and dx. But this confuses everybody, and so they warn that one should first eliminate fractions and irrationals (where indeterminate things are involved). It is also clear that our method extends to transcendent lines—those which cannot be reduced to an algebraic calculus, that is to say, have no definite degree—and thus holds in a most universal way, without recourse to any particular suppositions, which are not always satisfied. This is so provided only that one holds in general that to find a *tangent* is to draw a line which joins two points on a curve an infinitely small distance apart, that is, the produced side of an infinitangular polygon, which for us is equivalent to a curve. That infinitely small distance, however, can always be expressed by a known differential, such as dv, or by a relation to it, that is by a certain known tangent. In particular, if y is a transcendent quantity, for example, the ordinate of a cycloid, and it enters into a calculation by means of which z, the ordinate of another curve, would be determined, and we want to find dz, that is to say, the tangent through it of this latter curve, dz would have to be determined by dy, because the tangent of a cycloid is known. But if this tangent of the cycloid is feigned to be unknown, it may be found in a similar way be a calculation from the given property of the tangents of the circle. ...

Text 22. From *Extract from a Letter from M. L[eibniz] on a General Principle, Useful for the Explanation of the Laws of Nature by the Consideration of Divine Wisdom; to Serve as a Reply to the Response of Rev. Father M[alebranche]*[3]

I have seen in the *Nouvelles de la Republique des lettres* what the Reverend Father Malebranche replies to the remark I made about some of the laws of nature which he had established in *La Recherche de la Vérité*. He seems quite willing to abandon them himself, and this frankness is highly commendable; but as he gives reasons and restrictions which would lead us back into the obscurity from which I believe I have delivered this subject, and which conflict with a Principle of General Order which I have observed, I hope he will be so kind as to allow me to use this occasion to explain this principle, which is of great use in reasoning, and which I do not yet find used, nor well enough known in its full extent. It has its origin in the *infinite*; it is absolutely necessary in geometry, but it is also effective in physics, because the sovereign wisdom, which is the source of all things, acts as a perfect geometer, and in accordance with a harmony to which nothing could be added. That is why this principle often serves me as a proof or test for revealing the defect of an ill-conceived opinion at the outset and from the outside, before even entering into an internal examination. It can be stated as follows: When the difference between two cases can be diminished below any given magnitude *in datis* or in what is supposed, it must also be diminished below any given magnitude *in quaesitis* or in what results from it. Or, to speak more familiarly: When the cases (or what is given) continually approach and finally disappear one into the other, it is necessary that the consequences or results (or what is sought) do so also. This again depends on a more general principle, namely: As the given items are ordered, so also are the items sought [*Datis ordinatis etiam quaesita sunt ordinata*]. But to understand this, examples are needed. It is known that the case or supposition of an ellipse can approach the case of a parabola as much as one wants, in such a way that the difference between the ellipse and the parabola can become less than any given difference, provided that one of the foci of the ellipse is far enough away from the other one, because then the rays coming from this distant focus will differ from the parallel rays as little as one wants, and consequently all the geometrical theorems which are verified of the ellipse in general, may be applied to the parabola, considering it as an ellipse one of whose foci is infinitely distant or (to avoid this expression) as a figure which differs from a certain ellipse by less than any given difference. The same principle holds in physics. For example, rest can be considered as an infinitely small speed, or as an infinite slowness. That is why everything that is true with respect to slowness or speed in general must also be true of rest taken in this

---

[3] *Nouvelles de la République des lettres*, Juillet 1687, 744–753; GP III, 51–55. There exists a Latin version of this article in manuscript, *Principium quoddam Generale non in mathematicis tantum sed et physicis utile, cujus ope ex consideratione sapientiae divinae examinantur naturae leges, qua occasione nata cum R. P. Mallebranchio controversia explicatur, et quidam errores notantur*, [Fall 1688], (A VI, 4, 2032-33 = GM VI, 129–130).

way, so that the rule of rest must be considered as a particular case of the rule of motion: otherwise, if that does not succeed, it will be a sure sign that the rules are badly formulated. In the same way equality can be considered as an infinitely small inequality, and one can make the inequality approach equality as much as one wants.

Text 23. The Lemmas on Incomparables, from *An Essay on the Causes of the Celestial Motions* (the *Tentamen*) [February 1689][4]

5) In demonstrating these things I have assumed *incomparably small quantities*, for example, the difference between two common[5] quantities incomparable with the quantities themselves. Now if I am not mistaken, such things can be lucidly expounded as follows. Thus if someone does not wish to employ *infinitely small* quantities, one can assume them as small as one judges sufficient for them to become incomparable, and to produce an error of no importance, indeed, an error smaller than any given. Just as the Earth is taken as a point, or the diameter of the Earth is taken as an infinitely small line with respect to the heaven, in the same way it can be demonstrated that if the sides of an angle have a base that is incomparably smaller than them, [85/86] the angle comprised will be incomparably smaller than a right angle, and the difference between the sides will be incomparable with those very differences. Likewise the difference of the whole sine, of the sine of the complement, and of the secant, will be incomparable with the differences, and likewise the difference of the sine,[6] the chord, the arc, and the tangent. Therefore since these quantities are themselves infinitely small, their differences will be *infinitely many times infinitely small*, and the versed sine will also be infinitely many times infinitely small, and so incomparable with a right angle. And there are infinitely many degrees of the infinite, and just as many degrees of the infinitely small. And one can use common triangles that are similar to those *unassignable* ones, which are of the greatest use in finding tangents, and maxima and minima, and for working out the curvature of lines; and likewise in almost every application of geometry to nature. For if motion is expressed by a common line that a moving body traverses in a given time, impetus ôr velocity is expressed by an infinitely small line, and the element of velocity, such as the solicitation of gravity or centrifugal endeavour, is expressed by an infinitely many times infinitely small one. And I reckoned that these things ought to be noted down here as *lemmas* for *our Method of incomparable quantities* and *analysis of infinites*, and as it were *Elements* of this new doctrine.

---

[4]*Tentamen de motuum coelestium causis*, Acta eruditorum, February 1689, 85–86; GM VI, 168/Leibniz (2023), pp. 99–100.

[5]By "common" Leibniz means finite, *assignable* as opposed to unassignable.

[6]Here "the sine" should be deleted, as Leibniz notes in 1706.

Text 24. *From a Letter of Mr. Leibniz to Mr. Huygens*, 1690[7]

I conceive very easily, Sir, that you have a Method equivalent to that of my calculus of differences. For what I call dx or dy you can designate by some other letter. So nothing prevents you from expressing things in your own way. However, I imagine that there are certain truths which do not come as easily as by my expression, and it is almost as if instead of roots or powers one would always want to substitute letters, and instead of xx or x^3, to take m or n after having declared, that these must be the powers of some magnitude. See, Sir, how cumbersome this would be. It is the same for dx or ddx and the differences are no less affections of magnitudes, indeterminate in their places, than the powers are affections of a magnitude taken apart. It seems to me, therefore, that it is more natural to designate them in such a way that they immediately indicate the magnitude of which they are the affections.

11.2 (for 5.2) Mathematical Foundations: Quantity, Magnitude and Quasi-Minima

Text 25. From *On Quantity* [1680s?][8]

Those things are *determinants* which simultaneously only belong to one thing alone, as the two extrema A and B only belong to one straight line.

Those things are *coincident* which are clearly the same and only differ in denomination, as the path from A to B does from the path from B to A.

Those things are *congruent* which, if they are different, can only be distinguished with respect to external things, like the squares C and D, namely because they are at the same time in a different place or situation, or because one, C, is made of golden matter, the other, D, of silver. Thus a pound of gold and a pound of lead are congruent, and so are yesterday and today. Any point is congruent to any other, as is also an instant to an instant.

Those things are *equal* which are either congruent ... or can be rendered congruent by a transformation ... . Thus those things can be defined as *equal* each of which can be resolved into different individual parts congruent to the individual parts of the other.

Those things are *similar* in which it cannot be discovered by considering them in themselves one by one how they are distinguished, such as the spheres or circles A and B, (or two cubes or two perfect squares). For instance, if we feign only an eye, without any other members, now to be inside sphere A, now to be inside sphere B, it cannot distinguish them; but it could if it saw both at the same time, or if it brought inside with it another member of the body or some other measure, which it applied now to one body, now to the

---
[7] A III, 4, 620.
[8] GM VII, 29–32.

other. Thus in order for similar things to be distinguished, it is necessary either that they be compresent with each other, or that a third body is successively compresent with each of them. [But in dissimilar things, when some proportion of parts is observed in one that is not observed in the other, it suffices for distinguishing them individually, concerning which more will be said later.]

Those things are *homogeneous* which are either similar or can be rendered similar by a transformation. Two straight lines are homogeneous, because similar; but a straight line and the arc of a circle are homogeneous things, because a circle can be extended into a straight line. We can also define *homogeneous* things as those things which agree in some respect, in which respect other things that can be indefinitely assumed in each of them, also agree.

If there are several things, such as *A* and *B*, and one, *C*, and they agree in some respect, and in these things, there is something homogeneous, which is common to *A* and *B*, and if all the homogeneous things in *A* and *B* are common to *C*, then those several things are called *integrant parts*, and the one is called the *whole*. That is *smaller* which is equal to a part of the other (the *greater*).

Quantity is that which belongs to a thing insofar as it has all its parts, ôr, on account of which it is said to be equal to, greater than, or less than, another thing (with each homogeneous with the other), ôr can be compared with it. The quantity of a thing, for example of the area *ABCD*, is expressed by a number, for example, four, supposing another thing, such as the square foot *AEFG* is taken as the primary measure ôr the real unit. For *ABCD* is four square feet. If on the other hand one assumes a different unit *AHIK*, which is the square of the half-foot *AH*, the quantity of the area *ABCD* would be 16. Thus the number will turn out to be different for the same quantity, according to the unity assumed. And therefore quantity is not a definite number, but the material basis for a number, ôr an indefinite number that is to be defined by assuming a certain measure. So quantity is expressed either by definite numbers, like 1 or 2, or indefinite ones, ôr letters or other characters, *a b*, ☉, ב.

Number is homogeneous with unity, and so it can be compared with unity by adding to it or subtracting from it. And it is either an aggregate of unities which is called an *integer*, such as 2 (ôr 1 + 1) ..., or an aggregate of aliquot parts of unity, which is called a *fraction*, as when a unit, for example, a foot, is divided into four parts, then the thing, for example the line *BH*, which has three quarters of a foot or thrice $1/4$, is expressed thus $3/4$ ...; or finally number is determined in some other way by relation to unity, which relations may even be infinite, but are most usually determined through roots. For example, let there be a number 4 (for the square *ABCD*), and its square root (ôr its side *AB*) is sought, that is the number which, when multiplied by itself, makes 4; this number will be 2, and so when 2.2 ôr *aa* is 4, $\sqrt{4}$ ($\sqrt{aa}$) is 2 (*a*). And in this case the root can be reduced to a common ôr *rational* number. But sometimes this reduction does not succeed. For example, let the number be sought which when multiplied by itself makes 2, this is neither an integer (for otherwise, since it is necessarily less than 2, it will be one, but one multiplied by itself makes 1), nor is it a fraction, because every fraction multiplied by itself produces another fraction, as $\frac{3}{2}$ produces $\frac{9}{4}$ ôr $2 + \frac{1}{4}$; and so it is not a number but an *irrational* , as they call it, ôr rather ineffable, ἄλογος, *surd*, which is written thus $\sqrt{2}$, or $\sqrt{q\,2}$, or $\sqrt[2]{2}$, that is the quadratic root of 2; for supposing this number to be y, then its square yy ôr y^2 will be 2. And in order to show that there is such a number in the nature of things, let us draw the diagonal AF of the square $AEFG$. Let AG be 1, namely one foot, whose square is 1 (namely the space $AEFG$ ôr one square foot), then AF, which we have called y, will be $\sqrt{2}$; for its square yy, $AFBM$, is 2 (namely two square feet), for the square $AFBM$ is twice the square $AEFG$, since its half, the triangle AFB, is equal to the whole $AEFG$. Therefore, since we have defined number to be homogeneous to unity, there must certainly be a number whose relation to unity is that of the straight line AF to the straight line AG, that is to say, assuming AG is 1, there must be a number which expresses the quantity of AF, which is called $\sqrt{2}$; while AB will be 2.

...

Text 26. From *On Magnitude and Measure* [1690s?][9]

(1) *Magnitude* is that which is expressed in a thing by the number of parts congruent with a given thing, which is called *Measure*.

Scholium. For example, the magnitude of a line is expressed by the number of feet or inches, that is, by the number of parts each of which is congruent with a foot or an inch in some actually given material (whether brass or wood). Thus the magnitude of an arm-span (how far a man can extend his arms) is judged to be designated to a certain something (as if by aversion)[10] is expressed numerically by six feet, or seventy-two inches, since there are twelve inches in a foot. The magnitude of a cubit is one and a half feet, or one foot and six inches, or eighteen inches. On the other hand, we suppose the magnitude of a foot or an inch to be actually given in an instrument. Whence it is clear also that the same magnitude of the same thing can be expressed by different numbers, as the measure is varied, indeed, different measures can be combined together, as when a cubit is designated by feet and inches.

(2) Those things are *homogeneous* whose magnitudes can be expressed by numbers by assuming the same measure as unity for all of them.

Scholium. Thus if a foot is taken as the unit, an inch will be as $^1/_{12}$, a cubit as $^3/_2$, an arm-span as 6. But if the inch is taken as the unit, a foot will be as 12, a cubit as 18, and an

---

[9] GM VII, 35–40.

[10] Leibniz's *"per aversionem"* is a a legal term, "applied to that kind of sale where the goods are taken in bulk, and not by weight or measure, and for a single price; or where a piece of land is sold as containing in gross, by estimation, a certain number of acres." The idea is that the buyer averts his or her face, without subjecting the items purchased to a detailed examination. Leibniz uses the same expression in the dialogue *Pacidius Philalethi*: "Thus action in a body cannot be conceived except through a kind of aversion [*per aversionem quandam*]" (A VI, 3, 566/ LLC 211).

arm-span 72. And in this way the length of each straight line can be expressed by a number, by an integer, in fact, if when the measure is subtracted a number of times—for example, by subtracting a foot three times—nothing is left, it will then be three feet long. But if when the measure or foot is subtracted as many times as possible something is left over, for the measuring of which there could also be assumed a certain part of a foot, for example, a tenth, which in turn can be subtracted as many times from this remainder—for example, seven times, with the foot assumed as unit—then the number of the quantity subtracted will be the fraction 3 and $7/10$, or $37/10$. And in this way, if the thing to be measured is exhausted so that there is nothing left over, this number will correspond to the thing and will express its magnitude. If something is still left over then either we can again assume a new part as measure, for instance, a hundredth, and subtract this as many times as it can be; and if the *error* with a hundredth part is sufficiently small not to seem important to us, we can be content with this through a measure and a tenth or a hundredth of a measure by an *approximation*.

We are accustomed in practice to use a *scale*, that is, a certain constant division of a measure made of brass or another durable material, and even in tenths, and tenths of tenths or hundredths, and thousandths since in this way fractions can be explicitly treated decimally after the fashion of integers, which we are accustomed to express in decadic progression, that is, by means of ones, tens, hundreds, thousands, myriads. In this way Ludolph of Cologne discovered by producing a long calculation, that with the diameter of a circle being 1, its circumference is $3 + 1/10 + 4/100 + 1/1000 + 5/10000 + 9/100000 + 2/1000000$, or (allowing /37/ decimals) by combining them at once into one fraction, $3141592/1000000$, etc. up to <…> a seat.

But approximations of this kind, even though sufficient in practice, do not give an exact knowledge of the magnitude sought. And so we continue in a scientific process until it is clear that the series is proceeding to infinity; and to this end we do not apply decimals indiscriminately, or any other constant divisions of a scale, but we accommodate fractions to the nature of the thing, so that it is easier for us to arrive at the law of progression. And so in this way I discovered that if the diameter is $1/8$, then the circumference will be $1/3 + 1/35 + 1/99 + 1/195$ and so on to infinity, with the numerator of the fractions supposed as unity, whereas the denominators will be made from two odd numbers, 1 and 3, 5 and 7, 9 and 11, 13 and 15, 17 and 19, and so on, multiplied together. And by this method not only can all approximations that can be done by continuing be expressed at the same time, but this can even be done in such a way that the error is smaller than any given. For it can be shown that if the circumference is said to be $1/3$, the error will be smaller than $1/5$; if it is said to be $1/3 + 1/35$, the error will be smaller than $1/9$; if it is said to be $1/3 + 1/35 + 1/195$, the error will be smaller than $1/13$, and so on, always taking the former by the combination of the nearest odd numbers. Now the whole infinite series exactly expresses the nature of the circle.

But we derived the latter by *reasons* from the inner nature of the circle; *instrumentally*, on the other hand, the magnitude of the circumference of a circle or another curved line can

11.2 (for 5.2) Mathematical Foundations: Quantity, Magnitude and Quasi-Minima

be obtained by means of a thread accommodated to the rigid curved line,[11] and then applied to an extended straight line and scale, or when a rigid curved line is rotated in a plane, although that revolution is to be regulated by securing the thread or chain, lest a traction gets mixed up with it. The result is also obtained by motion, when two moving bodies having a uniform velocity traverse a straight line and a curve. For the lines completed in the same times will be as the velocities of the moving bodies. But if the thing to be measured is a surface, another surface can be assumed as measure, for example, a square foot, which, or the determinate parts of which, are subtracted from a plane surface as many times as possible. But if the surface is not a plane, it must be seen whether it can be conveniently transformed into a plane. For a solid measure another solid is assumed, for instance a cubic foot, and one proceeds in the same way. Also the magnitudes of solids can be compared by immersing them in liquid, and measuring how much that rises in a vase. But also by weighing, if both are made out of the same matter, and the same can also be transferred to lines and surfaces in its own certain way. And thus Galileo investigated the dimension of the cycloid by means of weights, even though he did not derive the true dimension afterwards discovered by the scientific method of other people. Also a surface and a solid are meanwhile conveniently measured by motion, as the trace of a line or a surface. Generally, though, by means of our definition of magnitude every estimation comes down to the repetition of a certain measure expressed by numbers, or to the ascribing of a number to a thing by supposing another given thing is taken as the unit. Also by the same method not only extensions and diffusions of parts beyond parts, as in space and time, but also intensions ôr degrees of qualities or actions, and by equal right, values, probabilities, perfections and other unextended things, are reduced to numbers, namely by discovering a measure congruent either with it or with the various parts of it that are found in the thing to be measured; then the estimation of magnitude will be formed from this measure, or from the repetition of that of its various parts. How important this consideration is, and how it contains the force of a true Mathesis Universalis ôr art of estimating in general, is shown in specimens of our dynamics.

(3) Those things are *commensurable* with each other for which a common measure can be found that exhausts them, by the repetition of which their magnitudes are constituted. But if less, they are called *incommensurable*, and the number that must be assigned to that which is incommensurable with the measure assumed as unity is called *surd* or irrational; whereas if they are commensurable with unity, they are called *rational*.

Scholium. If, that is, by subtracting a measure, or the various parts constituting the measure by repetition as often as can be done, one arrives at an exhaustion, a constituent common measure can always be had by means of repetition. For the whole magnitude that

---

[11] This way of measuring the length of a curve with a thread was given mathematical form by Christian Huygens with his notion of the involute of a curve, and further extended by Tschirnhaus in his *Medicina Mentis* (1687). See also Leibniz's article *Deux Problemes construits par M. De Leibniz, en employant la regle generale de la composition de mouvemens, qu'il vient de publier* in the *Journal des Sçavans*, September 1693, 423–424, given in translation in Leibniz (2023, pp. 174–175).

is to be measured is expressed either by integers or by a combination of integers and fractions. Now, any fractions whatever can be reduced to a common divisor, and thus to a common measure. Suppose we find a number $2 + {}^2/_3 + {}^1/_6$ expressing the magnitude of a line we are seeking, then by reducing it to a common denominator we will have $^{17}/_6$; and so if a foot is 1 or $^6/_6$, then of course the common measure of the estimated thing and the foot will be $^1/_6$, which quantity is contained in the estimated thing seventeen times multiplied by a sixth of a foot. But if a sixth of a foot or a line of two inches is assumed as unity or measure, a foot will be 6 and the line to be estimated 17, and so a foot and the line will be commensurable. But if the fractions proceed to infinity, and cannot be collected into one assignable whole number or fraction by adding them, then the magnitude to be estimated will be incommensurable with that which we have assigned unity, or with its various parts (that is, with those making it up by repetition). For instance, if a line is what consists in one foot, and two tenths of a foot, and three hundredths of a foot, and four thousandths, and five ten thousandths, and so on to infinity, then with a foot is taken as unity, the line will be $^1/_1 + {}^2/_{10} + {}^3/_{100} + {}^4/_{1000} + {}^5/_{10000} + {}^6/_{100000} + {}^7/_{1000000}$ etc. or $^{1234567}/_{1000000}$ etc. or in decimals, 1.234567 etc. For in this way the line that is to be measured will never be exhausted, and yet its magnitude is considered to be expressed exactly. For whenever a number is rational, as they call it, ôr commensurable with unity, it is known that it is expressed periodically by decimals, so that the same characters always recur to infinity (as we will show elsewhere), which a construction shows does not happen in this case. Moreover this method for investigating the common measure is provided by mathematics, as we will explain in its place, so that the smaller is subtracted from the greater as many times as possible, then the remainder is subtracted as many times as possible from the smaller that was subtracted before, and the second remainder from the second that was subtracted, i.e. from the first remainder, and similarly the third remainder from the second. And so either we necessarily come to an exhaustion, and the last thing subtracted in exhausting will itself be the greatest common measure contained as many times in the first magnitude, that is, compared to the greater, as unity is contained in the product of all the quotients multiplied together; or if remainders are left over to infinity, the first two quantities will be incommensurable, just as all the remainders will be too. But this series of quotients, if it follows a certain law which expresses how the smaller can be subtracted from the preceding one as often as you wish, will give a scientific comparison of the two magnitudes.

Meanwhile by a certain fiction we can conceive all homogeneous quantities to be as if commensurable with each other, namely by feigning some infinitieth or infinitely small element. The calculus of logarithms is founded on such a fiction, by the establishing of a certain logarithmic element. A similar fiction takes place in geometry, by conceiving the matter just as if all lines are made up out of infinitely many infinitely small straight linelets, and so just as if curved lines were polygons having infinitely many sides, or just as if surfaces were made up of infinitely many planelets, that is, just as if concave or convex solids were all polyhedra with infinitely tiny faces. In the same way it can be feigned that all solids are made up of equal elementary corpuscles infinite in number and infinitely small in magnitude. And this fiction cannot introduce any error since (if you proceed correctly according to the hypothesis) the error can never become greater than some of the elementary particles which have no comparison with the whole, as is seen in our Lemmas on Incomparables. Whence if instead of the fictitious ôr infinitely small elementary particles we assume true assignable things however small, it can be shown that the error admitted in the reasoning can be seen to be smaller than any given error, that is, none can be assigned. Yet although it can be conceived that these infinitesimal ôr infinitely small elements are equal to one another for the purpose of imitating commensurability, it is nevertheless still preferable for it to be feigned as another useful way of proceeding for the purpose of assisting reasoning. These things will be made clearer by the more profound part of the doctrine of magnitudes or Universal Mathesis, that, namely, in which is contained the Science of the Infinite.

Text 27. From *Specimen Geometriae Luciferae* [1695?][12]

Also the method of indivisibles and infinites, ôr rather of the infinitely small or infinitely large, ôr of infinitesimals and infinituples, is of the utmost use. For it contains a certain resolution as if into a common measure, albeit one that is smaller than any given quantity, ôr a means by which it is shown that by neglecting other things which make an error smaller than any given, and thus null, one of two things that are comparable can be transformed into the other by a transposition. It must be recognized, however, that a line is not composed of points, nor a surface of lines, nor a body of surfaces; rather, a line is composed of indefinitely small linelets, a surface of indefinitely small surfacelets, and a body of indefinitely small corpuscles; that is, it can be shown that two extensa can be compared by resolving them into equal or congruent particles as small as we please, as if into a common measure, and the error will always be smaller than any one of these particles, or at least it will be in a finite constant or decreasing ratio to the common measure; whence it is clear that the error in such a comparison is smaller than any given. The Method of Exhaustion also belongs here; it is somewhat different from the preceding, although ultimately they agree at root.

There it is shown how a certain series of magnitudes is infinite if the first and last can be known, and they continually approach a certain proposed magnitude in such a way that the difference finally becomes less than a given difference, and is therefore ultimately null ôr exhausted. Thus the last magnitude of this series (which we have said is known) is equal to the proposed magnitude; but this method has seemed to pertain only in this case.

...

Next, in almost the same way as equal things are generated from congruent ones, so also homogeneous things are generated from similar ones, which is worth noting. For as equals

---

[12] GM VII, 273, 282–283.

are either those that are congruent or can be rendered congruent by transforming, so homogeneous things are those that are either similar (the homogeneity of which is obvious per se, as two squares are with one another, or two circles) or they can at least be rendered similar by being transformed. This transformation occurs, moreover, if nothing is taken away or added and yet something else happens, where a certain transformation is made by conserving certain parts, as when we cut the square $ABCD$ into the two triangles ABD and BCD, and join them up again in a different way (for example, by transferring ABD into BCE) and from this we form the triangle DBE. On the other hand, there is a kind of transformation that does not conserve the parts, as when a straight line is transformed into a curve, a gibbous surface into a plane, and anything rectilinear into something curvilinear, and conversely. In that case, therefore, only the minima are conserved, and the transformation is when one thing is made from another with at least the same minima remaining, and so is conserved in a perfect real transformation through something flexible or liquid. But in a mental transformation instead of minima we can employ quasi-minima, that is, indefinitely small parts, in order to make a quasi-transformation, since instead of something curvilinear we can employ something quasi-curvilinear, namely a rectilinear polygon having as great a number of sides as you wish; and if then the quasi-transformation we seek in this way should succeed, that is, the error ôr difference between a quasi-transformation and a true one will always come out smaller and smaller, so that finally it becomes smaller than any given, then it can be inferred to be a true transformation. And since equals are those things one of which can come out of the other by being transformed, it is clear that homogeneous things are those that are similar to each other, or are at least those things to which similar things are equal.

It is also clear that homogeneous things are those which are generated by a continuous increase or decrease of the same thing, at least excepting minima and maxima, ôr extremes. And so if we suppose a path ôr line to grow continuously by the motion of a point, the lines described from one point are homogeneous with one another, but also if the lines are generated from different points, for although they are dissimilar, it is clear that that dissimilarity arises from certain particular impediments which cannot change homogeneity. And it is the same concerning those things that are described by the motion of a line or a surface. It must be understood, however, that the motion by which a point describing one of them does not fall through the trajectory of point describing the other. But we can also imagine homogeneous things as continuously arising from one another, as by continuous transmutation a circle can transition into one ellipse after another, through infinite ellipses of all possible kinds. And in general the following axiom holds for homogeneous things: what makes a continuous transition from one extreme to another passes through all intermediate ones; which, however, does not apply to the angle of contact, which is in fact not an intermediate thing, but is of a different and plainly heterogeneous nature.

11.3 (for 5.3) Further Thoughts on the Existence of Infinitesimals

Text 28. *Quicunque Attentius ea Considerabit...* [1677–1682?][13]

Whoever will consider more carefully the things that have so far been known about the dimensions of curved lines will readily admit to me that they are discovered by feigning the curve to be a rectilinear polygon, with infinitangular sides infinite in number but infinitely small in magnitude; and the same things are afterwards demonstrated by an inference to absurdity, by assuming a finitangular polygon having many enough and small enough sides in which, it is found in some fashion, the same things hold as for the infinitangular polygon previously, by showing that the difference between the assumption of a curve and that of a finitangular polygon is that much less the smaller the sides of the polygons we have assumed, that is, that the error is smaller than any assignable, and is therefore null. For if someone says it is so much, then, by assuming more points on the curve closer to each other, and thus by having described a polygon with more sides, we will show the error to be still smaller, and thus none could be assigned at all. Euclid already used this method, for once he had shown that a regular polygon is as its circumscribed square, it could easily have come into his mind that a circle could be considered as a regular infinitangular polygon, so that the circles will be as the circumscribed squares, that is, as the squares of their diameters. But even if this is clear and contains the origin of the theorem discovered, Euclid, however, preferred not to use the method of *reductio ad absurdum* in demonstrating. Archimedes then cultivated the same method. And, indeed, I do not believe that there is anything else in the nature of things that is not resolved into this, so long as we proceed by congruences, not similitudes. Therefore, for finding the dimensions of curved lines, it is necessary to use either the indirect method leading to absurdity, or fictitious infinitely small quantities.

[*cancelled passage*: But the fact that infinitely small quantities are fictitious can be shown as follows (Fig. 11.2):

Let there be a truly infinitely small line [*AB*],[14] with *CD* an ordinary finite line. Now we seek the mean proportional *EF* between *AB* and *CD*, and this will also be infinitely small. For if it is an ordinary finite line, then of course the third proportional *AB* to the ordinary finite lines *CD* and *EF*, will also be an ordinary finite *AB*, contrary to hypothesis. Let there now be found a third proportional *GH* to *CD* and *EF*; this will be infinite. For it cannot be infinitely small, since it is greater than the finite *CD*. But it is greater than any finite, otherwise, if it were an ordinary finite line, then *EF*, the third proportional to ordinary finite *GH* and *CD*, would be an ordinary finite line, contrary to what we have shown. Therefore

---

[13] LH 35, 8, 12, fol. 37. Transcription in Pasini (1985-86, App., pp. 35–39). An earlier transcription was given by Robinet, which is reproduced in Bassler (2008, pp. 136–137).

[14] Here Leibniz had mistakenly written '*AD*'. The mismatch between the capitals in the text and the lowercase letters in the figure is his.

Fig. 11.2 The scale of infinites in LH 35, 8, 12

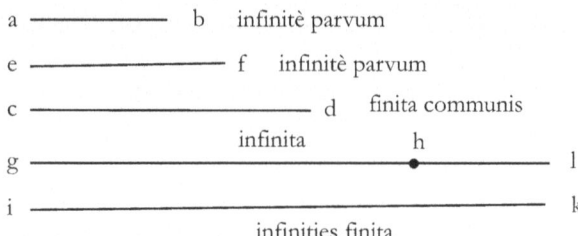

GH is infinite. Now let us seek a third proportional *IK* to *CD* and *GH*, and this will be greater than *GH* —in fact, infinitely greater, because, since *GH* (infinite) is infinitely greater than *CD* (finite), also *IK* (which is to *GH* as *GH* is to *CD*, will be infinitely greater than *GH*. Let *IK* be applied to *GH*, by transposing it onto *GL*, so that *G* is their common beginning. It is necessary that *IK* or *GL*, being extended much farther beyond *GH*, is greater than it, and so will have a part *HL* beyond *GH*. Therefore *GH* is finite, that is, it has a bound, namely the point *H*, common to itself and to that by which the greater *GL* exceeds it. But it is absurd for a straight line bounded on each side by the points *G* and *H* to be infinite in magnitude.—end of deleted passage]

Moreover, even if these quantities are fictitious, they are not therefore any less useful for reasoning, just like imaginary roots in algebra, and just as in the general contemplation of conic sections we can conceive a circle as an ellipse whose two foci are an infinitely small distance apart, that is, coincide in the centre, and we can conceive a parabola as a hyperbola whose other focus is infinitely far away; and in this way many properties of the hyperbola can be transferred in some fashion to the parabola, and many truths can be discovered which would not have come to mind without this comparison. Thus for all conic sections I have a unique general equation; for, supposing a side *a* and a transverse line *b*, with *x* the part of the axis cut off from the vertex (the abscissa), and *y* the ordinate to the axis, we will have $2ax \pm \frac{a}{b}xx = yy$, where + is for the hyperbola, and—for the ellipse and the circle, and in the circle $a = b$. Now if *b*, the transverse side, is alone taken to be infinite, the quantity $\frac{a}{b}xx$ becomes infinitely small and vanishes, and what remains is only $2ax = yy$, which is the equation for the parabola. But if both *a* and *b* are infinitely small, being equal or having a certain ratio, then $2ax$ will vanish. But $\frac{a}{b}xx$ becomes an ordinary quantity, for one infinitely small divided by another one, whether equal to it or in a certain ratio, gives an ordinary quantity, so $\frac{a}{b}$ is an ordinary quantity, and so there will remain $\frac{a}{b}xx = yy$, which is the equation for a *straight line*. Finally, if *a* alone is infinitely small, it will be $yy = 0$, which is the locus of a *point*. It is the same if *a* is infinitely small and *b* infinite.

Also, those who are versed in calculus are not ignorant of the fact that infinite or infinitely small lines often occur. For example, if in some problem a straight line is to be drawn from a given point to a straight line given in position, and one tries to find the point where the straight line to be drawn should meet the line given in position, or the angle by which the straight line should be drawn to the given one: then in certain cases it happens that the point of intersection is at an infinite distance, ôr that its distance from the given point is infinite, which signifies that a straight line should be drawn through the given point that is parallel to the one given in position, and thus the problem will be solved. And so this very analytical calculation also often usefully leads us to imaginary quantities even in the problems of ordinary geometry, as could be shown by many examples. Since, therefore, it has been proposed to me that I treat the key to discovery and the very beginnings of the analytic art in that more sublime geometry which is not subjected to algebra, I will be that much less afraid of using imaginary quantities. And this I believe to be not only advisable, but also necessary. For while it can be doubted whether there truly are, or could be given in nature, any curves which do not consist of fractional rectilinear portions, that there are such can certainly only be demonstrated with difficulty. It will be evident that our reasonings will apply if there are curves whose sides are so subtle that they could not be distinguished. For the more they approach a true curve, the smaller the error will be, but if they finally terminate in it, the error will be null. This is why, content to expound the true art of discovery, I will not waste time with apagogical demonstrations, for anyone will easily be able to get them right in imitation of the things that are found in Archimedes and others.

Text 29. From Annotations on Thomas White's *Euclides Physicus* [Summer 1689?][15]

[p. 19 sq.] Prop. 20. Both the parts of a locus and the parts of a moving body to which they are applied are beyond number, that is, infinite, because the locus and the located things are measurable lines, and any line is divisible into as many other parts as it is cut, that is, into just as many other lines as can be added up to its length.

[p. 20 sq.] Prop. 21. Neither the locus nor the located thing can be infinite.

For if from some point there are infinitely many spans, some part will be infinitely distant (+ this does not validly follow, rather, there will always be a part more distant +), but nothing can exist which is not in itself determinate to itself (+ N.B. I like this; it seems to follow from it that those infinitely small lines cannot really exist, for there is no principle for determining the exact proportion. But this needs examining. +)

[p. 21 sq.] Prop. 22. The parts of the locus or of the located thing cannot all be actual.

For if there is an infinite multiplicity, a transition needs to be made at some stage from the finite to the infinite. (+ If one infinite is greater than another, and straight lines can be assigned proportional to any quantities you please, it will follow necessarily that there are bounded straight lines of infinite magnitude, for every straight line than which there is a greater is bounded. Hence it seems to follow that there are infinitely small lines, for it is certain that there is an infinity of other things in any body, and so an infinite multiplicity of things. Corresponding to which the number of spans falls on an unbounded line, but an infinite one; that is, one which is greater than any given one; but I am always brought up

---

[15] A VI, 4, 2092–2093.

short by the difficulty that if an infinite number is once admitted, or an infinite body, there is not for me an essential mark for distinguishing the finite from the infinite, except with respect to us, except, that is, that we cannot explicate it by any definition. +) (+ A new difficulty: if at some time, that is, at a bounded time, it turns from being a finite line into an infinite one, it is necessary that at some time it ceases to be finite and starts to be infinite, which cannot be conceived. Therefore these things are imaginary: this to me seems demonstrative. Then from this it now follows that an infinite number is not something whole, nor can as many spans be added to it as there are parts in some body, for the infinite is not forged by addition, but is already present. +) Hence there are not actually parts of the continuum, since they would be infinite. (+ I do not know whether that validly follows, since there are actually infinite parts, but not for that reason a number corresponding to them, ôr a number of spans that can be added to it. Should we not rather infer from this that bodies are not real beings, but then what about possibility? —for the difficulty is the same. And it seems to follow at least that infinite things are substances. Or, it must be denied that there is any substance aside from a unique one, which is absurd, that is, it is not really one thing, ἄτοπον. +)

Text 30. From Reflections on Libert Froidmont's *Labyrinthus Continui* [1693–94?][16]

For the rest, I wish to assign this for the sake of a more profound meditation, for it is of no concern to geometry, that if we understand instants to exist now, since only they exist, time will consist of instants.[17] Won't a line then also consist of points? since when a sphere meets a plane, at any instant one point of a line is produced, and nothing further; and yet in this way it produces a line. But it must be said that in motion something happens other than contact, for there is a beginning of progress along with contact, in any instant there is already velocity, ôr endeavour, ôr a beginning of motion. And so not even a line is produced in an instant, but still a point which afterwards involves a certain law of continuation; that is, more than a point of such a kind as touch, or being at rest, designates. Whence a line is not composed from bare points, nor time from instants, except insofar as they involve a continuation; certainly they will not be parts of the continuum, nor will a continuum be aggregated from them—even though they are immediate congenious requisites. Now I call those things *congenious* one of which can arise from the other by a certain continuous change, as, for instance, a plane, by being continuously contracted, can be understood to be converted finally into a line <...> ratio between similar things and

---

[16] LH 37, 4, fol. 57–58; transcribed and translated in Arthur and Ottaviani (forthcoming); translation used by permission of the authors.

[17] *Marginal note:* "If in fact we understand by an indivisible nothing other than the negation of further progress, then all these [difficulties] cease; to exist at an instant, or to touch at a point, will be nothing other than to begin to exist when something is ceasing or beginning."

11.3 (for 5.3) Further Thoughts on the Existence of Infinitesimals

assimilable things, indeed a plane may be feigned to undergo a continuous conversion into a point and even into a sphere.[18]

From this it follows that there is something lying between a line and a point; which is true to the extent that it can be said that there is an infinite difference between them, not indeed between [the point and line] themselves, nor while they are comparable to a line; for there is a endeavour which is to impetus as an instant is to time, like the endeavour which is in a heavy body at the beginning of its descent or in that which moves by centrifugal force, where the endeavour at the beginning is to the impetus of progress that is acquired in the time as an instant is to time; and so, that endeavour of the heavy body that is initially present in an instant does not have a ratio in any way to that which is present after the descent has continued for some time. Yet either of them is the middle term between a simple point, or a point of contact of something at rest, and the line itself. But if we take these rudiments ôr beginnings of a line to be lines, albeit infinitely small ones, it follows from this that since a third proportional to any two given straight lines may be assigned, there can be an infinite straight line, but one bounded at both ends. And such we feign certain asymptote lines to be, as if, that is, the straight line and curve finally meet, but at an infinite interval from here. For no such thing is intelligible in reality, and nor can any such similar straight line be assigned, but since every straight line is similar to a straight line, it follows that there are no infinitely small straight lines either; then what finally will be that by which we distinguish the infinite from the finite? For why do we call asymptotes infinite rather than call them our ordinary straight lines, so that the appellation of finite things is transferred onto those things that were formerly called infinitely small? Thus there would be no principle for distinguishing the finite from the infinite; rather, they would consist in a comparison, and would be called infinite only with respect to ours. But these things are not to be admitted; for see what absurdity follows from this: an infinite but bounded straight line of this kind is traversed in a time that is also an infinite bounded one; at any rate, at the beginning of the motion, traversing the point, it will move for some time in such a way that it will have an infinite ratio to the space traversed in completing the whole. That is because, at the beginning only so many feet have been completed, but afterwards the completed part of what was the whole <before> will at some time be the completion of half the whole, or something having an assignable ratio to the other. Therefore at some time the ratio of what is traversed to the unassignable whole must necessarily begin to be assignable, but such a point or instant in which that transition comes about is impossible. Therefore such a

---

[18] *Marginal note*: "Things are *congenious* when one of them is similar to that which can arise from the other by continuous variation, by which means, assuming many things exist, one of the congenia of a mutual generation contains the other; that is, the other is contained in it. But accidents are those things which are in a thing, not as ingenia, but so that by the fact that when this one is posited, something else is posited, so that is not necessary for this to posit something congenious which is not in it. Hence generally: those things which are not in another congenious thing unless something can arise from another thing, without supposing many things, are in that to which the posited things are understood to exist."

motion, and so such a line, must also be held to be impossible, and there is no infinite yet bounded straight line. Whence it also follows that infinitely small straight lines are chimeras, although nonetheless useful, like imaginary roots, just as <...>.

Text 31. *A Marvellous Axiom* [1695][19]

A marvellous and very beautiful axiom, of great use in mathematics and equally in geometry, and recognized and often used fruitfully in ordinary mechanics: *If the given [values] of* A *vanish in* B, *so also what is sought or the resultant [values] of* A *vanish in [the values] sought in* B. Or, more distinctly, *if a certain hypothesis passes away into the incomparably small by transforming it into another hypothesis, [then] the assignable [values] (arising from the incomparably small) of the former also pass away into assignable [values] of the latter, that is, they agree with the former.* I have used this axiom not only in discovery, but also in demonstration—although demonstration is also a kind of discovery of truth, for making known *a priori* what is otherwise only established by observations. I will omit now how much I have been served by discovering a method that flows from this Axiom, and which consists in *a new use for the infinitely small*, combined, so to speak, with the Law of Continuity: and I will only explain how from it I have demonstrated in mechanics what certain learned men have been able to accept, even though they have seen it confirmed in marvellous ways by experiments and in agreement with reason, but not yet, as it were, demonstrated *a priori* mathematically, because they have seen it as conflicting with their prejudices.

Text 32. From the Leibniz–Johann Bernoulli Correspondence [1698–99]

(a) From Leibniz to Johann Bernoulli, 7/17 June, 1698[20]

I gather from his[21] response that you have written certain profound and ingenious things to him about infinitely varied bodies. I seem to understand your meaning, and I have often deliberated about these matters, but not yet dared to pronounce on them until now. Perhaps the infinite and infinitely small that we conceive are imaginary, but appropriate for determining real things, just as roots are also usually considered imaginary. They are those things in ideal reasons by which real things are regulated as if by laws, even though they are not among the parts of matter. But if we posit infinitely small real lines, it follows that we must also posit straight lines bounded at both ends, which are nonetheless to our

---

[19] LH 35, 8, 29, fol. 1. Transcription in Pasini (1985-86, App., p. 47).
[20] A III, 7, 796–797.
[21] The "he" is Pierre Varignon, whose progress in the "new method", and acknowledgement of debt for this to Bernoulli, Leibniz was discussing in the previous paragraph.

ordinary lines as infinite to finite; assuming this, it follows that there is a point in space which cannot ever be reached from here in any assignable time by an equable motion; and similarly we will need to consider a time bounded at both ends which is nonetheless infinite, and thus there will be a certain kind of eternity, so to speak, that is bounded; ôr, it would be possible for someone to live in such a way that he would die after a number of years that is not ever assignable, and yet he would die at some time—all of which I would not dare to admit, unless forced to do so by indubitable proofs. The real infinite is perhaps the absolute itself, which is not made up of parts, but comprises parts by having an eminent reason and, as it were, a degree of perfection. If there were something perfectly rigid, and perfectly equable, they would be held to be just as we conceive them in our geometry, but I am afraid nature does not allow this. Meanwhile I praise the force of your genius, willing to root out the most abstruse things. If we get the opportunity to speak together, perhaps you would hear from me many marvels more concerning the universe and principles that I take as demonstrated. Now, farewell and prosper.

* * *

(b) From Johann Bernoulli to Leibniz, 5/15 July, 1698[22]

You have come near enough to my opinion on the variety of infinities. And I do not assert for certain that there are infinite degrees of infinities, rather I only adduce conjectures, by which I have stated the matter to be possible and probable. And indeed I have offered a reason for this to be so, that there is no reason why God would have wished there to exist only this degree of infinity ôr this kind of quantity which constitute our objects and are proportionate to our intellect; when, however, I can easily conceive that there could exist in the least speck of powder a world in which everything would be in proportion to this large one, and on the other hand our world could be nothing other than a speck of powder in another infinitely larger one; and that this conception could be continued in ascending and descending without end, whence our kind of quantities would make only one of an infinity of degrees. But there is nothing which might persuade me that this should exist rather than that, for whatever reason could be adduced for one would be applicable to any other degree. So, for example, if I were to conceive in a drop of air a formed world having parts proportional to those in ours, sun, stars, planets, the Earth with its inhabitants, and all other quantities in the same ratio, then what is for us the tiny time of one second, for them would be a series of many ages, and so on for the others. Meanwhile these people could use the same arguments to prove that only they exist, that their world is infinite, and that nothing exists outside it. But I should break off, for if I wished to recount here even to myself all my sweet deliriums by which I sail through these infinities, many pages would not be enough for me.

* * *

---

[22] A III, 7, 810.

(c) From Leibniz to Johann Bernoulli, 12/22 July, 1698[23]

From the actual division it follows that in however small a part of matter there is a kind of world consisting of innumerable created things; but that still leaves open to question whether there will be any portion of matter which has an unassignable ratio to another portion; ôr, whether there will be a straight line bounded at both ends, but which nonetheless has a ratio to another straight line that is infinite or infinitely small. In the calculus we usefully assume this, but from this it does not follow that they can exist in nature. Therefore the matter needs to be investigated further.

* * *

(d) From Johann Bernoulli to Leibniz, 23 July/2 August, 1698[24]

I am surprised that you ask "whether there could be any portion of matter which would have an unassignable ratio to another portion; ôr, whether there could be a straight line bounded at both ends, but which nonetheless has a ratio to another straight line that is infinite or infinitely small", since you nonetheless admit the actual division of matter into parts infinite in number. For if a finite body has parts infinite in number, I always believed and still do believe that the least of these parts must have an unassignable ôr infinitely small ratio to the whole. And there is no need for an actual division, it suffices that such a particle co-exists with the whole it is in, just as a mathematical line co-exists with a surface or the surface with a body, or whatever differential with its integral; or, as I say more fittingly, just as according to Harvey and others, though not according to Leuwenhoek, in innumerable animals there are little eggs, in any egg an animalcule or several, in every (female) animalcule in turn innumerable little eggs and so on to infinity. But whatever my thoughts may be on the infinity of worlds, I have not meant to sell them as certain and demonstrated, but rather as mere probable conjectures; having tried with this fundamental principle, that their existence implies no contradiction, because our thought, inasmuch as it is only relative concerning the infinite as well as the finite, because it is nothing in itself, and as neither large nor small, so neither infinite nor finite, and because finally there is no argument against the infinity of worlds which could not equally well be used by the inhabitants of another world to demonstrate that they alone exist. But perhaps an occasion will arise for me to explain these things more thoroughly.

* * *

(e) From Leibniz to Johann Bernoulli, 29 July/8 August, 1698[25]

Also it seems to me that you have solved very correctly the extremely ingenious and elegant objection of the most distinguished gentleman against the infinitesimal calculus,

---

[23] A III, 7, 827–828.

[24] A III, 7, 847–848

[25] A III, 7, 855–858.

11.3 (for 5.3) Further Thoughts on the Existence of Infinitesimals

Fig. 11.3 Quadrature of a "simply analytic figure"

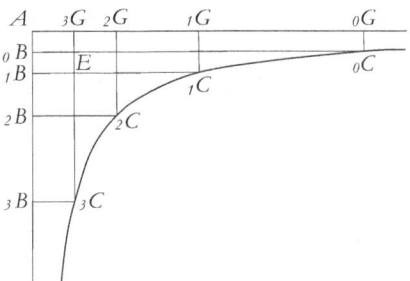

namely that in reality the infinitely small is very different from zero. And when the equation of the hyperbola is $x^2y = a^3$, it is clear that, when x is supposed to be infinitely small, namely of the first degree, then y is to a, as a^2 is to x^2, and so y is an infinity, not simply of the first degree, but of the next higher or second degree; which is different from the case of a simple hyperbola, where y is to ordinary a as the same ordinary a is to an infinitely small of the first degree. I myself once formulated a cognate objection to this very one in the Scholium of Proposition 22 of the unpublished treatise which I composed in France on my Arithmetical Quadrature, shortly after its discovery. There the objection appeared to apply not only to our calculus, but also with equal right to the geometry already accepted up to that time.

I had demonstrated, namely, (see Fig. 11.3) that in a *simply analytic figure* (this is what I called those figures whose equation containing an ordinary relation between ordinate and abscissa, consist of only two members, such as Paraboliforms and Hyperboliforms, ôr those where certain powers of the abscissas are as certain powers of the ordinates) "the zone $_1C_1B_2B_2C_1C$ is to the conjugate zone $_1C_1G_2G_2C_1C$, *as the exponent of the powers of the ordinates BC is to the exponent of the powers* of the powers proportional to them *of the conjugate ordinates, that is, of the abscissas GC or AB.*" Whence in a hyperbolic conic the zones are equal; whereas in a hyperboloid, which may be called an antiparabola, where the ordinates are reciprocally as the squares of the abscissas, the zone will be to the conjugate zone as 1 to 2, and so forth. …

…

And in fact you might communicate to Mr. De Volder, if you see fit, certain excerpts from these as well as preceding letters concerning the estimation of forces and the nature of body, as well as concerning the infinitesimal calculus. But between ourselves, I would also add this, that I also once wrote in the said unpublished manuscript that it is possible to doubt whether there could be infinitely long straight lines that were nonetheless also in fact bounded. Meanwhile, it suffices for the calculus that they be taken as fictions [*fingantur*], like imaginary roots in algebra. For it is always the case that what is concluded by means of the infinite and infinitely small can be evinced by a *reductio ad absurdum* by my method of incomparables (the Lemmas for which I gave in the *Acta*). So you also shouldn't wonder that I doubt whether there is an infinitely small thing, or an infinitely great one bounded at both ends. For even though I concede that there is no portion of matter that is not actually cut, one does not on that account come to uncuttable elements, or minimum portions, nor indeed to infinitely small things, but only to ones perpetually smaller, and yet ordinary; similarly in increasing one comes to perpetually greater ones. Thus also I easily concede that there are always animalcules in animalcules, and yet that it is necessary that there are not infinitely small animalcules, still less ultimate ones. If I were to concede such things as the infinite and the infinitely small that we are discussing to be possible, I would also believe them to exist.

* * *

(f) From Johann Bernoulli to Leibniz, 16 /26 August, 1698[26]

I do not remember ever having said that in the division of matter it is possible to arrive at uncuttable elements or to minimum portions. But here it is not a question of how far I can reach in a division, whether an actual or a mental one, it is a question of how far it has already reached. You concede a finite portion of matter already actually divided into parts infinite in number, and yet you deny that it is possible for one of these particles to be infinitely minute, how is that coherent? For if there is no infinitely minute one, then the individual particles are finite, if they are all finite then taken together they constitute an infinite magnitude, contrary to hypothesis. Conceive a determinate magnitude divided into parts in this geometrically descending progression, $\frac{1}{2}, \frac{1}{4}, \frac{1}{8}, \frac{1}{16}, \frac{1}{32}$, etc.: whenever the number of terms is finite I admit that the individual terms will also be finite, but if all the terms *actually* exist, the infinitieth and all subsequent terms will certainly be of infinitely small magnitude. But in any body, because of the infinite division actually made, not about to be made, all the terms of the progression *really* and *actually* exist. Therefore, etc. Moreover, a body which by being moved describes a line, at any rate *actually* exists at each of the individual points I can conceive in that line, and therefore also in two points that I conceive infinitely close to one another; and so that infinitely minute interval or particle is traversed. And finally, although such an infinitely small particle could not exist separately, nonetheless it coexists with the whole. But I am surprised that you say that if you were to concede that such infinite and infinitely small things as we are discussing were possible, you would concede that they also exist. I wish then that you would demonstrate their impossibility to me. For just as I do not simply attribute to myself the possibility of proving their existence, so on the other hand I am most persuaded that its impossibility cannot be shown by any arguments.

* * *

(g) From Leibniz to Johann Bernoulli, 22 August/1 September 1698[27]

Like the learned De Volder, Grégoire de Saint–Vincent once said somewhere that the axiom that the whole is greater than the part does not hold in the infinite. But it seems to me

---

[26] A III, 7, 873–874.
[27] A III, 7, 884–886.

11.3 (for 5.3) Further Thoughts on the Existence of Infinitesimals

that one of two things must be said, either that the infinite is not really one whole, or that, if the infinite is a whole and yet is greater than its part, then this is something absurd. Indeed, many years ago I demonstrated that the number or multiplicity of all numbers implies a contradiction, if it is taken as one whole. It is the same with the greatest number or the least number, or with the least of all fractions. And the same should be said about these as about the fastest motion, and similar things.

Also the Universe is not one whole, nor should it be conceived as an animal whose soul is God, as the ancients held. But just as there is no numerical element, ôr least part of unity or minimum in numbers, so there is no least line, ôr lineal element; for a line, like Unity, can be cut into parts or fractions. On the other hand, I accept that since the maximum is different from the infinite, and the minimum from the infinitely small, this does not immediately refute the possibility of our infinitely small things. And they can at least be used in calculation and reasoning, which, as I have already observed, is not permissible with the maximum and unbounded, or with the minimum.

When I said that if I believed infinitely small things were even possible, I would have conceded that they exist, I did not on that account say they were impossible, but left the matter somewhere still in the middle. When I denied that we arrive at minimum portions, it could easily be judged that I was speaking not [just] about our divisions, but also about those that actually occur in nature. So even though I hold for certain that any part of matter is again actually subdivided, I do not however think that it follows from this that there is an infinitely small portion of matter, and still less do I concede that it follows that there is any absolutely minimum portion. Anyone who wants to insist on consequence in form, will feel the difficulty.

But you say, "if there is nothing infinitely minute, then all individual things will be finite." (This I concede.) "If all individual things are finite, then all of them taken together will constitute an infinite magnitude." I do not concede this consequence. I would concede it if there were some finite thing which was smaller than all others, or certainly greater than no other; for then I agree that if one assumes a greater plurality of such things than any given number, there arises a quantity greater than any given. But it is well known that for any given part there is another smaller finite one. You would of course appreciate an example appropriate to the matter.

Let us suppose that in a line its $1/2$, $1/4$, $1/8$, $1/16$, $1/32$, etc. parts are actually given, and that all the terms of this series actually exist. You infer from this that there also exists an infinitieth term. I, on the other hand, think that nothing follows from this other than that there actually exists any assignable finite fraction, however small you please. Similarly in motion, even though it passes through all the points, it still does not follow that there are two points infinitely near to one another, and even less does it follow that they are next to one another. And in fact I conceive points, not as elements of a line, but as limits or negations of further progress, or as endpoints of lines.

...

My treatise on Arithmetical Quadrature was able to garner praise at the time when I wrote it, whereas now it would be more pleasing to beginners in our Method than to you.

Since your learned brother thinks that you have not given the right answer anywhere, it seems that he himself must have a different general solution.

* * *

(h) From Johann Bernoulli to Leibniz, 6 /16 September, 1698[28]

At this very hour I am about to leave town, so I cannot respond to the contents of your letter as extensively as I would wish. I will say this at least: I also believe that there is no maximum or minimum quantity; infinite and infinitely small things cannot be demonstrated to exist, but also that it is not possible to demonstrate their nonexistence; nonetheless, it is probable that they exist. If all the terms of this progression, $\frac{1}{2}$, $\frac{1}{4}$, $\frac{1}{8}$, $\frac{1}{16}$, $\frac{1}{32}$, etc., actually exist, *then an infinitieth term exists, as do all the terms which succeed it*, it seems to me one can rightly infer this from actual existence. Nor do I conceive points as elements of a line, but only as limits.

* * *

(i) Leibniz to Johann Bernoulli, 20/30 September 1698[29]

I come now to what in your most recent letter are metaphysical considerations. You infer as follows: "If all the terms of this progression, $\frac{1}{2}$, $\frac{1}{4}$, $\frac{1}{8}$, $\frac{1}{16}$, $\frac{1}{32}$, etc., actually exist, then an infinitieth term and those following it also exist." I reply that I approve of the inference, if it is conceded that there really is an infinitieth term, or post-infinitieth, but this is the very thing that I do not concede.

* * *

(j) From Johann Bernoulli to Leibniz, 8 /18 November, 1698[30]

I do not disapprove of your metaphysical considerations, and readily admit them and your dynamics, provided only that you provide me with a clear idea of them. Your responses on this score are too laconic, and are definitions rather than explanations. It seems to me a contradiction to say that all the terms of this progression, $\frac{1}{2}$, $\frac{1}{4}$, $\frac{1}{8}$, $\frac{1}{16}$, etc. exist, but that there really are no infinitieth terms: for if infinitieths do not exist, then the terms are only finite, therefore not all the terms exist, contrary to hypothesis. I see indeed where you are coming from, namely that we are not able to reach an infinitieth term, since however long we continue the progression, the terms are of finite magnitude. But it is not a question of how far we can reach, whether actually or conceptually, but how far has already been reached by nature itself. Yet you concede that all the terms exist at the same time, therefore surely also an infinitieth also exists, and really exists, or is real, for if it weren't it would not exist.

* * *

---

[28] A III, 7, 899.
[29] A III, 7, 908.
[30] A III, 7, 935.

(k) Leibniz to Johann Bernoulli, 18/28 November, 1698[31]

You say that these metaphysical considerations of mine are too laconic, although if I am not mistaken I gave the means for speaking accurately and elegantly. But if you have remaining doubts, I will endeavour to satisfy them by responding. You say that I had given definitions rather than explanations. But I wish definitions would always be provided, for explanations are virtually contained in them. As far as infinitieth terms are concerned, it seems to me not only that they cannot be reached by us, but also that they do not exist in nature, that is, they are not possible, otherwise I admit, as I already said, if I conceded that it is possible for them to exist, I would concede them to exist. Therefore it must be seen by what reasoning it could be demonstrated that it is possible (for example) for there to exist a straight line that is infinite and yet bounded at both ends.

* * *

(l) Johann Bernoulli to Leibniz, 6/16 December, 1698[32]

As for infinitieth terms either you do not understand me, or I you. I say that "if there are no infinitieth [terms] in nature, then certainly the number of terms will only be finite, therefore not all of them will exist, contrary to hypothesis." But here I present a dilemma: "the number of terms existing in nature is either finite or infinite, there is no third alternative; if it is finite, then not all of them exist, since there could be more; if infinite, then by that very fact an infinitieth terms exists, as do all those following it." You say, perhaps there are infinitely many terms and yet every one of them is of finite magnitude, just as is evident in this progression $\frac{1}{2}$, $\frac{1}{4}$, $\frac{1}{8}$, $\frac{1}{16}$, $\frac{1}{32}$, etc., where there are necessarily infinitely many terms of finite magnitude. For if there were only finitely many terms (*by number*) of finite magnitude then the number of terms would be determinate, which is absurd. But if I consider the progression from another aspect, insofar as there are infinitely many terms *by number*, an infinitieth necessarily exists. From this I conclude that there must necessarily be one infinitely smaller than a finite term, that is, there must be an infinitely small one.

* * *

(m) Leibniz to Johann Bernoulli, 17/27 December, 1698[33]

Concerning infinitieth terms it comes down to this, that the following proposition you make use of should be proved: If the terms in a series such as $\frac{1}{2}$, $\frac{1}{4}$, $\frac{1}{8}$, $\frac{1}{16}$, $\frac{1}{32}$, etc. are infinite in number, there exists an infinitieth term. But what if the distance to any of them from the first is finite, and by an assignable number of intervals? Nor do I see what prevents a series from being conceived as made up of terms not only finite in magnitude but infinite in number.

* * *

---

[31] A III, 7, 943.

[32] A III, 7, 956.

[33] A III, 7, 966.

(n) Johann Bernoulli to Leibniz, 7/11 January, 1699[34]

This proposition, "If the terms in a series such as $1/2$, $1/4$, $1/8$, $1/16$, $1/32$, etc. are infinite in number, there exists an infinitieth term", which you say it remains for me to prove, in order to complete the demonstration of the existence of an infinitely small quantity, I easily prove as follows: if there are ten terms, then there certainly exists a tenth; if a hundred, then the hundredth too; if a thousand, there will be a thousandth, etc. Therefore if they are infinite in number, there exists an infinitieth.

* * *

(o) Leibniz to Johann Bernoulli, 13/23 January, 1699[35]

P. S. I was almost forgetting the question of whether there exist infinitieths. It can be doubted whether this follows: Assuming ten terms, there is a tenth; therefore, assuming infinite terms, there is an infinitieth. For one says perhaps that an argument from the finite to the infinite is not valid here. And when it is said that there are infinite terms, it is not said that their number is bounded, but that there are more than any bounded number. And by the same right that among ten numbers there is a last one which is the greatest of them, it will then also be concluded that among all numbers there is a last one which is the greatest of all numbers, which kind of number, however, I believe implies a contradiction. Also, you do not respond to my objection, when I warned that it is possible to understand an infinite series consisting of only finite numbers. For it is evident, even if one supposes with you a series consisting of finite as well as infinite terms, then, with that supposed, it is possible to understand part of it consisting of only finite terms, with the remaining part comprising infinite ones omitted. But this series of only finite terms will indeed also itself be infinite, but it will have no infinitieth term.

* * *

(p) Leibniz to Johann Bernoulli, 24 February/6 March, 1699[36]

You do not respond to the argument I gave for why it does not follow that, given infinitely many terms, there is therefore an infinitieth: namely, that one may conceive an infinite series consisting merely of finite terms, or terms ordered in a decreasing geometric progression. I concede the infinite multiplicity of terms, but this multiplicity makes neither a number nor one whole. It means only that there are more terms than can be designated by a number; just as there is, for instance, a multiplicity or complex of all numbers; rather, this multiplicity is neither a number nor one whole.

* * *

---

[34] A III, 8, 33.
[35] A III, 8, 40.
[36] A III, 8, 66.

Texts for Chapter 6, *Leibniz's Mature Justifications of the Calculus*

12.1 (for 6.1) The Infinite as Syncategorematic

Text 33. From *Towards a Science of the Infinite* [c. 1698][1]

§8 [*Apart from the absolute, there is no infinite thing, nor is a collective infinite valid, except insofar as it can be explained in a distributive sense.*]

And in general it can be said that an infinite number, an infinite line, a series composed from infinitely many terms, or an aggregate of an infinite multitude of things, is in metaphysical rigour not one thing, since they always involve the greatest number, which is impossible. But in mathematical matters they are taken as one thing as an abbreviation of speech, since they have a foundation in reality. In this way, when I say that the infinite series of fractions $1, \frac{1}{2}, \frac{1}{4}, \frac{1}{8}$ etc., is equal to 2, I mean that if each of these fractions is assumed and none besides, then neither more nor less is assumed than what is in 2. And in this sense it is understood that the whole infinite series is equal to 2, so that what in fact is called a collective whole, is understood to be a distributive one; the sense is the same when the infinite space contained under the asymptote is said to be equal to a finite one, namely because none of the parts it contains can be understood to which a corresponding equal part cannot be assumed in the finite one, one to one.[2] And when we say that in a plane a straight

---

[1] LH 35, 7, 10, Fol. 8r-8v. Editors' title. This selection is excerpted with permission from the translation of Osvaldo Ottaviani and Richard Arthur in (Ottaviani and Arthur, forthcoming).

[2] See Leibniz's discussion of this case of the area contained under a hyperbola between one axis (as an asymptote) and a line parallel to it, and how to understand the infinity involved, in Propositions XXI and XXII of the *DQA* (Leibniz 1993, pp. 64–67). As he tells Johann Bernoulli in a letter dated July 29, 1698, in that unpublished manuscript he also wrote that "it suffices for the calculus that [bounded infinite lines] be taken as fictions [*fingantur*], like imaginary roots in algebra. For it is always the case

line *B* that is not parallel to a given straight line *A* will meet it when <*B*> is produced to infinity, it is not necessary to think of one certain thing which is an unbounded line. Rather, the sense is that it is impossible that the straight line *B*, when produced sufficiently far, would not at some time meet any of the straight lines lying in the direction of the given one *A*, that is to say, any of the lines reducing to it that have been continued sufficiently. For example, I once discovered that, with the radius taken as one, and the tangent called *t*, and the arc of the circle *y*, it results that $y = \frac{1}{1}t - \frac{1}{3}t^3 + \frac{1}{5}t^5 - \frac{1}{7}t^7$ etc. (where the 'etc.' will signify for me that it is always to be continued, ôr that the series is designated infinite). But we can, however, also understand by such an expression an indefinite finite series, so that when *y* signifies a whole rational number and *x* signifies ten, and *a*, *b*, *c*, etc. numbers less than 10, namely 0, 1, 2, etc. up to 9, any whole rational number whatever could be expressed by this equation

$$y = a + bx + cx^2 + dx^3 + ex^4 \text{ etc.,}$$

where the 'etc.' signifies that it is perhaps to be continued, even if it does not state how far. So the general value of the said whole number will be designated by an indefinite finite series. But if it is understood to continue without end, the 'etc.' will signify an infinite series. For example, let *x* signify $\frac{1}{10}$, then $[\frac{5}{9}]^3$ could be designated thus:

$$5x + 5x^2 + 5x^3 + 5x^4 + 5x^5 \text{ etc.to infinity,}$$

which according to the decimal system would be written $[\frac{5}{9}] = 0.55555$ etc. For in this way in the equation $y = a + bx + cx^2 + dx^3 + ex^4 + fx^5$ etc., *y* signifies $[\frac{5}{9}]$ and *x* signifies $\frac{1}{10}$, and *a* would be 0, and *b*, *c*, *d*, *e*, *f* and the remaining letter coefficients to infinity signify 5. But if they signified 9, then *y* would signify unity, for 0.99999 etc. to infinity is the same as 1, that is to say, $\frac{9}{10} + \frac{9}{100} + \frac{9}{1000} + \frac{9}{10000} + \frac{9}{100000}$ etc., if they are understood to be continued *to infinity*, make 1. But the same series taken *indefinitely finitely* would signify $\frac{99999}{100000}$, or $\frac{999999}{1000000}$, or $\frac{9999999}{10000000}$, and so on, by which fact a fraction less than one is always designated.[4] Whence it is clear how great a difference there is between these two. And we can even express irrational numbers by means of infinite series, which cannot be done by means of indefinite finite series. But these same irrational numbers, even though they cannot be expressed by finite values of common whole numbers or fractions, can

---

that what is concluded by means of the infinite and infinitely small can be evinced by a *reductio ad absurdum* by my method of incomparables (the Lemmas for which I gave in the *Acta*)." (GM III 522–524; A III 7, 855–858)

[3] Instead of $\frac{5}{9}$, Leibniz has written $\frac{1}{2}$ here and in what follows, clearly an arithmetical error.

[4] The original conclusion of this paragraph was: "Meanwhile it should be noted that when we call a series infinite, this is a somewhat improper locution, or when we say that the above infinitely many fractions taken together equal unity, the sense is that in one entire foot there are contained $\frac{9}{10}$ of a foot, and $\frac{99}{100}$, and $\frac{999}{1000}$, and any other fraction of this kind."

Fig. 12.1 Sum of the Dichotomy Series

however be expressed through affected equations, that is, by finite equations of certain degrees of which they are the roots. But those *Numbers* which I have thought can conveniently be called *Transcendent* cannot even be expressed in this way; instead one needs to employ an equation composed of infinite terms. The use of indefinite finite equations or other formulas, however, is important for discovering theorems that are general and common to any degree.

§9 [*An infinite series sometimes equals a finite quantity, and sometimes exceeds every finite quantity.*]

Lest someone should think, however, that an infinite series involves a maximum number, and suffers the difficulties mentioned above, when we say that the infinite series $9/10 + 9/100 + 9/1000 + 9/10000$ etc. equals 1, the sense is only that in an entire unity, as in a foot, there are in it $9/10$ of a foot, and furthermore $99/100$ of a foot, and $999/1000$ of a foot, and no matter what other finite fraction of this kind, but nothing further. Nor, therefore, do we hold that there is some number of all fractions of this kind, which there is not; in the same way as we acknowledge any unity whatsoever or any created thing whatsoever taken one at a time, even though we deny that an infinite thing is compounded from them. Whence our series also, even if it is written with the signification of continuing without bound, is not an infinite thing. But how the innumerable individual parts are in it and not anything further, and in this way constitute or exhaust a finite whole, can be understood more easily if the imagination aids the intellect in a figure set before the eyes (Fig. 12.1).

From the straight line *AB* cut off the half *AC*, there remains *CB*; from it again cut off the half *CD*, that is ¼ of the whole, there remains *DB*; from this again cut off the half *DE*, that is $1/8$ of the whole, there remains *EB*; from which again cut off the half *EF*, ôr $1/16$ of the whole, there remains *FB*, and so on. Supposing then that whatever remainder is bisected, and that the prior part is added to the preceding ones, it is evident that the whole line itself will be constituted, in which each of these fractions, $1/2$, $1/4$, $1/8$, $1/16$, in decreasing geometric progression is contained; and yet nothing further can be conceived to be in it which does not map onto these parts. And this very infinite series, $1/2 + 1/4 + 1/8 + 1/16 + 1/32$ etc. equals unity. But not every infinite series compounded from innumerable finite fractions less than unity is finite; for the most simple of all, $1/1 + 1/2 + 1/3 + 1/4 + 1/5$ etc., when it is continued sufficiently far, contains more than can be expressed by any given number, since however great a number is taken, the series can be continued far enough that even if the number is finite, it exceeds that number, as we will show in the appropriate place. But it must not be said that when the series is continued without end it can therefore constitute a truly infinite quantity, or that an infinite thing compounded from innumerable parts is found in nature. …

Text 34. From *New Essays on the Understanding*, Book II, Chapter XVI [1704]⁵

THEOPHILUS: Properly speaking it is true that there is an infinity of things, that is to say that there are always more of them than can be assigned. But there is no such thing as an infinite number or line or other infinite quantity, if we take them to be true wholes, as is easy to demonstrate. The Scholastics wanted to say this, or ought to have done so, in admitting a syncategorematic infinite, as they call it, and not the categorematic one. The true infinite, in all rigour, is only in the *absolute*, which is prior to any composition, and is not formed by the addition of parts....

The idea of the *absolute* is internal to us, like that of being: these absolutes are nothing other than the attributes of God; and it may be said that they are no less the source of ideas than God himself is the principle of beings. The idea of the absolute, in relation to space, is nothing but the idea of the immensity of God, and thus of other things. But we are mistaken in wanting to imagine an absolute space that is an infinite whole composed of parts; there is no such thing, it is a notion that implies contradiction; and these infinite wholes, and their opposites, the infinitely small, have no place except in the calculation of geometers, just like the imaginary roots in algebra.

Text 35. From the Leibniz-Des Bosses Correspondence

(a) From Des Bosses to Leibniz, 2 March 1706⁶

You indicated elsewhere in the *Mémoires de Trévoux*, when you were discussing the differential calculus, that it is not necessary that the infinite be taken rigorously; and in your *Specimen Dynamicum*, after speaking of infinite degrees of impetus, you say: "I do not mean, though, that these mathematical entities are really to be found as such in nature, but only that they are useful for making accurate calculations by mental abstraction".⁷ From what you say in these two places I would have been led to conjecture that the infinite you advance can be confined to the syncategorematic; for what would prevent us from transposing what you say about the degrees of impetus to the multitude of substances? Or do you believe that the plenitude of the world, the uniform divisibility of matter, and the laws of varying motion, cannot be explained without assuming a strictly actual infinite?

---

⁵ A VI, 6, 157–158.
⁶ GP II, 302/LDB 27.
⁷ Des Bosses is referring to the "Mémoire de Mr. Leibnitz touchant son sentiment sur le calcul différentiel," *Mémoires de Trévoux*, November 1701 (GM V, 359); and Leibniz's *Specimen of Dynamics*, Part I, published in the *Acta eruditorum*, April 1695, 145–157 (GM VI, 234–246); of which a new translation can be found in (Leibniz 2023, pp. 176–189). The quoted sentence is from p. 149 (GM VI, 238; Leibniz 2023, p. 181).

(b) From Leibniz to Des Bosses, 11 March 1706[8]

Arguments against an actual infinity assume that if this is admitted, there will be an infinite number; likewise, that all infinites are equal. But it must be recognized that an infinite aggregate is not one whole or endowed with magnitude, and is not consistent with number. And, accurately speaking, instead of 'infinite number' it should be said that there are more than can be expressed by any number, and in place of 'infinite straight line', that it is a line produced beyond any magnitude that can be assigned, so that there is always a longer and longer straight line. It is of the essence of number, of line, and of any whole whatsoever, to be bounded. Consequently, even if the world were infinite in magnitude, it would not be one whole... . It is therefore an abbreviation of speech when we say "one" when there are more things than can be comprised in one assignable whole, and when we treat as a magnitude something which does not have its properties. For just as it cannot be said of an infinite number whether it is even or odd, so it cannot be said of an infinite straight line whether it is commensurable with a given straight line or otherwise; so that these are only improper ways of speaking of infinity, as though of one magnitude, based on some analogy, but which, if you examine them more carefully, cannot be sustained. Only the true and indivisible infinite has a true unity, namely God. And this, I think, is enough to satisfy all arguments against the actual infinite, which should also be applied to the potential infinite in its own way. For it cannot be denied that there really are natures of all possible numbers, at least in the divine mind, and so the multitude of numbers is infinite.

Speaking philosophically, I maintain that there are no more infinitely small magnitudes than there are infinitely large ones, that is, no more infinitesimals than infinituples. For I hold both to be fictions of the mind through an abbreviated way of speaking, suitable for calculation, as imaginary roots in algebra are too. Meanwhile I have demonstrated that these expressions are very useful for abbreviating thought and thus for discovery, and cannot lead to error, since it suffices to substitute for the infinitely small something as small as one wishes, so that the error is smaller than any given, whence it follows that there can be no error. The Reverend Father Gouye, who objects, seems not to have sufficiently understood me.

To pass now from the ideas of geometry to the realities of physics, I hold that matter is actually fragmented into parts smaller than any given, that is to say, that there is no part that is not actually subdivided into other parts undergoing different motions. This is demanded by the nature of matter and motion, and by the whole frame of the universe, for physical, mathematical and metaphysical reasons.

(c) From a "Supplementary Study" Included in a Draft of Leibniz's Letter to Des Bosses of 1 September 1706[9]

There is a Syncategorematic infinite, that is, a passive power having parts, namely the possibility of further progress in dividing, multiplying, subtracting, or adding. *There is* also

---

[8] GP II, 304-05/LDB 31-35.

[9] GP II, 314-15/LDB 52-3. This study was crossed out, and not sent with the letter. See Look and Rutherford's editorial note 8, LDB 409.

a Hypercategorematic infinite, or potestative infinite, an active power having parts, as it were, eminently but not formally or actually. This infinite is God himself. But *there is not a Categorematic infinite*, that is, one having actually infinite parts formally.

There is also an actual infinite in the sense of a distributive whole, not a collective one. Thus, something can be stated of all numbers, though not collectively. In this way it can be said that for every even number there is a corresponding odd number, and vice versa; but it is not therefore accurately said that there is an equal multitude of even and odd numbers.

Text 36. From *Essays on Theodicy* [1710], Preliminary Discourse[10]

We embarrass ourselves in the same way with *number series* which go to infinity. We conceive a last term, an infinite number, or an infinitely small one; but all this is nothing but fictions. Every number is finite and assignable, every line is so likewise, and the infinite or infinitely small signify only magnitudes that one may take as big or as small as one wishes, to show that an error is less than that which has been assigned, that is to say, that there is no error at all: or else by the infinitely small we understand the state of vanishing or of beginning of a magnitude, conceived in imitation of magnitudes already formed.

12.2 (for 6.2) Mature Justifications of the Use of Infinitesimals

Text 37. A Memoir by Mr. Leibniz Concerning His Opinion on the Differential Calculus (From a Letter to Pinsson)[11]

One of the issues of the *Journal de Trévoux* contains some of Mr. Jacob Bernoulli's method, and mixes in some reflections on the calculus of differences, in which I have taken so much part. The author of these reflections seems to find the route through the infinite and the infinitely infinite insufficiently secure, and too far from the method of the ancients. But he will have the goodness to consider that while the discoveries are considerable, the novelty of the method enhances their beauty. With regard to the security of the route, the Marquis de l'Hôpital's book will give him satisfaction. To what this illustrious mathematician says of it, I would even add that there is no need to take the infinite here in all rigour, but only as when one says in optics that the rays from the sun come from a point infinitely far away, and so are considered parallel. And when there are several degrees of the infinite or infinitely small, it is as when the globe of the Earth is judged to be a point

---

[10] GP VI, 90.

[11] This memoir was included by Leibniz in his letter to François Pinsson of 29 August, 1701, and published in the *Memoires de Trévoux* of November/December, pp. 270–272 (pp. *223–*234 in the 1702 reedition). It is reproduced also in GM IV, 95–96.

with respect to the distance of the fixed stars, and a ball that we handle is still a point in comparison with the radius of the Earth, so that the distance of the fixed stars is an infinitely infinite or infinity of infinity in relation to the diameter of the ball. For instead of the infinite or infinitely small, we take quantities as great or as small as is needed for the error to be less than the given error; so that we differ from the style of Archimedes only in the expressions, which are more direct in our method, and more in keeping with the art of discovery.

Journal editor's note : Some Geometers who have carefully examined the Marquis de l'Hôpital's *Analyse des Infiniment Petits*, and who even profess to follow his method, say that one must take the infinite in all rigour, and not as Mr. Leibniz explains here. What was said in the May and June issues of the *Mémoires* was in relation to their opinion.

Text 38. From the Correspondence with Varignon [1702]

(a) Extract from a Letter from M. Leibniz to M. Varignon, Containing the Explanation of What Was Reported by Him in the *Mémoires de Trévoux* in the Issues of Last November and December[12]

I am greatly obliged to you, Sir, and to your learned colleagues,[13] who have done me the honour of reflecting upon what I had written to one of my friends on the occasion of what was published in the *Journal de Trévoux* against the calculus of differences and sums. I do not recall exactly what expressions I may have used, but my intention was to point out that it is unnecessary to make mathematical analysis depend on metaphysical controversies, or to affirm that there are lines in nature which are infinitely small in all rigour in comparison to ours, or accordingly, that there are lines infinitely greater than ours [and yet bounded; inasmuch as it has seemed to me that the infinite, taken in all rigour, must have its source in the unbounded; otherwise I see no way of finding an adequate ground for distinguishing it from the finite].[14] This is why, to avoid these subtleties, I believed that, in order to make my reasoning sensible to everybody, it would suffice here to explain the infinite by the incomparable, that is, to conceive of quantities incomparably greater or smaller than ours. This would provide as many degrees of incomparability as we may wish, since it is useless for that which is incomparably smaller to enter into account with respect to that which is incomparably greater than it; it is in this sense that a particle of magnetic matter which passes through glass is not comparable to a grain of sand, nor that grain of sand to the terrestrial globe, nor that globe to the firmament. And it was to this effect that I once gave in the Leipzig *Acts* some Lemmas on incomparables, which can be understood as one wishes,

---

[12] *Journal des sçavans*, March 20, 1702, 183–186. The letter was dated Hanover, February 2, 1702 (GM IV, 91–95=A III, 9, 21–26); Leibniz instructed Varignon to make a couple of minor corrections to the version that he had sent him earlier for its publication in the *Journal des sçavans*.

[13] —literally, "savants"—perhaps a humorous allusion by Leibniz to the title of the journal.

[14] In the margin Leibniz has noted that the passage set in brackets is left out of the extract published in the *Journal des sçavans*: "ausgelaßen was in [ ]" (A III 9, 23).

either as infinites in all rigour, or only as magnitudes that do not enter into account in relation to the other magnitudes.[15] But at the same time one must consider that these common incomparables themselves, which are not at all fixed or determined, and which can be taken as small as one wishes in our geometrical reasonings, have the effect of rigorous infinitely small things, since whenever an opponent tries to contradict this statement, it follows from our calculus that the error will be less than any error he could assign, it being in our power to take this incomparably small as small enough for this purpose, inasmuch as one can always take a magnitude as small as one wishes. Perhaps that is what you mean, Sir, by talking of the inexhaustible; and it is certainly in this that the rigorous demonstration of the infinitesimal calculus consists. It has the advantage of giving directly and visibly, and in a way fitted to exhibit the source of the invention, what the ancients, like Archimedes, gave in a circuitous way with their reductions *ad absurdum*; not being able to arrive, in default of such a calculus, at clear truths or solutions, although they possessed the foundation of the invention. Whence it follows that even if someone does not admit infinite and infinitesimal lines in all metaphysical rigour and as real things, he can still use them safely as ideal notions which shorten the reasoning, similar to what we call imaginary roots in the ordinary analysis (such as, for example, $\sqrt{-2}$); all of which, even though they are called imaginary, remain useful and even necessary for expressing real magnitudes analytically. It is impossible, for example, to express the analytic value of a straight line necessary to trisect a given angle without the intervention of imaginaries, just as it is impossible to establish our calculus of transcendents without using differences which are on the point of vanishing, by taking the incomparably small all at once in place of that which can be assigned smaller and smaller to infinity. It is in this same way that one can also conceive of dimensions beyond three, and even of powers whose exponents are not ordinary numbers—all in order to establish ideas fitting to shorten our reasoning, and which are founded in realities.

Yet we must not imagine that the Science of the Infinite is debased by this explanation and reduced to fictions, for there always remains a syncategorematic infinite, as they say in the Schools, and it remains true, for example, that 2 is as much as $\frac{1}{1}+\frac{1}{2}+\frac{1}{4}+\frac{1}{8}+\frac{1}{16}+\frac{1}{32}$ etc., which is an infinite series in which all the fractions whose numerators are 1 and whose denominators are in a double geometric progression are comprised at once, although only ordinary numbers are used in it, and no infinitely small fraction, or one whose denominator is an infinite number, ever occurs in it. Furthermore, imaginary roots likewise have a *fundamentum in re*, so that when I communicated to the late Mr. Huygens that $\sqrt[2]{(1+\sqrt{-3})} + \sqrt[2]{(1-\sqrt{-3})}$ is equal to $\sqrt[2]{6}$, he found this so remarkable that he replied that there is something here incomprehensible to us. So it can also be said that infinites and infinitesimals are founded in such a way that everything in geometry, and even in nature, takes place as if they were perfect realities. Witness not only our Geometrical

---

[15] In the margin, Leibniz has written "N.B." [note well] by this passage.

Analysis of Transcendents, but also my law of continuity, by virtue of which, it is permitted to consider rest as an infinitely small motion (that is, as equivalent to an instance of its contradictory), coincidence as an infinitely small distance, equality as the last of [a sequence of] inequalities, etc. This law I once explained and applied in Mr. Bayle's *Nouvelles de la république des lettres*, to the example of the rules of motion of Descartes and Father Malebranche. I have since observed, by the second edition of the latter's rules which has since appeared, that the entire force of this law has not yet been properly considered.

Nonetheless, one can say in general that all continuity is something ideal, that there is never anything in nature with perfectly uniform parts, but that in recompense, the real never ceases to be governed perfectly by the ideal and the abstract, and that the rules of the finite are found to succeed in the infinite, as if there were atoms (that is, assignable elements of matter), even though there are none, matter being actually subdivided[16] without end. And conversely, the rules of the infinite succeed in the finite, as if there were metaphysical infinitely small things, even though there is absolutely no need for them, and the division of matter never does proceed to infinitely small particles. This is because everything is governed by reason; otherwise there could be no science and no rule, which would not at all conform with the nature of the sovereign principle.

For the rest,[17] when what I read in the *Journal de Trévoux* led me to write something on what was said there against the calculus of differences, I assure you that I was not thinking of the controversy that you, Sir, or rather those who use the calculus of differences, are having with Mr. Rolle. Also, it was not until I received your last letter that I learned that the Abbé Gallois, whom I always greatly honour, has taken part. Perhaps his opposition comes only from his belief that we base our demonstration of this calculus on metaphysical paradoxes, from which I myself hold it can readily be disengaged. Although I can't imagine this learned Abbé capable of believing that this calculus is as faulty as it seems Mr. Rolle says it is, according to what I have learned from you, I have never yet seen the works published by this author.

(b) Extract from the Letter of Mr. Varignon to Mr. Johann Bernoulli[18]

I have given Mr. Leibniz's letter to be inserted in the *Journal des Sçavans* as an explanation of the paper I sent you from the *Mémoires de Trévoux*. Awaiting this letter to be published, I did not miss the chance of letting it be seen by Father Gouye, who was very much surprised to see in it that the infinite in all rigour, which Mr. Leibniz said was of no use in his calculus, is only a real and existent infinite, and not the ideal or mentally inexhaustible infinite, as this Father had first thought in opposing this to us in his *Memoires*, and after

---

[16] Here the editor of the journal has wrongly corrected Leibniz's "sousdivisée" to "sous-divisible", which spoils the intended sense. We have restored the original.

[17] The remainder of this letter did not appear in the *Journal des sçavans*.

[18] GM IV, 97.

(c) From Leibniz to Varignon, 14 April, 1702[19]

I learned from what Mr. Bernoulli of Groningen communicated to me that you have received the letter which will be used in the *Journal des sçavans*, but that according to Father Gouye's opinion, I explain myself there in a different manner than in the memoir which was made public in the *Journal de Trévoux*. I admit that I said more in this letter, but it was necessary, since it was a matter of explaining the memoir, but I don't think there is any opposition. If this Father finds some and makes it known to me, I will endeavour to relieve it. At least there is not the least thing in it which could make one believe that I meant a quantity that is truly very small, but always fixed and determined. Moreover, I wrote some years ago to Mr. Bernoulli of Groningen that the infinite and infinitely small could be taken as fictions, similarly to imaginary roots, without that having to do a wrong to our calculus, these fictions being useful and founded in reality.

(d) From Leibniz to Varignon, 20 June, 1702[20]

Between ourselves, I believe that when Mr. Fontenelle, who has a gallant and beautiful mind, said that he wanted to create the metaphysical elements of our calculus, he wanted to mock it. To tell the truth, I am not too persuaded myself that we must consider our infinites and infinitesimals otherwise than as ideal things or as well founded fictions. I believe that there is no created thing below which there is not an infinity of created things, but I do not believe that there are any infinitely small ones, nor even that there can be any, and this is what I believe I can demonstrate. It is only simple substances (that is, those which are not beings by aggregation) that are truly indivisible, but they are immaterial, and are only principles of action.

Text 39. From *Quaestio De Jure Negligendi Quantitates Infiniti Parvas* [Autumn 1702-1703][21]

The question concerning by what right infinitely small quantities can be neglected in comparison with ordinary ones, or ordinary ones with respect to infinite ones, and any

---

[19] GM IV, 97–98.

[20] GM IV, 110.

[21] LH 35, 8, 21 fol. 1–2. First transcription in Pasini (1985-86, App., pp. 40–47). A new transcription and French translation may be found at https://eman-archives.org/philiumm/justification-of-the-differential-calculus/quaestio-de-jure-negligendi-quantitates-infiniti-parvas (accessed 14 July 2024).

12.2 (for 6.2) Mature Justifications of the Use of Infinitesimals   239

that are infinitieths with respect to those that are infinite multiples of them, has recently engendered many disputes in the Royal Academy of Sciences, and this affair has led to me being requested to give an opinion of some sort as well.

Indeed, the matter seems to reduce to this: one needs to allow for pure mathematics to be preserved intact from metaphysical controversies. We can do this if, caring not at all whether the infinite and infinitely small in quantities, numbers and lines are real, we use the infinite and infinitely small as apt expressions for condensing thought. Thus it is granted that they are imaginary. These quantities can nevertheless be employed like the imaginary roots in algebra.

Hence it follows that this abbreviation is never to be used without an explanation of what is substituted, by which the matter reduces to a rigorous demonstration in the style of Euclid or Archimedes. And this is the method I gave in the *Acta eruditorum*, when I explained the Lemmas on Incomparables.

For it can be said that the various degrees of the infinite and infinitely small behave in the way in which I conceive the distance away of the fixed stars to be incomparably greater than the globe of the Earth, and the globe incomparably greater than this grain of sand, and the grain of sand incomparably greater than a corpuscle of light, so that one can be taken for a point in comparison with the other; even if no such definite quantity is employed in our method, but only one as small as one wishes.

Therefore, when it can be shown that in the assertions of the geometers the error is smaller than any assignable, that is, null, we say by way of abbreviation that the error or discrepancy is infinitely small, and we employ it in our calculus as a quantity, and neglect it in comparison with that of which it is the discrepancy. For the reasoning is immediately made rigorous by changing the indefinitely small into one determinately small, but less than an assigned quantity, which a demonstration shows can be held to be zero, so that the matter reduces in a way to that peculiar kind of demonstration, examples of which are to be found just about everywhere, where from the supposed inequality of two quantities we directly conclude their equality.

And this is the principle of the differential calculus, when we suppose discrepancies or differences incomparably smaller than these very differences, and here we recognise various degrees, constituting a certain law of homogeneity of a new kind.

And these indefinite quantities are useful, even for [making] universal propositions in determinate [quantities]. For example, theorems about converging straight lines are verified for parallels taken as a species of converging lines where the point of concurrence is infinitely distant. In this way everything that is demonstrated about ellipses is also verified in a certain way for the parabola, if one conceives the parabola as an ellipse whose other focus is infinitely distant.

From this there arises that Law of Continuity whose great utility in physics I showed some time ago in the *Nouvelles de la République des Lettres*,[22] by conceiving equality as a vanishing inequality, and rest as an infinitely small motion, and other things of this kind; where, when one of the given things vanishes into the other, as motion into rest, so also in the same case the rule for motion must vanish into the rule for rest. Otherwise the rules are inconsistent.

And so by an admirable artifice of nature it happens everywhere that everything proceeds in reasoning as if these quantities were as real as can be; and otherwise the order and connection of things would not be consistent with itself.

As regards certain difficulties arising from infinite series, I say that it is not proper to treat nothing as opposite to infinity, but rather nothing is opposite to everything, and the opposite of the infinite is the infinitely small. And so '$\frac{1}{0}$ = infinity' is not a true proposition, unless 0 signifies an infinitely small number, which, as a substitute for that same 0, can certainly be dropped, for the reasons stated above—as if I were to say that $\frac{aa}{dx} = f$, so the ordinary quantity a is the mean proportional between the infinite f and the infinitely small dx.

Hence one cannot consider twice 0 to be something different from 0, as I have heard maintained by certain ingenious men, who have thus confused the infinitely small with 0 in some fashion.

When one investigates infinite series by division or by extraction or by some other means, the gist of the truth consists in the fact that the discrepancy is shown to be less than any quantity whatever. Thus, if someone says that there is a discrepancy between the circle and this quantity found by me, $1 - \frac{1}{3} + \frac{1}{5} - \frac{1}{7} + \frac{1}{9} - \frac{1}{11}$ etc. (assuming a diameter of 1), I will show that this discrepancy is less than anything anyone could assign. Because in this series, wherever you stop the progression, the error will always be smaller than the fraction immediately following, and indeed if you progress far enough it is smaller than any given quantity whatever.

This is the reason why continuous division only gives a series equal to a fraction when the terms are decreasing. For in order for the error to be shown to be less than that given, the series must be decreasing, and must even decrease sufficiently.

Thus[23] in the fraction $\frac{a}{b+c}$, if you divide a by $b + c$ it produces

---

[22] In an apparent lapse or abbreviation Leibniz has written "*rei litterariae Novellis*" instead of the usual "*Rei publicae litterariae Novellis*". He means the *Nouvelles de la république des lettres*, in which he published in 1687 his "Extrait d'une lettre de M. L. sur un principe géneral, utile à l'explication des loix de la nature, par la considération de la sagesse divine ; pour servir de réplique à la réponse du R. P. M."

[23] In the manuscript Leibniz provides a schema showing his calculations.

12.2 (for 6.2) Mature Justifications of the Use of Infinitesimals

$$\frac{a}{b} - \frac{ac}{bb} - \frac{acc}{b^3} - \frac{ac^3 : b^3}{b+c}$$

so that it is permissible to stop at any point, provided that one must write underneath the last remainder, here $-ac^3{:}b^3$, the divisor $b + c$, so that the fraction $\frac{ac^3:b^3}{b+c}$ appears, which, when added to the series, makes it equal to the given fraction $\frac{a}{b+c}$.

Thus, if you have $\frac{a}{b+c} = \frac{4}{3+1} = \frac{4}{4} = 1$, this will give $\frac{4}{3} - \frac{4}{9} + \frac{4}{27} - \frac{4:27}{3+1}$, that is, $\frac{4}{3} - \frac{4}{9} + \frac{4}{27} - \frac{1}{27}$ $= \frac{4}{3} - \frac{4}{9} + \frac{3}{27} = \frac{4}{3} - \frac{4}{9} + \frac{1}{9} = \frac{4}{3} - \frac{3}{9} = \frac{4}{3} - \frac{1}{3} = 1$.

And up to whatever point you continue the series, there will be $\frac{4}{3} - \frac{4}{9} + \frac{4}{27} - \frac{4}{81}$ etc. up to $\pm$ (+ or -) $\frac{4}{3+1} \frac{1}{3^e}$, which can be made less than any given. But if it were $\frac{4}{1+1}$, then wherever you stop there will always remain (+ or -) $\frac{4}{1+1} \frac{1}{1^e}$, but $\frac{1}{1^e} = 1$, so there will always remain $\frac{4}{2}$, which will not decrease and will not become less than any given.

What, then, pertains to a series such as $\frac{2}{2-2} = 1 + 1 + 1 + 1$ etc. $= \frac{1}{0}$,[24] or such $\frac{2}{1-1} = 2 + 2 + 2 + 2$ etc. $= \frac{2}{0}$, which ought to be equal to one another, yet one is twice the other? It must be said that there is no equality here between the infinite series and the fraction. It must also be said that absolutely nothing does not divide anything at all, nor does it multiply it, except to remove it completely. But this $\frac{1}{0}$, when 0 is taken for perfectly nothing, must be absolutely infinite, which has no extension ôr quantity. Now this absolutely infinite, namely God in calculable quantity, is certainly deprived of parts, whereas everything, which really has parts, is not a true whole, that is to say, cannot constitute a quantity that can be changed by multiplication or division.

We must respond in the same way to someone who, assuming 0 as a quantity, and using these two accepted axioms, (1) those things equal to a third are equal to each other, and (2) equals multiplied or divided by the same quantity remain equal—without any reference to our infinitesimal calculus—will argue as follows: 1 times $0 = 0 = 2 - 2 = 2$ times $1 - 1 = 2$ times 0. Therefore 1 times $0 = 2$ times 0, therefore $1 = 2$; which is absurd.

One must reply that the true 0, ôr 0 taken absolutely, cannot be subjected to such a calculation, i.e. if something is multiplied by 0, it is not restored by dividing the product again by 0, since anything multiplied by 0 will give the same thing, namely nothing.

But, you will insist, it can at least be shown that $\frac{1}{0} = \frac{2}{0}$, ôr that something is twice itself. For $\frac{1}{0} = \frac{1}{1-1} = \frac{2}{2-2} = \frac{2}{0}$. It must be responded that $\frac{1}{0}$ cannot be subjected to such a calculation, that is to say that $\frac{1}{0}$ is not doubled by being multiplied by 2, so that it is not permissible to say "twice $\frac{1}{0}$", except in the same way as we say that twice something is adding it to itself, as if we were repeating the same truth, since $\frac{1}{0}$ signifies the opposite of nothing, i.e. everything, the number of all unities. And if something multiplies everything, it will produce nothing new, since everything cannot be increased—just as if someone divides nothing by some number, say, into two, three, etc. parts, this will produce nothing new, because nothing cannot be diminished. In turn, just as nothing cannot be doubled, so

---

[24] Here the original had $\frac{2}{0}$, which appears to be an error.

one cannot take half the number of all things, i.e. the number of all unities. From these things it is understood that nothing and everything are not changed by being multiplied or divided. But it is different with the supposed infinite and infinitesimally small which we use in the calculus; this is as small or as large as is needed for the error to be shown to be smaller than the one given.

Text 40. From a Letter to the Abbé Gallois [July, 1705][25]

Hanover, July 1705

Sir, I have learned more than once that the great number of years that have elapsed since we were last in communication has by no means diminished the kindness you have shown me on other occasions. And although I was told some time ago that you are not entirely happy with the use of a new Calculus which I first proposed, I did not believe that this marked a change in your feelings towards me. [...]

And I would never have said anything about those objections that have just been refuted in the *Journal des sçavans* of the 13th of April of this year, if I had not been asked to examine them by persons whom I esteem, and if I did not believe that the Académie Royale des Sciences had some interest there.

I am told, Sir, that you are favourable to the person who made them, but I believe that it is because of other talents that he may have, and other difficulties that he may have made, and not in relation to these objections, which seem to me to be the least excusable, since they amount to refusing to those who handle the new Calculus the liberty of making use of the axioms and most accepted operations of ordinary Geometry. [...]

After having spoken enough now of these objections, I return to the new Calculus in general. I do not know if you have seen, Sir, what I said on this subject on the occasion of the difficulties that the Rev. Father Gouye had made, against the infinites and infinities of infinites. It is easy to see that assignable magnitudes (for example, the ordinates of curves, or the motions that correspond to them) have their continuous or unassignable increments which make them grow without making any leap, and these increments are what are called infinite[simal] differences. Now, since these increments may still be continuously unequal (as the ancients already recognized in uniformly difform motions or changes) they will still have their continuous increments or second differences. And since thus the unassignable increments can be considered as proportional to the assignable magnitudes, it is manifest, that they can be treated here <--> and that thus these degrees of differences can go down to infinity and be used as required. It may be said that motion through some assigned space gives that which corresponds to the ordinary magnitude; that the velocity which makes this

---

[25] Gotha FB A 448–449, Fol. 34–35. A transcription of this text was kindly provided to us by Charlotte Wahl from the Leibniz Archiv, and a digital version of the original may be found at https://dhb.thulb.uni-jena.de/rsc/viewer/ufb_derivate_00015564/Chart-A-00448-449_00075.tif.

space grow continuously gives that which corresponds to the first differences; and that the centrifugal solicitations, or those of gravity and everything that makes the velocity itself grow or decay continuously, give that which corresponds to the second differences. It may be said further that the *ordinates* of the curves are assignable magnitudes; that the differences of the ordinates which determine the *direction* of the curves (or which give the tangents) are differences of the first order (regularly), and that the angle of contact or the *curvature* of the lines is determined by the second differences. And finally the *angles of the osculations* which are of an infinite kind depend on the following differences.

But in order that one should not be troubled by the heterogeneity of these degrees of magnitude and by their elisions which the new calculus requires for abbreviating, by neglecting the magnitudes or differences that are inferior in relation to those which are superior; and in order that one may see the infallibility of this procedure; I usually advise that one should conceive the one in relation to the other, not at first as infinitely less, but as incomparably less: having presented for this purpose on another occasion an Essay on the Lemmas on Incomparables. That is to say, more or less, that the ordinary magnitudes are like the distance from the sun to some fixed star; that the first differences are like the diameter of the great orbit which the earth describes around the sun, which may be taken as a point with respect to this distance from the fixed stars; that the diameter of the earth is like a difference of the second order, which may be taken as a point with respect to the diameter of the great orbit; that the diameter of a grain of sand is like a third difference; and that the diameter of a particle of the matter which flows from the magnet is like a fourth difference, or like a magnitude of the fifth degree. However, all these magnitudes of various degrees are of one and the same dimension, that is to say, lines in the case of which I speak, and never indivisible points. Now by neglecting what is inferior in relation to the superior, we see that the error is inconsiderable. But as the same reasoning remains however great one may take the interval from one degree of magnitude to another, and as one may take an inferior magnitude as small as one wants in relation to the superior one, it is evident that the error will be less than any given magnitude, and consequently that there is no error whatsoever, which renders the reasoning entirely rigorous, just as Archimedes' reasoning was, without going through the detours that he took. In the beginning, when this calculus was not yet sufficiently known, I saw some very skilful people commit paralogisms in trying to use it. But this was always for want of making use of the precaution which I have just indicated, and by confusing the magnitudes of different degrees, which I call violating the laws of the homogeneity in the Analysis of the infinitesimals, which has this peculiarity, that it puts a heterogeneity between the magnitudes of the same dimension; for example by comparing lines with each other according to what I have just said. Instead of this, in ordinary analysis there is only heterogeneity between magnitudes of different dimensions; for example between lines and surfaces.

I do not know, Sir, if what I have just said will give you any satisfaction. It would put me at ease to justify my calculus in the mind of a person whom I honour as much as you, and whose cooperation with us will be of no little effect for the advancement of science. And wishing you many more years to contribute more and more to this advancement, which is one of the main objects of my passion; I am entirely your most humble and obedient servant, Leibniz.

Text 41. *An Observation That Ratios or Proportions Do Not Hold for Quantities Less Than Zero, and on the True Meaning of the Infinitesimal Method*[26]

When in Paris that great gentleman Antoine Arnauld once communicated to me his new Elements of Geometry, and in these same Elements he testified that he wondered how it could be that 1 to –1, is as –1 to 1; which seemed to him to be proved from the fact that the product of the outer terms is the same as that of the middle terms, since both produce +1. I then said that it seemed to me that those are not *true ratios* in which the antecedent or consequent is a quantity less than nothing, even though on this calculation they could be employed safely and usefully as *imaginary*. And certainly the foundation of the identity of true ratios is similarity, which brings it about that, for example, when we assume similar segments of different circles there is everywhere the same ratio of the chord to the radius, that is to say, that the chord of the smaller one bears the same relation to the radius of the smaller one as does the chord of the larger one to the radius of the larger one. But it is clear on the other hand that there is no similarity in the above analogy. For if –1 is less than nothing, then 1 to –1 will certainly be the ratio of the larger to the smaller. But on the other hand the ratio of –1 to 1 is the ratio of the smaller to the larger. How, then, will the ratio in each case be the same? But that those ratios are imaginary, I will also prove by another argument, that is, by logarithms. For a ratio to which there is no corresponding logarithm is not a true ratio. Moreover, supposing the logarithm of unity is 0, that of the ratio of –1 to 1 is the same as the logarithm of –1. But there is no logarithm of –1. For it is not positive, since every such logarithm is that of a positive number greater than unity. But yet it is also not negative, since the logarithm of every such logarithm is that of a positive number less than unity. Therefore since the logarithm of –1 is neither positive nor negative, it remains that it is not true, but rather imaginary. And so the ratio to which it corresponds will also be *not a true one, but imaginary*. I prove the same thing like this: if there were a true logarithm of –1, or of the ratio of –1 to 1, then half of its logarithm would be the logarithm of $\sqrt{-1}$, but $\sqrt{-1}$ is an imaginary quantity. So there would be a true logarithm of an imaginary quantity, which is absurd.

And so the distinguished John Wallis made something of a human error in a few places in his *Geometry*, when he said that the ratio of 1 to –1 is greater than infinity; and this is rightly rejected (even if according to other considerations) by the most celebrated Varignon. On the other hand, I do not want to deny with the latter that –1 is a quantity less than nothing, provided this is understood in a sound sense. Such statements are *true*

---

[26] *Acta eruditorum, Leipzig,* April, 1712, 167-69; GM V, 387–389.

within a tolerance, as I am accustomed to say with that great man Joachim Jungius; the French would call them *passables*. They cannot be sustained in all rigour, yet they have great utility in calculation, and are valid for the art of discovery and universal concepts. Such was Euclid's way of speaking when he said that the angle of contact is smaller than any rectilinear angle; and such are many other things in Geometry, in which there is a kind of figurative and cryptic way of speaking. Yet they do indeed have, so to speak, *tolerability*. Moreover, just as I deny the reality of a ratio one of whose terms is a quantity less than nothing, so I also deny that properly speaking there exists an infinite or an infinitely small number, or an infinite or an infinitely small line—even though Euclid often speaks of an infinite line, but in a sound sense. The *infinite*, whether *continuous* or *discrete*, is properly neither one, nor a whole, nor a quantity [*quantum*]; and if we take it as such by a certain analogy, it is only, so to speak, a way of speaking; when, namely, there are more things than can be comprised by any number, yet by analogy we attribute to these things a number, which we call infinite.

Thus I once declared that when we say that an error is infinitely small, it is understood as smaller than any given, and really nothing; and when we compare the ordinary, the infinite, and the infinitely infinite, it is exactly as if we compare in ascending order the diameter of a particle of powder, the diameter of the Earth, and the diameter of the sphere of the fixed stars, or things arbitrarily greater or smaller (by degree) than these; and in the same sense in descending order, the diameter of the sphere of the fixed stars, the diameter of the Earth, and the diameter of a particle of powder, can be compared with the ordinary, the infinitely small and the infinitely times infinitely small, but in such a way that any of these things is understood to be conceivable as arbitrarily greater or smaller in its own kind. When, however, by making a leap to the ultimate, we talk of the infinite itself or infinitely small, we make use of a convenient expression or a mental abbreviation, but we say only *tolerably true things* that can be given a *rigorous* explanation. And this is also my opinion of those areas under the asymptotes of hyperbolas, which are said to be infinite and infinitely infinite; that is to say, such things cannot be true rigorously speaking, yet are tolerated in a sound sense in various ways. And in fact these considerations can serve not only to end the controversies of those most distinguished gentlemen Varignon and Grandi, but also to guard against certain chimerical concepts, and also, finally, to crush objections against the method of the *infinitely similar*.

Text 42. From a Letter to Grandi [September 1713][27]

Furthermore, my opinion, very often expounded, is that infinitely small as well as infinite quantities are indeed fictions, but useful for reasoning compendiously and at the same time safely. And it suffices that they are understood to be truly as small as is necessary in order

---

[27] GM IV, 218–219.

for the error to be smaller than any given; from which it is shown that there is no error. I have indubitable arguments for this opinion, but which would be too prolix to expound now. Meanwhile we conceive infinitely small things, not as simply and absolutely nothings, but as *relative nothings* (as is well known), that is, as indeed vanishing into nothing, yet retaining the character of that which is vanishing. Such quantities multiplied by a quantity that is also infinitely modified we conceive as producing an ordinary quantity. The affair of Creation has been quite elegantly illustrated by you from this, where an absolutely infinite force makes something from absolutely nothing. Certainly, in our Analysis we conceive a modified infinite straight line, such as $aa{:}dx$, multiplied into the straight line dx that is disappearing into nothing—or, what is the same thing, disappearing into a state of annihilation of the continually decreasing straight line x—as producing an ordinary rectangle aa. In my view, magnitudes that are infinite in number (i.e. more than any number whatever) never compose one infinite whole, and true infinitude does not occur except in an infinity of power, lacking every part; and therefore neither eternity nor an infinite straight line, even though expressed by one name, is one whole, and those extraordinary quantities of our calculus are fictions. But they are not for this reason to be spurned or rejected, by analogy with those things which I do not at all deny can be advantageous to the true religion; since in the calculus it is exactly as if they were true quantities, and have a foundation in reality and a certain ideal truth, like imaginary roots, which were incorrectly said by Prestet, the French Analyst, to produce a contradiction. For just as imaginary roots are necessary for maintaining equations that contain possible cases on a par with impossible ones, so extraordinary quantities are necessary for maintaining general rules that include intermediate values together with extreme ones, for example, as parallelism is comprised under convergence as an extremum of convergence. And Nature prescribes to things an inviolable *law of continuity*, once expounded by myself in the *Nouvelles de la République* [*des Lettres*] formerly published in the Netherlands, so that the use of these [general rules] even in physics never fails, even if there it is not established by a rigorous demonstration, but by an agreement of reasons, so that it should be said that God himself had respected them. And it can be said quite appropriately that the very case of the modified infinite multiplied into the infinitely small modified by continuously increasing and decreasing, finally evades the combination of the absolute infinite with absolute nothing, that is, creation. And by rejecting these subtleties of philosophy and this, so to speak, metaphysics of geometry (which Caramuel would call "Metageometry"), the Scholastic philosophers are ignorant of many otherwise ingenious and well known matters that would prove more useful than those with which they are commonly concerned in the schools.

Text 43. *Regula de Transitu per Saltum Non Admittendo* [After 1710?][28]

The Rule of not Admitting a Transition through a Leap seems to suffer a certain amazing exception, but this arises from fictitious expressions, useful in calculations, which are nevertheless, as Jungius used to say, true within a tolerance. It is known that n^0 is 1, whatever the number n may be. So this should also hold when $n = 0$, so we will have $0^0 = 1$. So that 0^0 is more than 0^1 or 0^2 or 0^3, etc. This is paradoxical enough, that a smaller exponent in rational wholes makes a greater power. But now we come to the counterexample to the rule concerning not admitting a leap. So perhaps it is not permissible to apply powers to 0.

Let there be 0^{a-x}. Here if x is 0, 1, 2, 3, etc. or any other number, provided it is less than a, (which I take to be positive ôr greater than zero) we will have $0^{a-x} = 0$. But at the very moment that x reaches a, so that 0^{a-x} becomes 0^0, it passes from 0 to 1. And it will jump over all the numbers between 0 and 1, that is, the fractions smaller than unity. But if x continues to increase beyond a, there will at once be a still more amazing leap, from unity to infinity, with nothing interposed. For now let $x = a + y$, giving $a - x = - y$ and $0^{a-x} = 0^{-y} = \frac{1}{0^y}$. But this is an infinite quantity, for if y is a positive quantity greater than nothing, we will have $0^y = 0$ et $\frac{1}{0^y} = \frac{1}{0} =$ infinity. And so Nothing, Unity, and Infinity will follow one another immediately, with nothing between. And 0^{a-x} is 0 if x is less than a, it is unity if x coincides with a; and it is infinite if x exceeds a (Fig. 12.2).

Let us exhibit this leap by way of a figure (Fig. 12.2). Let there be a straight line BC, from point B to point C, and with BC produced indefinitely far, let us assume a line AB from a point A taken between B and C. Next, assuming any point ξ in the straight line BC, let $B\xi = x$. Now let a dotted line $F\nu HIKLM$ be drawn[29], from any point of which is drawn a perpendicular to the straight line BC, such as $FB, \nu\xi, HD, KC, ME$, and let these be called ν, and let $0^{a-x} = \nu$. It is obvious that $FB, \nu\xi, HD$ will be equal to nothing, that is to say, up until the dotted line proceeds through BC itself, but at that moment when it reaches A, it will shoot out far from it and KA will become $= a$, assuming a to represent unity, but by

Fig. 12.2 Counterexample of Transition through a Leap

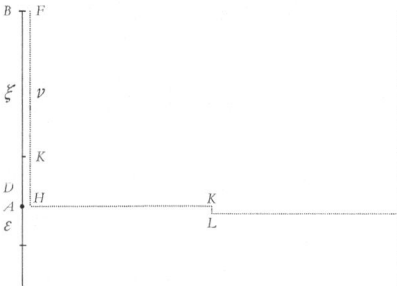

---

[28]LH 35, 4, 12 fol. 4. *Mathesis* N. 139.

[29]The letters missing in the figure (such as C, I, M) are already missing in Leibniz's manuscript.

progressing below *C*, from *K* or *L* it will at once shoot much further outwards than this, indeed to infinity at *M*, so that the perpendicular from *M* to the straight line *BC*, namely *ME*, is infinite. So three of the *v* will be indefinitely close to each other: *HD*, 0; *KA*, 1; and *ME*, infinite; will be immediately next to each other.

But where there is a leap in any asymptote, such as that of a Hyperbola, it is not contrary to the Rule. For there, before an infinite ordinate may be understood to occur, the ordinate grows to a width greater than any given, and so through all straight lines smaller than an infinite one. Here, on the other hand, it suddenly passes from 0 to 1, but not through all numbers greater than nothing and yet smaller than unity. And similarly it passes at once from unity to infinity, but not through numbers greater than unity and smaller than infinity. I do not know another example of this.

Text 44. From a Letter to Christian Wolff[30]

Moreover, this is consistent with the Law of Continuity that I once first proposed in Bayle's *Nouvelles Lettres*, and applied to the Laws of Motion from which it follows that the exclusive extremum in continua can be treated as inclusive, and so the ultimate case, even though different in its whole nature, lies hidden in the general law of the other cases. And at the same time, by a certain paradox of reason and, so to speak, philosophico-rhetorical figure, rest can be understood as included in motion as a special case in contradistinction to a general one, just as a point is an infinitely small ôr evanescent line, or rest is an evanescent motion, and other things of that kind. Joachim Jungius, a most profound gentleman, called these things true within a tolerance; they contribute a great deal to the art of discovery, even if in my opinion they involve something fictional and imaginary, which, however, by a reduction to ordinary expressions, is easily rectified in such a way that no error can intervene: and otherwise, Nature, proceeding always in order and not by a leap, cannot violate the law of continuity.

12.3 (for 6.3) Justifications of the Differential Algorithm

Text 45. From the Response to Nieuwentijt [July 1695][31]

Response to Some Difficulties Proposed by Mr. Bernard Nieuwentijt concerning the Differential or Infinitesimal Method.

As concerns the first objection, the most distinguished author puts forward this proposition in his preface of Considerations, which he avers to be of the most limpid truth: *Only*

---

[30] *Acta eruditorum, Suppl.* Vol. 1713; GM V, 385.
[31] GM V, 321–323.

12.3 (for 6.3) Justifications of the Differential Algorithm

those quantities are equal whose difference is null ôr equal to nothing. And in the analysis of curves, under Axiom 1, p. 2: *Whatever cannot be taken so many times, that is, cannot be multiplied by such a number* (even an infinite one, for this is how he understands it), *that it has a value equal in magnitude to any given quantity, however small, is not a quantity, but in geometry, a mere nothing.* Hence, given that in investigating equations for tangents and for maxima and minima (which he attributes to the learned author Barrow, although, if I am not mistaken, Fermat employed it first), there remain quantities that are infinitely small, yet their squares or rectangles are rejected; he deduces the reason for this state of affairs from the fact that the infinitely small or infinitesimal quantities themselves are something, since multiplied by an infinite number they make a given (that is, an ordinary or assignable) quantity; whereas it is otherwise with their rectangles or squares, which according to the axiom premised, are therefore merely nothing.

For my part, I acknowledge that for me it is very important to be diligent with those who claim to accurately demonstrate everything all the way to first principles, and I have also often taken interest in such matters; but it is also very important not to persuade in such a way as to make an obstacle to the art of discovery by excessive scrupulousness, or so that on such a pretext it is best that we should reject discoveries and deprive ourselves of their fruit, which Father Gottignies and his disciples also once insisted on with their scruples concerning the principles of algebra. For the rest, I believe those things to be equal, not only whose difference is absolutely nothing, but also those whose difference is incomparably small; and even though this difference should not be said to be nothing at all, there is however no quantity comparable with them whose difference it is. Just as you do not increase the quantity of a line if you add the point of another line, nor a surface if you add a line, so, in the same way, you do not increase a line if you add a line that is incomparably smaller than it. Nor is there any construction by which such an increase can be exhibited. That is to say, only those homogeneous quantities are comparable, I hold with Euclid, Book 5, Definition 5, one of which can be made greater than the other when multiplied by a finite number. And those things which do not differ by such a quantity I hold to be equal, as Archimedes also assumed, as have all others after him. And this is the very thing which is said to be a difference smaller than any given difference. And by a process that is indeed Archimedean, the matter can always be confirmed by an inference *ad absurdum*. But since the direct method is easier to understand and more useful for discovery, once this way of reducing is known it suffices that afterwards the method is applied in which incomparably smaller things are neglected, which certainly also brings with it its own demonstration according to the Lemmas communicated by me in February 1689. And if someone rejects such a definition of equality, he is only disputing a name. For it is sufficient for it to be intelligible and useful for discovery, since that which can be discovered by another method that is more rigorous (in appearance), is necessarily always produced by this method no less accurately. Thus I assume not only infinitely small lines, such as dx, dy, as true quantities of their kind, but also their squares or rectangles $dxdx, dydy, dxdy$, and I believe the same about cubes and other higher powers, especially since I find them useful for reasoning and discovery. Nor do I see at all how the most learned author could bring it into his mind to hold that a line ôr side dx is a quantity, but that the square ôr rectangle of such lines is nothing. For although these infinitely times infinitely small quantities, when multiplied by an infinite number of the first degree, do not produce a given ôr ordinary quantity, they still make one when they are multiplied by an infinitely times infinite number; and there is no right to reject this if you admit infinite number, for it is produced by an infinite number of the first degree multiplied into itself. On the other hand, concerning the fact that in Fermatian equations one discards terms containing such squares or rectangles, but not those terms containing simple infinitesimal lines, the reason for this is not that the latter are something whereas the former are nothing, but that ordinary terms are themselves eliminated, hence there remain not only terms containing simple infinitesimal lines, but also those containing their squares or rectangles: but when these terms are themselves incomparably smaller, they are discarded. But if the ordinary terms have not vanished, then also the terms of infinitesimal lines no less than the squares of these, should be discarded. One can add certain Lemmas of mine, serving for foundations of the differential calculus, from the *Acta Eruditorum* of Leipzig, February 1689, which the author professed to have come across only after publishing the Considerations in his Preface to the *Tractatus Analytici*, where the consideration of incomparables was already brought forward then in order to prevent these difficulties.

Text 46. From a Letter to Wallis [30 March 1699][32]

In my view, the form of the Characteristic Triangle on a curve can be correctly explained by the degree of declivity, but for the calculation it is useful to feign infinitely small quantities, ôr as Nicolaus Mercator called them, infinitieths [*infinitesimas*]: which are of such a kind that, since a ratio between them is sought that is certainly assignable, it is already illicit to hold them to be nothings. Meanwhile they are rejected whenever they are added to incomparably greater ones, according to the Lemmas on Incomparables once proposed by me in the *Acta* of Leipzig; which foundation is also used by the Marquis de l'Hospital. Thus, if $x + dx$ stands by itself, then dx is rejected. It is different if one seeks $(x) - x$ ôr $x + dx - x$; for then the assignable quantity vanishes. And by the same right, xdx and $dxdx$ (that is, $x + dx$ multiplied into dx) cannot stand together. Hence if one must differentiate xy, and one writes $(x)(y) - xy$, assuming $(x) = x + dx$ and $(y) = y + dy$, the assignable rectangle vanishes, and there remains the rectangle formed from an assignable and an elementary [line] of the first degree, and the rectangle formed from the two elementary [lines] must be rejected, that is to say: $x + dx, y + dy = xy + xdy + ydx + dxdy$, as you rightly advise, when xy is subtracted, there remains $xdy + ydx + dxdy$. But this $dxdy$ must be rejected, as it is incomparably smaller than $xdy + ydx$, and this gives $d, xy = xdy + ydx$, so that, if someone wished to translate the calculation into the Archimedean style, it is always evident that,

---

[32] GM IV, 63 = A III, 8, 91.

when the thing is done using assignables, the error which could result from this is smaller than any given. But when it comes down to second differentiation, then those rectangles formed from an assignable and an elementary [line] also vanish, and the rectangles formed from the two elementary [lines] are left, and (what is memorable) something homogeneous with these is made by multiplying an assignable into a differentio-differential [line]. Nothing, therefore, is neglected by us except in its proper place, nor is anything else considered as nothing, except comparatively. Nor do we need another postulate. So dd, xy is $2dxdy + xddy + yddx$—which calculus has been used for osculation and innumerable other things of that kind. What you say is simpler, I grant, that a multiple of nothing is nothing, but that does not have the use that we ourselves propose. [Meanwhile I do not dispute whether unassignable quantities are true or fictitious; it suffices that they serve for the abbreviation of thought, and always can be removed by changing the style of demonstration; thus I have noted that if someone substitutes the incomparably or sufficiently small for the infinitely small, this does not oppose me.]

Text 47. *Justification of the Infinitesimal Calculus by that of Ordinary Algebra* [Supplement to a Letter to Pinsson of April 1702][33]

Let there be two straight lines *AX* and *EY* intersecting at *C*, and then from the points *E* and *Y*, let us drop the perpendiculars *EA* and *YX* onto the straight line *AX*. Let us call *AC*, *c* and *AE*, *e*; *AX*, *x* and *XY*, *y*. Then because of the similarity of the triangles *CAE*, *CXY*, we will have that $x - c$ is to y as c is to e, and consequently if the straight line *EY* approaches closer and closer to the point *A*, always keeping the same angle at the variable point *C*, it is evident that the lines *c* and *e* will always diminish, but that the ratio of *c* to *e* will nevertheless remain the same, which we will suppose here to be other than a ratio of equality, and that the said angle is less than half a right angle (Fig. 12.3).

Now let us suppose the case where the straight line *EY* comes in this way to fall on *A* itself; it is obvious that the points *C* and *E* will also fall onto *A*, the straight lines *AC* and *AE*, or *c* and *e*, will vanish, and the proportion or equation $\frac{x-c}{y} = \frac{c}{e}$ will become $\frac{x}{y} = \frac{c}{e}$. Then in the present case we will have $x - c = x$, if we suppose the case is included under the general rule. And nonetheless *c* and *e* will not be absolute nothings, since together they preserve the ratio of *CX* to *XY*, or the ratio between the entire sine or radius, and the tangent of the angle at *C*, which angle we have assumed to have remained always the same while *EY* approaches the point *A*. For if *c* and *e* were absolute nothings in the calculation in this case, where the points *C*, *E*, and *A* are reduced to a coincidence, then, since one nothing

---

[33] GM IV, 104–106. The original letter (sent to Pinsson on April 21, 1702) is now lost, but the paper appended to it was meant to be transmitted to Varignon and was received by him, as testified by the PS to his letter to Leibniz from May 1702. In it, he says that he will try to have it published in the *Journal de Trevoux* (A III, 9, 107)—an attempt which did not in the end succeed.

Fig. 12.3 Including the Limiting Case under the General Rule

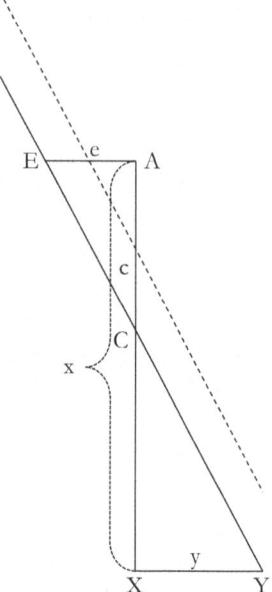

equals another, c and e would be equal, and the equation or proportion $x{:}y = c{:}e$ would become $x{:}y = 0{:}0 = 1$; that is to say, we would have $x = y$, which is an absurdity, since we have supposed that the angle is other than half a right angle. Thus c and e are not taken as nothings in the algebraic calculation except comparatively in relation to x and y but c and e still have a ratio to one another, and so they are treated as infinitesimals, exactly like the elements that our calculus of differences recognizes in the ordinates of curves, that is to say, as momentaneous increments or decrements. Thus one finds in the calculations of ordinary algebra traces of the transcendent calculus of differences, and those same singularities about which some scholars have scruples. And even the algebraic calculus cannot do without them if it is to retain its advantages, one of the most important of which is the generality that is due to it in order for it to comprise all the cases, even that in which some given straight lines vanish. It would be ridiculous not to want to do this, and so to deprive ourselves voluntarily of one of its greatest uses. All analysts skilled in the ordinary Specious have profited from it in order to render their calculations and constructions general.

And this advantage, when again applied to physics, and particularly to the laws of motion, reduces in part to what I call the Law of Continuity, which has served me for a long time as a principle of discovery in physics, and also as a very convenient test to see if some of the rules that have been given are correct. Several years ago I published a sample of this in the *Nouvelles de la Republique des Lettres*, taking equality to be a special case of inequality, rest to be a special case of motion, and parallelism to be a case of convergence, etc., supposing not that the difference between magnitudes that become equal is already nothing, but that it is in the act of vanishing, and likewise in the case of motion, that it is not yet absolutely nothing, but that it is on the point of being nothing. And anyone who is not satisfied with this can be made to see, in the style of Archimedes, that the error is not assignable, and cannot be given by any construction. It is in this way that we answered a certain mathematician, and a very capable one too, who, based on scruples similar to those raised against our calculus, found fault with the quadrature of the parabola. For he was asked whether he could by some construction *assign* a magnitude smaller than the difference he claimed there to be between the parabolic area given by Archimedes and the true one, as can always be done when a quadrature is false.

However, although it is not rigorously true that rest is a species of motion, or that equality is a species of inequality, any more than it is true that a circle is a species of regular polygon, it can still be said that rest, equality, and the circle terminate the motions, inequalities and regular polygons which by a continuous change arrive at them by vanishing. And although these terminations are exclusive, that is to say, are not comprised in all rigour in the variations which they bound, nonetheless they possess the same properties as if they were so comprised, in accordance with the language of infinities and infinitesimals, which takes the circle, for example, as a regular polygon with an infinite number of sides. Otherwise the law of continuity would be violated, that is to say, when one passes from polygons to the circle by a continuous change and without making a leap, it is also necessary not to make a leap when passing from the affections of the polygons to that of the circle.

Text 48. From Leibniz's Responses to Jenisch's Criticisms [October 1702][34]

Response (a)
Jenisch "As for most of those who attack the differential method, the most elegant discovery of our age, when immediately on its very threshold they see some of its *suppositions*[2] or *postulates*, they are so offended that they disdain to put a foot forward; ...";

Leibniz [2] (a) I do not make such postulates or suppositions.

Response (b)
Jenisch "so they are very often observed to hold this scruple as the worst, that the *infinitieth part*[3] in relation to the whole of which it is said to be a part, ôr what comes down to the same thing, the last term of an infinite series of terms decreasing in continuous proportion, can be safely neglected and held to be nothing....";

---

[34] LH 35, 7, 17 fol. 1–8; A III, 9, 212–223. Leibniz wrote his responses, numbered (a) to (m), in a separate column opposite Jenisch's criticisms.

Leibniz ³ (b) Whenever there is a need for the infinitesimal calculus, ôr method of differences, we show two quantities to be equal to each other, then tacitly apply a rigorous method, as if we were to say: Let there be a difference between them, from which very thing we will prove that there is none, ôr that no error can be assigned. For whatever [error] is assumed, and however small it is, we will show that this is still too much, that is, it is found to be smaller than assumed, and so that difference from the true one which was assumed as some quantity, is really nothing. Hence it is clear why in the calculus these differences ôr quantities however small not only could be rejected by comparison with those of which they are said to be the differences, but must be: because the difference assigned is really nothing, even though it is assumed as something by our adversary and by us in examining this argument. And since it can be assumed to be as small as you like, by an abbreviation of speech we have immediately assumed ôr feigned it to be as small as possible ôr infinitely small, even though it should really be interpreted as indefinitely small, ôr small enough to come out smaller than some given quantity. And this is one of those really marvellous kinds of demonstrations which, having assumed something, like the difference we have assumed here, conclude the opposite, namely that there is no difference, ôr equality; because, that is, by continuously decreasing into equality, it finally vanishes into that equality.

Response (c)
Jenisch "... and by the inverse operation, if from the given sum is subtracted the last *as yet unknown* ⁴ [term], and this is to be indicated by x, the remainder multiplied by the difference between the first and the second, and the product divided by the first, I will obtain the value of the difference between the first and the last, and thus the last, as was plain....";

Leibniz ⁴ (c) In my judgement, the demonstration does not proceed sufficiently rigorously when we conceive a certain term x as the last one in a geometrically decreasing series; for in fact there is no last term when it is continued to infinity, therefore through the fiction of a last ôr infinitesimal term (to reduce things to a rigorous demonstration), one has to conceive an error smaller than a given one. For it can be shown that whatever error is assigned, it cannot be so great but that if it were so, it would be much smaller; namely by continuing a finite series so far that whenever there is a remainder (which comprises the error ôr interceding difference, and that which is said to be equal to it) that remainder is smaller than the error assigned. Whence it follows that since no definite error can ever be assigned or exist, none is in fact committed.

Response (d)
Jenisch "And these things turn out very much for the best, provided only that an infinite series of this kind, arising from the continuous division of a fraction, can be shown to exactly *recover*⁶ its value so evidently that nothing would be made of this business by those who are more scrupulous. For indeed it is not beyond reason that the following could

12.3 (for 6.3) Justifications of the Differential Algorithm

persuade that it should still be called into doubt: 2/(2 − 2), that is *without doubt*[7] = 2/(1− 1), seeing as each equally = 2/0."

Leibniz [6] There ought to be a *recovery* [of the exact value], if namely an infinite series which is supposed equal to a fraction, multiplied by the denominator of the fraction, produces the numerator of the same fraction. And this will always happen as often as the supposition is true.

(d) [7] Such things here do not proceed as in a calculus of quantities. For neither 0 nor 1/0 ought to be considered as quantities. And 0, ôr nothing, is not the opposite of infinity, but of the most absolute, ôr everything, and what is properly opposite to infinity is not nothing, but infinitesimal ôr infinitely small; moreover, 0 and 1/0, namely nothing and everything, are not quantities, for twice 0 is nothing other than 0, and to multiply by 0, is the same as to eliminate. If 0 really were a quantity, it would follow that the whole would be equal to the part, the double to the single, namely twice 0 to 0 itself. But if now 0 is not a quantity, also 1/0 will not be a quantity, and this is also confirmed by the present calculus. For let 1/0 = 1/(1 − 1) = 2/(2 − 2) = 2/0. Therefore 1/0 = 2/0, the whole to the part, ôr the double to the single, which is absurd. And so everything ôr 1/0 is not a quantity. Furthermore, multiplying $1/x$ by x produces 1. Therefore multiplying 1/0 by 0 produces 1. But it is known from elsewhere that multiplying by 0 is to eliminate or reduce to nothing; 1/0 multiplied by 0 will be 0. Therefore 1 = 0. Which is absurd. Here it is clear that absurdities are produced without following any method of ours or any use of infinite series or of the differential calculus, and thus proceed from the nature of the matter, ôr from an impossible supposition, as if such things were quantities.

Response (k)
Jenisch "It is *postulated* that 1/0 = ∞."

Leibniz [14] (k) I do not make this postulate. For 1/0 is in fact the opposite of 0, but infinity is not the opposite of this same 0, the true opposition is between the infinite and the infinitely small, and likewise between everything and nothing. Since therefore 1/0 is the opposite of 0, it is clear that 1/0 does not properly signify the infinite, but, so to speak, the most absolute infinite, ôr everything; which, we showed above, is not a quantity; and the same thing can be recognized here. That is, if 1/0 were a quantity, its double 2/0 would also be a quantity, but since 1/0 signifies everything, it is clear that it cannot be doubled, otherwise it would be increased, which cannot happen, since more things could be assumed than everything, that is to say, there would be something beyond everything. With equal reason it could be said that as 0 is to 2, so 2 is to everything. Therefore this infinite will be greater than before, but it is absurd for something to be greater than everything.

Response (l)
Jenisch "Let there therefore be instituted on each side of the equation a division by y = 0, giving $\frac{ax-xx}{y} = \frac{0}{0} (= 1) = y (= 0)$."[18]

Fig. 12.4 Not Nothings, but Infinitely Small Quantities

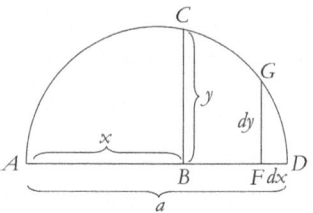

Leibniz [18] (l) I respond that, when we calculate by our method, we ought not to use nothings, but infinitely small quantities. In order to show this, let there be a semicircle *ACD* of diameter *a*, with abscissa *AB* or *x*. *BC* or *y* is applied (Fig. 12.4). When indeed *B* has approached *D*, and *C* has approached the same *D*, in such a way that they must coincide in *D*, that is to say, when *B* reaches *F* and *C* reaches *G*, at an infinitely small ôr evanescent interval from *D*; then for $a - x$ ôr $AD - AB$, one should not take 0, otherwise all calculation must cease, but instead the last interval *dx* ôr *FD*, and so *BC* ôr *y* will disappear into a last interval *dy* ôr *FG*, and from the equation $ax - xx = yy$ we will get $adx - dxdx = dydy$. Now since *dxdx* is incomparably smaller than *adx*, it is rejected in the quantity $adx - dxdx$, and we get $adx = dydy$, which is a sign that here *dx* is a quantity infinitely many times infinitely small, ôr incomparably smaller than *dy*; otherwise *dy* could not be the middle proportional between *a* and *dx*. Of course it often happens that two quantities such as *x* and *y*, whether in their very origin or on occasion, are incomparable with each other, even though they are comparable in process. Certainly, in the circle a nascent or evanescent chord is incomparably greater than the nascent or evanescent sagitta, and you will observe this in definite quantities, the smaller you assume *FD*, *FG*, for the ratio of *FG* to *FD*, if you approach near enough to *D*, will be greater than any given. Hence it is clear that in our method it is not possible for *x* and *y* always to be assumed indiscriminately as nothings at the same time; certainly, if you had wished to use nothings, you would have had to suppose one nothing, *dy*, as evanescent with the other, *x*, evanescent, infinitely greater, and the calculation would have had to suppose $x = 0$ and $y = \underline{0}$, whence the equation $ax - xx = yy$ would have given $a0 - 00 = \underline{0.0}$, that is, $a.0. = \underline{0.0}$., that is to say $\underline{0}$ would be the mean proportional between *a* and 0. But it is absurd to talk in this way about nothings, and so instead of nothings, infinitely small quantities should be used, and whatever is said concerning these, will always be verified by a rigorous method, by means of the infinitely small, that is, smaller than any given.

Response (m)

Jenisch "It seems it could be responded that, just as two infinites are not necessarily equal, so *neither are two nothings*[19] *equal*…".

Leibniz [19] (m) No, two nothings are always equal, and just as everything is equal to everything ôr the most infinite with the most infinite, so nothing coincides with nothing, but it is otherwise with infinities and the infinitely small, which are conceived as having different magnitudes. Whether truly or fictitiously so conceived, is nothing to mathematics, which has no need of metaphysical controversies, and it suffices that everything which we suppose in our method be verified straightaway by the common method by means of things smaller than any assigned.

Text 49. *Defense du Calcul des Differences* [1702?][35]

I learn that talented people are opposed to the Calculus of Differences, because it seems that in it one necessarily proceeds by infinitesimals, or by infinitely small magnitudes, and because they believe that in it one makes elisions at pleasure. One can always show them that everything that is concluded by this calculus can be proved by a *reductio ad absurdum* in the style of Archimedes, and by using the Lemmas on Incomparables proposed in the Leipzig *Acta*; and that it is always easy to recognize what one can neglect with impunity without any error arising from it, so that the elisions are made according to certain rules, and not as we see fit, except in that it is permitted to give to the continuously variable magnitudes differences as one wishes by choosing the progression one finds appropriate; this means that one can choose a series whose differences are constant, and whose second and other differences vanish. But without here making use of these Archimedean-style demonstrations, which are extremely long and not really suitable for enlightening the mind, I want to propose here two things: a very palpable means of justifying our method of calculation by means of ordinary algebraic calculus; and then an interpretation of the Calculus of Differences where instead of infinitesimals one understands only very small magnitudes, and this does not stop one from reaching a conclusion. As to the first point, I will show, with regard to the Ordinary Calculus of Algebra, that without noticing it enough, we have already long practiced the method that is contested with regard to infinitesimals, when we have applied a general calculation to some particular case, where certain magnitudes vanish. Here is a very easy example, where even those who are not very well versed in Geometry and Algebra, may understand what we mean (Fig. 12.5).

Let two perpendiculars *IA* and *VE*, coming from opposite sides, be drawn to the straight line *AE*. Now let there be a straight line *LM* that cuts the straight line *AE* in *H* so that the angle *EHC* is equal to *EAV*, and other than half a right angle; and let the same *LM* cut *IA* in *B*, and *VE* in *C*. Then let it be conceived that while *IAEV* remains motionless, the line *LM* will move from point *H* to point *A*, always keeping the parallelism during its movement; so that being transported into *lm*, and cutting *IA* in *b*, and *VE* in *c*, one always finds *lm* parallel to *LM*, and the angles in *b*, *h* and *c* equal, respectively, to the angles in *B*, *H* and *C*. To come now to some calculation, let us call *AB* or *Ab*, x; *AH* or *Ah*, z; *EC* or *Ec*, y; and *AE*, f and *HE* or *hE* will be $f - z$. Thus, as the triangles *HAB* and *HEC* (or *hAb*, *hEc*) are similar, *HE* to *EC* will be as *HA* to *AB* (or *hE* to *Ec* as *hA* to *Ab*) so $f - z$ to y will be as z to x or, changing the

---

[35] LH 35, 6, fol. 22, fol. 1–2. 1; Pasini (1988, pp. 705–708).

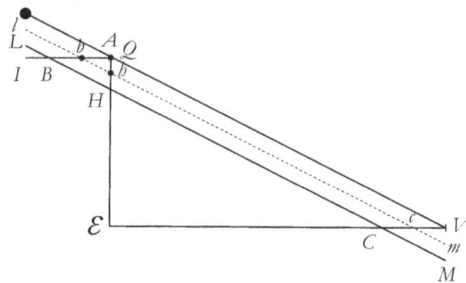

Fig. 12.5 Illustration of Applying the Calculus of Differences

equation into a proportion, we will have $\frac{f-z}{y} = \frac{z}{x}$. Now since the motion of the straight line *LM* is always continued with the same parallelism until it falls on *A*; it is obvious that points *b* and *h* will also fall on *A*, and point *c* on *V*; and that the lines *Ab*, *Ah*, or the values *x* and *z*, will vanish, so that $f - z$ will be equal to *f*. However, although *z*, since it is vanishing, is no longer comparable with *f*, and cannot increase or decrease it, it is not an absolute nothing in this calculation any more than is the magnitude *x*: for although in the previous equation or proportion it is necessary for this case to put $\frac{f-0}{y}$ or $\frac{f}{y}$ instead of $\frac{f-z}{y}$, it is nonetheless not allowed in the remaining cases, namely $\frac{z}{x}$, to put 0 instead of the vanishing magnitudes *z* and *x*—meaning that *bA*, *hA* always keep the proportion of inequality between them until the end; otherwise if they were absolutely nothing, instead of $\frac{f-z}{y} = \frac{z}{x}$, one will get $\frac{f-0}{y} = \frac{0}{0}$, i.e. $\frac{f}{y} = \frac{0}{0}$ or *f* is to *y* as 0 is to 0, hence *f* would be equal to *y*, i.e. *AE* to *EV*, contrary to hypothesis. For since the angle at *h* was supposed not to be a right angle, the sides *he* and *Ec* will never be equal, even in the case where *h* falls on *A* as well as *c* on *V*. For *AE* and *EV* are unequal. Thus, provided that we agree with all the algebraists up to now, that the particular case when *h* falls on *A* is included in the general calculation, we can only make this fit by following the method of the Calculus of Differences, considering *z* as 0 with respect to *f*, and as something with respect to what it is homogeneous with, which is *x*.

This is what I proposed to show with regard to the first point. I would like to add, however, that the foundation of all of this can be explained by taking *z* and *x*, i.e. *ba*, *hA*, *in the very act of vanishing* and falling on *A*. It is like a nascent motion, since an instant of motion is different from an instant of rest, and in this element of time, there is an element of nascent progress, which is more than nothing. But even if the instant of this act of vanishing or emerging is in metaphysical rigour only a fiction (in order not to sink into the labyrinth *de compositione continui*), it suffices that no error could arise from it, and that these fictions could always hold in place of truths in the calculus, in a like manner to imaginary roots. Since by rejecting all the infinitely smalls, and employing in their place only magnitudes as small as one wishes, one will always show that the error would be less than any given error. That is to say that there is none.

This is what made me speak on other occasions about incomparables, because what I say about it holds whether we understand infinitely small magnitudes or whether we use magnitudes of an inconsiderable smallness and sufficient to make the error less than the one given. This serves to combine the convenience of calculation with the rigour of the demonstration. Here is an essay on it.

Text 50. *Cum Prodiisset* [1702?][36]

When[37] my infinitesimal Analysis, which comprises the calculus of sums and differences, had come out and been widely disseminated, certain people began to raise ancient scruples—like those that the Sceptics once opposed to the Dogmatists, such as appear in the work of [Sextus] Empiricus *Against the Mathematicians* (i.e. the dogmatists), and that Francisco Sanchez, author of the book *Quod Nihil Scitur* [That Nothing is Known], sent to Clavius; and those objections which his opponents made to Cavalieri, and those which Thomas Hobbes in his Geometry made to everyone, and those which the renowned Detlef Clüver recently made even to Archimedes' quadrature of the parabola. Thus when our method of infinitesimals, which has come to be called the calculus of differences, began to be promulgated—first in certain essays of my own, and then in those of the exceptional Bernoulli brothers, and especially in the elegant writings of that illustrious Frenchman, the Marquis d'Hospital—recently a certain learned mathematician, writing in the *Journal de Trévoux* with his name suppressed, appeared to find fault with this method. But, writing in his own name even before this, there arose against me in Holland Bernard Nieuwentijt, certainly well equipped with learning and ability, but who has up to now preferred to become known by revising our method rather than promoting it. And whereas I had introduced not only first differences but also second, third and higher differences, unassignable or incomparable with the differences themselves, he wished to appear content with first differences; not considering that there are the same difficulties in first differences as in subsequent ones, and that wherever they might be overcome in the former, they also cease to exist in the later. To say nothing of how a very learned young man, Hermann from Basle, showed that in this matter the subsequent differences were avoided by Nieuwentijt in name only, and not in fact, but also that he did not succeed in demonstrating the legitimate use of those first differences—by accomplishing which he might at least have done something worth his while—having been forced to fall into doctrines admitted by no one; such as that something different is obtained by multiplying 2 by m than by multiplying m by 2; that in any case in which the former is possible, the latter is impossible; and also that the square or cube of a quantity is not a quantity, [but] nothing.

There is, however, something certainly worthy of praise in the fact that he wishes the infinitesimal calculus to be fortified by demonstrations in order to satisfy scruples. And this

---

[36] Gerhardt (1846, pp. 39–49).

[37] Leibniz makes the marginal comment: "All this must be edited very carefully so it could be published, omitting what is harsher in contradicting others. It is to be joined by my method for the law of continuity shown by the drawing of lines, and also by the tract I had sent the Parisians in order to show that in the common example a ratio between nothings is feigned to be something."

task he would already have procured from me more easily, if there had not appeared from his constant fault-finding everywhere a spirit somewhat alien to the custom of those who seek the truth rather than praise and fame.

It has been proposed to me many times that I should firm up the foundations of our calculus by demonstrations, and immediately below I have indicated sources for that plan, in order that anyone who has the leisure could take over this work. Yet up till now I have not seen anyone who would do it. For what the most learned Hermann began to do in the writing he published on my behalf against Nieuwentijt is not yet finished.

Apart from the infinitesimal mathematical calculus, though, there is also a method employed by me in physics which I once illustrated by an essay in the *Nouvelles de la République des Lettres*; and both of these I include under the *Law of Continuity*; by the application of which I showed that the Rules of Motion of those most renowned philosophers Descartes and Malebranche were in contradiction with themselves.

Moreover, I assume the following postulate: *In any proposed continuous transition ending in some terminus, one may institute a common reasoning in which the ultimate terminus is included.*

For example, if there are two [things] A and B, the former greater and the latter smaller, and with B remaining [the same], it is supposed that A is continuously diminished until A and B become equal, then it will be permissible to institute under a common reasoning the prior cases in which A was greater, as well as the ultimate case where, with the difference vanishing, A and B become equal. Similarly, if two bodies A and B collide with one another, and it is supposed that, with the motion of B remaining the same, the velocity of A is continuously diminished until it vanishes altogether, that is, the velocity of A becomes zero, this case may be included in one reasoning with the case of the motion of B.

We do the same in geometry, when two lines are taken, however produced, one VA (Fig. 12.6) in a given position, that is, always remaining in the same situation, the other BP passing through the given point P and varying its situation while the point P remains fixed; and first, converging with the line VA and meeting it in the point C, and then, as the angle of inclination BCA is continuously diminished, meeting it in the more remote point (C), until

Fig. 12.6 Including the Limiting Case

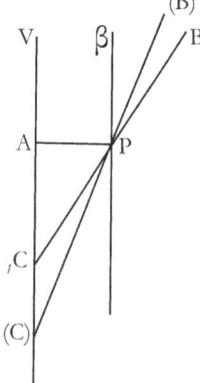

12.3 (for 6.3) Justifications of the Differential Algorithm

finally as it passes from *BP* through (*B*)*P* it reaches β*P*, where the straight line passing through *P* no longer converges towards *VA*, but is parallel to it, and the point *C* becomes impossible ôr imaginary. Granting these things, one may include under one reasoning not only all the intermediate cases such as (*B*), but also the ultimate one, β.

And hence it also happens that we may include ellipses and the parabola under one reasoning, as when *A* is considered to be one focus of an ellipse (with a given vertex *V*), which focus remains fixed, and the other focus *C* is variable as it passes from one ellipse to another, until finally (in the case where the straight line *BP* makes a variable focus by its intersection with the straight line *VA*) that focus *C* vanishes ôr becomes impossible, in which case the ellipse vanishes into a parabola. And so by our postulate the parabola may be included with ellipses under one reasoning. Geometers are also accustomed to use this method in constructions, when they include different cases under a general construction, noting that in a certain case the converging straight line passes away into a parallel one, with the angle between it and another straight line vanishing.

From this postulate, however, there arise certain ways of talking which, when applied freely for the sake of convenience, seem to entail an absurdity, but which absurdity ceases when a [different] meaning is substituted: for instance, when we speak of an imaginary point of contact as if it were something real, while in algebra it is accepted that imaginary roots are employed. And hence, preserving this analogy, we say that the straight line *BP*, when it ends up parallel to the line *VA*, is convergent with it, ôr makes an angle with it, but an infinitely small one, exactly like when we say that the motion of a body, when it ceases in rest, has a velocity, but an infinitely small one; and that a straight line, when it becomes equal to another, is unequal, but with an infinitely small difference; and that a parabola is an ultimate ellipse, which has a focus an infinite distance from a given focus nearer to a given vertex, or in which the ratio of *PA* to *AC*, ôr the angle *BCA*, is infinitely small.

Granted, it is true that those things are absolutely equal whose difference is absolutely null, and that parallel lines are those that never meet, since the distance between them is supposed everywhere equal; and that a parabola is not an ellipse, and so on. Yet a very state of transition, or vanishing, can be feigned[38] in which equality, or rest, or parallelism, have not yet come about, but yet in which there is a passing into such a state; which is assumed to be so close that the difference is smaller than every assignable difference; and that in this state there remains some difference, some velocity, some angle, but an infinitely small one; and that the distance of the point of intersection, ôr variable focus, from the fixed focus will be infinite, so that the parabola may be included under the denomination of ellipse (just as by a different reasoning it may be included under the denomination of hyperbola), since any difference between the things that are discovered about such a parabola, and what can

---

[38] A reminder that here (and throughout these translations) we have translated the verb *fingere* as "to feign" (rather than "suppose" or "imagine", as in Child) in order to preserve the intended semantic connection with its cognate, *fictio*, "fiction".

be rigorously affirmed of a parabola, is not assignable through any construction whatsoever.

And certainly it is credible that Archimedes, and someone who seems to have surpassed him, Conon, discovered their very beautiful theorems by the help of such notions. They confirmed these theorems by *reductio ad absurdum* demonstrations, which made their certainty manifest, but at the same time concealed the techniques used. And so Descartes noted very elegantly somewhere that Archimedes utilized a kind of metaphysics in geometry (Caramuel would call it "metageometry"), which technique scarcely any of the ancients promoted (save those who treated quadratures); in our time Cavalieri has revived the Archimedean Method, and has given others the opportunity of going further. As indeed did Descartes himself, when on one occasion he feigned the circle to be a regular polygon with an infinity of sides, and used the same reasoning when he treated the cycloid; and Huygens too, in his work on the pendulum, while he was accustomed to confirming his theorems by rigorous demonstrations, nonetheless sometimes used the infinitely small for the sake of avoiding too much prolixity, as also more recently did the renowned La Hire.

Meanwhile, whether this state of momentaneous transition from inequality to equality, from motion to rest, from convergence to parallelism, or the like, can be sustained in a rigorous and metaphysical sense, ôr whether infinite extensions each greater than the other, or infinitely small things each smaller than the other, are real, I admit can be called into doubt. But whoever wishes to discuss these matters will fall into metaphysical controversies about the composition of the continuum on which there is no need for geometrical matters to depend. In my opinion, it is certainly possible to conceive an unbounded line in some way, and if it is unbounded on one side, it is possible that something bounded on both sides can be added to it. But whether such a straight line is one whole of the kind that can be referred to computation, that is to say, whether it can be reckoned among quantities which can be used in calculation, is another question that there is no need to discuss here.

And so it will suffice that when we call things infinitely large (that is to say, infinite in the stricter sense) and infinitely small (that is to say, the infinitieths of quantities known to us), we understand indefinitely large and indefinitely small, that is, as large as anyone wishes and as small as anyone wishes, so that the error which anyone might assign is smaller than that which they assigned. And since it is clear in general that when an error however small is assigned, it can be shown that the error must be still smaller, it follows that the error is absolutely nothing. A roughly similar kind of argumentation to this was used in various places by Euclid, Theodosius and others, to whom it seemed somewhat amazing, although it could not be denied to be perfectly true, that from the very thing that is assumed as the error, it is inferred that the error is nothing. And so the indefinitely small (or infinitely large) is understood to be something however large (or however small) that behaves as a kind of thing of that genus, though not as some ultimate thing in that genus. If someone understands there to be such an absolutely ultimate thing, or at least a rigorously infinite thing, they can do this, even without deciding the controversy about the reality of extended, or in general continuous, infinite or infinitely small things, indeed, even if they

Fig. 12.7 Calculating the Tangent to a Parabola

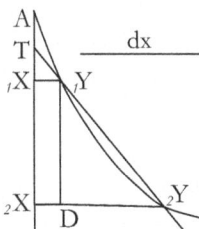

think such things impossible; for it will suffice that they can be usefully employed in a calculation, just as the algebraists employ imaginary roots with great profit. For they contain an abbreviation of reasoning, which it is clearly established, is always rigorously verified by the method already stated.

But it seems right to show the matter a little more distinctly, so that the algorithm (as it is called) of our differential calculus, proposed by me in 1684, may be confirmed to be perfectly true. First of all, when it is said that the elements of y are dy, the sense in which this is to be taken is best understood by considering some line AY referred to a straight line AX as axis (Fig. 12.7).

Let the curve AY be a parabola, and let the assumed axis AX be the tangent to the parabola at the vertex A. If AX is called x, $[XY]$ is called y, and the *latus rectum* is a, the local equation of the parabola will be $xx = ay$, and this obtains at each of its points. Now let A_1X be x and $_1X_1Y$ be y, and from the point $_1Y$ let a perpendicular $_1YD$ be dropped onto some greater succeeding ordinate $_2X_2Y$. Let $_1X_2X$, which is the difference between A_1X and A_2X, be called dx; and, similarly, let D_2Y, which is the difference between $_1X_1Y$ and $_2X_2Y$, be called dy.

Then, since $y = xx:a$, by the same law we will have[39]

$$y + dy = xx + 2xdx + dxdx, : a,$$

and subtracting y from one side and $xx:a$ from the other, there will remain

$$dy : dx = 2x + dx, : a$$

which is the general rule expressing the ratio of the difference in ordinates to the difference in abscissas. That is, if the chord $_1Y_2Y$ is produced until it meets the axis in T, then the ratio of the ordinate $_1X_1Y$ to T_1X, the intercepted part of the axis between the point of intersection and the ordinate, will be as $2x + dx$ to a. Now, since by our postulate it is permissible to include under one reasoning also the case where the ordinate $[_2X_2Y]$, having been moved closer and closer to the fixed ordinate $_1X_1Y$ until it finally coincides with it, it is clear that in

---

[39] In the following Leibniz has used the notation X + Y,:Z for (X + Y)/Z, although he sometimes uses the modern notation; and he has used ratios and quotients interchangeably.

this case dx will be equal to nothing ôr should be omitted, and so it is clear that, since in this case T_1Y is the tangent, $_1X_1Y$ to T_1X is as 2x to a.

Hence it may be seen that in all our differential calculus there is no need to call equal those things that have an infinitely small difference, but those things can be taken as equal that have no difference at all, provided that the calculation is supposed to have been made general, applying equally to the case where the difference is something and to where it is nothing; and only when the calculation has been purged as far as possible by means of legitimate subtractions and ratios of non-vanishing quantities, until at last application is made to the ultimate case, is the difference assumed to be nothing.

Similarly, if $x^3 = aay$, then we would have

$$x^3 + 3xxdx + 3xdxdx + dxdxdx = aay + aady,$$

that is, subtracting from each side,

$$3xxdx + 3xdxdx + dxdxdx = aady, \text{ or}$$

$$3xx + 3xdx + dxdx, : aa = dy : dx = {}_1X_1Y : T_1X$$

Hence, when the difference vanishes we have

$$3xx : aa = {}_1X_1Y : T_1X$$

But if we want to retain dx and dy in the calculation in such a way that they denote non-vanishing quantities even in the ultimate case, let $(d)x$ be assumed to be any assignable straight line whatever; and let the straight line which is to $(d)x$ as y ôr $_1X_1Y$ is to $_1X\,T$ be called $(d)y$, representing the interval between $_1X$ and $_2X$ that it is equal to it, or that it is proportional to it in some fixed ratio, so that dy and dx will always be assignable to one another in the ratio D_2Y to D_1Y, which latter vanish in the ultimate case. Whence an error needs to be corrected in the *Acta Eruditorum*, p. 467, l. 10, (and I do not know how it came to be committed), for instead of *VB* (or *WC* or *YD* or *ZE*) one should put *XB* (or *XC* or *XD* or *XE*).

With these things assumed, all the rules proposed for our algorithm in the *Acta Eruditorum* for the month of October in the year 1684 will be demonstrated without much trouble.

Let the curves *YY*, *VV*, *ZZ* be referred to the same axis *AXX* (see Fig. 12.8); and to the abscissae $A_1X (= x)$ and $A_2X (= x + dx)$ let there correspond the ordinates $_1X_1Y (= y)$ and $_2X_2Y (= y + dy)$, and likewise, the ordinates $_1X_1V (= v)$ and $_2X_2V (= v + dv)$, and $_1X_1Z (= z)$ and $_2X_2Z (= z + dz)$. Let the chords $_1Y_2Y$, $_1V_2V$, $_1Z_2Z$, when produced, meet the axis *AXX* in the points *T, U* and *W*. Let $(d)x$ be any arbitrary straight line, always fixed, with the point $_1X$ remaining fixed while the point $_2X$ approaches it however closely, and let $(d)y$ be another

12.3 (for 6.3) Justifications of the Differential Algorithm

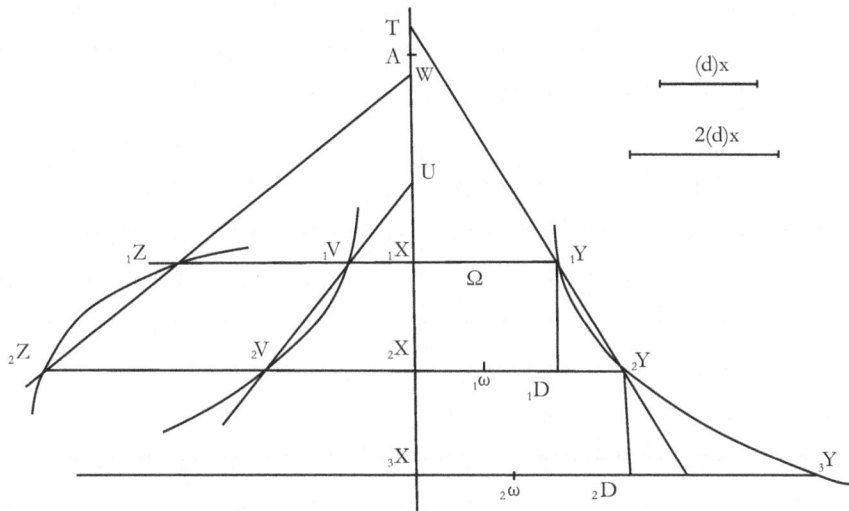

Fig. 12.8 Illustrating the Rules of the Calculus

line which is to $(d)x$ as y to $_1XT$, ôr as dy to dx; and similarly let $(d)v$ be to $(d)x$ as v to $_1XU$, ôr as du to dx, and let $(d)z$ be to $(d)x$ as z to $_1XW$, ôr as dz to dx, and $(d)x$, $(d)y$, $(d)v$, and $(d)z$ will be ordinary ôr assignable straight lines.

Now, *addition* and *subtraction* will go as follows: Let $y - z = v$, then $(d)y - (d)z = (d)v$, which I demonstrate thus: $y + dy - z - dz = v + dv$ (if we suppose that with y increasing, v and z also increase; otherwise, if we have decreasing quantities, such as z, then $-dz$ should be put instead of dz, which I note this once); then, subtracting the equals $y - z$ from the left hand side and v from the right, we obtain $dy - dz = dv$, and so also $(dy - dz):dx = dv:dx$. But $dy:dx$, $dz:dx$ and $dv:dx$ are equal respectively to $(d)y: (d)x$, $(d)z: (d)x$ and $(d)v: (d)x$. Similarly $(d)z: (d)y$ or $(d)v: (d)y$ are equal respectively to $dz:dy$ or $dv:dy$. So we have $(d)y - (d)z,: (d)x = (d)v: (d)x$. And so $(d)y - (d)z = (d)v$, as was proposed, ôr $(d)v: (d)y = 1 - (d)z: (d)y$.

This rule for addition and subtraction also follows from the postulate of a common calculation, when $_1X$ coincides with $_2X$, or when $_1YT$, $_1VU$, or $_1ZW$ are tangents to the curves TT, UU, and WW. Although we may be content with the assignable quantities $(d)y$, $(d)v$, $(d)z$, and $(d)x$, since in this way we can perceive the whole fruit of our calculus, namely a construction using assignable quantities, still it is clear from this that we may, at least by feigning, substitute for them the unassignables dx, dy by way of fiction even in the case where they vanish, since $dy:dx$ can always be reduced to $(d)y: (d)x$, a ratio between assignable or undoubtedly real quantities. And so even in the case of the tangents we have $dv:dy = 1 - dz:dy$, ôr $dv = dy - dz$.

Multiplication: Let $ay = xv$,[40] then we will have $a(d)y = x(d)v + v(d)x$.

Demonstration : $ay + ady = (x + dx)(v + dv)$

$$= xv + xdv + vdx + dxdv,$$

and, subtracting from each side the equals ay and xv, we will have

$$ady = xdv + vdx + dxdv$$

ôr $\quad \dfrac{ady}{dx} = \dfrac{xdv}{dx} + v + dv$

and transposing the case, as we may, to lines that never vanish, we have

$$\dfrac{a(d)y}{(d)x} = \dfrac{x(d)v}{d(x)} + v + dv$$

so that the only remaining term which can vanish is dv, since in the case of vanishing differences, since $dv = 0$, we have

$$a(d)y = x(d)v + v(d)x$$

as was asserted, or $(d)y: (d)x = (x + v):a$. Whence also, because $(d)y: (d)x$ always $= dy:dx$ one may feign this in the case of vanishing dy, dx and make $dy:dx = x + v,:a$, ôr

$$ady = xdv + vdx.$$

Division: Let $z:a = v: x$, then we will have $(d)z:a = v(d)x - x(d)v,:xx$.
  Demonstration: $(z + dz):a = (v + dv)/(x + dx)$
  And, getting rid of the fractions,

$$xz + xdz + zdx + dzdx = av + adv$$

and on subtracting the equals xz and av, and dividing the remainder by dx,

$$(adv{-}xdz)/dx = z + dz \qquad \text{ôr}$$

---

[40] Here the purpose of the constant a in these equations is to preserve their dimensional homogeneity.

12.3 (for 6.3) Justifications of the Differential Algorithm

$$(a(d)v-x(d)z)/(d)x = z + dz,$$

and so the only thing left that can vanish is dz. And in the case of vanishing differences, ôr when $_2X$ is coincident with $_1X$, then, because $dz = 0$, we will have

$$(a(d)v-x(d)z)/dx = z = av/x$$

Hence, as was proposed,

$$(d)z = (ax(d)v-av(d)x)/xx \qquad \text{ôr}$$

$$(d)z : (d)x = (a : x)(d)v/(d)x-av/xx,$$

and because $(d)z: (d)x$ is always equal to $dz:dx$ elsewhere, it will be possible to feign this also in the case of the vanishing quantities dz, dv, dx, and make $dz:dx = (axdv - avdx)/xx$.

For Powers let the equation be $a^{n-e} x^e = y^n$, and we will have $(d)y/(d)x = ex^{e-1}/ny^{n-1}$, which I will demonstrate in a little more detail than the previous ones, as follows.

$$a^{n-e}\left\{\frac{1}{1}x^e + \frac{e}{1}x^{e-1}dx + \frac{e(e-1)}{1\cdot 2}x^{e-2}dxdx + \frac{e(e-1)(e-2)}{1\cdot 2\cdot 3}x^{e-3}dxdxdx\right\}$$

(and so on until the factor $e - e$ ôr 0 is reached)

$$= \frac{1}{1}y^n + \frac{n}{1}y^{n-1}dy + \frac{n,n-1}{1\cdot 2}y^{n-2}dydy + \frac{n,n-1,n-2}{1\cdot 2\cdot 3}y^{n-3}dydydy$$

(and so on until the factor $n-n$ ôr 0 is reached);

Let $a^{n-e}x^e$ be subtracted from one side, and y^n from the other, since they are equal, and let the remainder be divided by dx, and then instead of the ratio $dy:dx$ between two quantities that are continuously diminishing, let there be put the ratio that is equal to it, $(d)y: (d)x$, that is, a ratio between two quantities, one of which, $(d)x$, always remains the same while the differences are diminishing, that is, while the point $_2X$ is approaching the fixed point $_1X$, and we will have

$$(e/1)x^{e-1} + (e, e-1)/(1, 2)\ x^{e-2}dx + (e, e-1, e-2)/(1,2,3)\ x^{e-3}dxdx + \text{etc.}$$

$$= (n/1)y^{n-1}\ d)y : (d)x + (n, n-1)/(1, 2)\ y^{n-2}(d)y$$

$$: (d)x\ dy + (n, n-1, n-2)/(1,2,3)\ y^{n-3}\ d)y : (d)x\ dydy + \text{etc.}$$

Since therefore (by the postulate), also included in this general rule is the case where the differences become equal to nothing, ôr where the points $_2X$, $_2Y$ become coincident with the points $_1X$, $_1Y$ respectively, therefore putting dx and $dy = 0$ in this case, we will have

$$(e/1)x^{e-1} = (n/1)y^{n-1} \; (d)y : (d)x$$

with the other terms vanishing, ôr

$$(d)y : (d)x = e \, x^{e-1} : ny^{n-1},$$

as was proposed. But as we have explained, the ratio $(d)y : (d)x$ is the same as the ratio of y ôr the ordinate $_1X_1Y$ to the subtangent $_1XT$, supposing T_1Y to touch the curve at $_1Y$.

This demonstration holds good whether the elevations are powers or roots whose exponents are fractions—although one may also remove fractional exponents from the equation by raising each side of the equation to some power, so that e and n then denote nothing but integral powers with rational exponents, which by the continual subtraction of the numbers 1, 2, 3, etc. are finally exhausted,[41] and there will be no need for a series proceeding to infinity. But anyway, by means of a fiction in the way explained above, one may also have recourse to the unassignables dx and dy, by making—in the case of vanishing differences, just as in all the other cases—the ratio of the vanishing differences dx and dy equal to that of the non-vanishing ones $(d)y$ and $(d)x$, since this fiction can always be reduced to an indubitable truth.

Thus far the algorithm has been demonstrated for differences of the first degree; now it must be shown that the same method is also valid for differences of differences. To this end let us assume three ordinates, $_1X_1Y$, $_2X_2Y$, $_3X_3Y$, of which $_1X_1Y$ remains fixed but $_2X_2Y$ and $_3X_3Y$ continuously approach it until both of them are coincident with it. This will occur if the speed with which $_3X$ approaches $_1X$ is to the [speed] with which $_2X$ approaches $_1X$ in the ratio $_1X_3X$ to $_1X_2X$. Also let two straight lines be assigned, $(d)x$ always the same wherever $_2X$ is situated, and $_2(d)x$ always the same wherever $_3X$ is situated, and let $(d)y$ to $(d)x$ always be as D_2Y is to $_1X_2X$, ôr as y (i.e. $_1X_1Y$) is to $_1XT$, so that with $(d)y$ remaining fixed, $(d)x$ is always changing while $_2X$ approaches $_1X$. Similarly, let $_2(d)y$ to $_2(d)x$ be as $_2D_3Y$ to $_2X_3X$, ôr as $y + dy$ (i.e. $_2X_2Y$) is to $_2X_2T$, so that with $_2(d)y$ remaining fixed, $_2(d)x$ is always changing while $_3X$ approaches $_1X$ (Fig. 12.9).

Moreover, let $(d)y$ always be taken in the varying straight line $_2X_2Y$, and let $_2X_1\omega$ be equal to $(d)y$, and similarly let $_2(d)y$ be taken in the varying line $_3X_3Y$, and let $_3X_2\omega$ be equal to $_2(d)y$. Thus while $_2X$ and $_3X$ continuously approach the straight line $_1X_1Y$, $_2X_1\omega$ and $_3X_2\omega$ will also continuously approach it, and finally coincide in it with $_2X$ and $_3X$. Further,

---

[41] The clause "which by the continual … exhausted" is missing from Gerhardt's transcription.

12.3 (for 6.3) Justifications of the Differential Algorithm

Fig. 12.9 Illustrating Second Differences

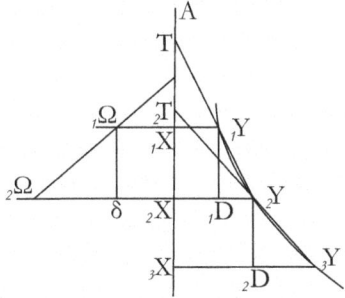

let us mark the point where $_1\omega$, continuously approaching the ordinate $_1X\ _1Y$, falls on it, and let that point be Ω. Then $_1X\Omega$ will be the ultimate value of $(d)y$, which is to the fixed line $(d)x$ as the ordinate $_1X\ _1Y$ is to the subtangent $_1XT$, supposing T_1Y touches the curve YY in $_1Y$, since then $_1Y$ and $_2Y$ will coincide. And since this can happen wherever the point $_1Y$ is assumed on the curve, it is clear that in this way a curve $\Omega\Omega$ will have been produced that is the *differentiatrix* of the curve YY, just as the curve YY is in turn the *summatrix* of the curve $\Omega\Omega$, as can easily be shown.

By the same method one can also demonstrate the calculus for differences of differences. Let there be three ordinates $_1X\ _1Y$, $_2X_2Y$, $_3X_3Y$, whose values are y, $y + dy$, and $y + dy + ddy$, and let $_1X\ _2X(dx)$ and $_2X_3X(dx + ddx)$ be any distances, and let the differences D_2Y and $_2D_3Y$ be dy and $dy + ddy$, respectively. Now the difference between $(d)y$ and $_2(d)y$, ôr between $[_1X_1\Omega]$ and $_2X_2\Omega$, is $\delta_2\Omega$, and that between $_1X\ _2X$ and $_2X_3X$, is ddx. Also let $(d)dx$ to $(d)x$ be as dx to $_2(d)x$,[42] and similarly let $(d)dy$ to $(d)x$ be as $_2\Omega\delta$ to $_1X\ _2X$, ôr $_1X_1\Omega$ to $_1XT$.

For the sake of example, let $ay = xv$. As shown above, we will have

$$ady = xdv + vdx + dxdv$$

and similarly,

$$ady + addy = (x + dx)(dv + ddv) + ((v + dv)(dx + ddx) + (dx + ddx)(dv + ddv)$$

$$= xdv + xddv + dxdv + dx \cdot ddv + vdx + vddx + dv \cdot dx + dv \cdot ddx + dx \cdot dv + dx \cdot ddv + ddx \cdot d + ddx \cdot ddv$$

and taking away ady from one side, and from the other $xdv + vdx + dxdv$, there will in any case remain

---

[42] Leibniz sometimes writes $_2(d)x$ for $(dd)x$, a potentially confusing notation.

$$\frac{ddy}{ddx} = \frac{x\,ddv}{a\,ddx} + \frac{v}{a} + \frac{2dx\,dv}{a\,ddx} + \frac{2dv}{a} + \frac{2\,dx\,ddv}{a\,ddx} + \frac{ddv}{a}$$

where it is evident that the ratio between ddy and ddx can be expressed by the ratio of the straight line $(d)dy$ to $[(d)dx]$, the straight line assumed above, which we have supposed to remain fixed as $_2X$ and $_3X$ approach $_1X$. So also $(d)dx$—since it bears an assignable ratio to $(d)x$, however nearly $_2X$ approaches to $_1X$, or however much dx, the difference between the abscissae, is diminished)—does not vanish even when, finally, dx, ddx, dv, and ddv, are all assumed equal to 0. In the same way, the ratio ddv:ddx may be expressed by the ratio of an assignable straight line $(d)dv$ to the assumed constant $[(d)dx]$; and even the ratio of $dv \cdot dx$ to $a \cdot ddx$ may be so expressed; for, since dv:$dx = (d)v$: $(d)x$, then $dv \cdot dx$:$dx \cdot dx = (d)v$: $(d)x$, we only need assume a new straight line, $(dd)x$, such that $a\,ddx$: $dxdx = (dd)x$: $(d)x$, and then the new straight line will remain assignable even if dx, ddx, etc., become evanescent. Since therefore $dv \cdot dx$:$dx \cdot dx = (d)v$: $(d)x$ and $dxdx$:$addx = (d)x$:$(dd)x$, it follows that $dv \cdot dx$: $a \cdot ddx = (d)v$:$(dd)x$, and so finally an equation will be produced that is purged as far as possible of those ratios that can vanish, namely[43]

$$\frac{(d)dy}{(d)dx} = \frac{x\,(d)dv}{a\,(d)dx} + \frac{v}{a} + \frac{2\,(d)v}{(dd)x} + \frac{2dv}{a} + \frac{2\,dx\,(d)dv}{a\,(d)dx} + \frac{ddv}{a}$$

Thus far all the straight lines have been considered to be assignable so long as $_1X$ and $_2X$ do not coincide; but in the case of coincidence, dv and ddv are zero, and we have

$$\frac{(d)dy}{(d)dx} = \frac{x\,(d)dv}{a\,(d)dx} + \frac{v}{a} + \frac{2\,(d)v}{(dd)x} + \frac{0}{a} + \frac{2\,dx\,(d)0}{a\,(d)dx} + \frac{0}{a}$$

or, omitting terms equal to zero,

$$\frac{(d)dy}{(d)dx} = \frac{x\,(d)dv}{a\,(d)dx} + \frac{v}{a} + \frac{2\,(d)v}{(dd)x}$$

Hence, if dx, ddx, dv, ddv, dy, and ddy, are by a certain fiction imagined to remain, even when they become evanescent, as if they were infinitely small quantities (for this may be

---

[43] Comparing the following equations with those given in Child's translation, we note that on several occasions he has written the variable y in place of v on the left (or vice versa on the right); see Child (1920).

12.3 (for 6.3) Justifications of the Differential Algorithm

done with no danger, since the whole matter can always be referred back to assignable quantities), then in the case of the points $_1X$ and $_2X$ coinciding we have the equation[44]

$$\frac{ddy}{ddx} = \frac{x\,ddv}{a\,ddx} + \frac{v}{a} + \frac{2dx\,dv}{a\,ddx}$$

---

[44] That would yield the unstated conclusion that $a \cdot ddy = x \cdot ddv + 2\,dv \cdot dx + v \cdot ddx$. The formula is incorrect, a result of Leibniz's having assumed that in the limit as $dx \to 0$, $(dd)y = (d)y$. As Henk Bos has pointed out, however, this is the case only if in that limit $(dd)x = (d)x$. This requires x to be an independent variable, so that dx is constant and $ddx = 0$. See (Bos 1974/5) and (Arthur 2013) for a corrected derivation using this assumption and Leibniz's method of finite surrogates, yielding the correct formula $a \cdot ddy = x \cdot ddv + 2\,dv \cdot dx$, in the case where x is the independent variable. (In this and all the preceding formulas, the use of the dot '·' to disambiguate the expressions is our own addition.)

Bibliography

Andersen, Kirsti. 1985. "Cavalieri's Method of Indivisibles", *Archive for History of Exact Sciences*, **31**, 4, pp. 291-367.
Antognazza, Maria Rosa. 2015. "The Hypercategorematic Infinite," *Leibniz Review* **25**: 5–30.
Arthur, Richard T. W. 2006. "The remarkable fecundity of Leibniz's work on infinite series," *Annals of Science* **63**(2): 221–225.
Arthur, Richard T. W. 2008. "Leery Bedfellows: Newton and Leibniz on the status of infinitesimals. In (Goldenbaum and Jesseph 2008), 7–30.
Arthur, Richard T. W. 2009. "Actual Infinitesimals in Leibniz's Early Thought", 11-28, in *The Philosophy of the Young Leibniz, Studia Leibnitiana Sonderhefte* **35**, ed. Mark Kulstad, Mogens Laerke and David Snyder. Stuttgart: Franz Steiner, 2009
Arthur, Richard T. W. 2013. "Leibniz's syncategorematic infinitesimals, smooth infinitesimal analysis, and second order differentials", *Archive for History of Exact Sciences* **67**: 553–593.
Arthur, Richard T. W. 2014. *Leibniz*. Classic Thinkers Series. Cambridge: Polity Press.
Arthur, Richard T. W. 2018. *Monads, Composition, and Force*. Oxford: Oxford University Press.
Arthur, Richard T. W. 2021. *Leibniz on Time, Space, and Relativity*. Oxford: Oxford University Press.
Arthur, Richard and Osvaldo Ottaviani. (forthcoming). *Leibniz on the Metaphysics of the Infinite*. In *BSHP New Texts in the History of Philosophy*. Oxford University Press: Oxford.
Bassler, O. Bradley. 2008. "An Enticing (Im)Possibility: Infinitesimals, Differentials, and the Leibnizian Calculus". In (Goldenbaum and Jesseph 2008), 135–152.
Beeley, Philip. 1996. *Kontinuität und Mechanismus. Zur Philosophie des jungen Leibniz in ihrem ideengeschichtlichen Kontext*. Stuttgart: Franz Steiner.
Beeley, Philip. 2015. "Leibniz, Philosopher Mathematician and Mathematical Philosopher", 23–48, in *G. W. Leibniz, interrelations between Mathematics and Philosophy*, ed. Norma Goethe, Philip Beeley and David Rabouin. Berlin/New-York: Springer.
Bell, J. L. 2006. *The Continuous and the Infinitesimal in mathematics and philosophy*. Milan: Polimetrica.
Bella, Sandra. 2022. *La (Re)construction française de l'analyse infinitésimale de Leibniz (1690-1706)*. Paris: Garnier Classiques.
Bernoulli, Johann. 1988. *Der Briefwechsel von Johann I Bernoulli. Band 2. Briefwechsel mit Pierre Varignon. Erster Teil: 1692-1702*. Edited and commented by P. Costabel and J. Peiffer. Basel: Springer.

Blay, Michel. 1986. "Deux moments de la critique du calcul infinitésimal : Michel Rolle et George Berkeley", *Revue d'Histoire des Sciences* **39** (3): 223–253.
Blåsjö, Viktor. 2017. "On what has been called Leibniz's rigorous foundation of infinitesimal geometry by means of Riemannian sums." *Historia Mathematica* **44** (2): 134–149.
Bos, H. J. M. 1974–75. "Differentials, higher-order differential and the derivative in the Leibnizian calculus. *Archive for History of Exact sciences* **14**: 1–90.
Bosinelli, Fabio C. M. 1991. "Über Leibniz' Unendlichkeitstheorie," *Studia Leibnitiana* **23** (2): 151–169.
Boyer, Carl B. 1949. *History of the Calculus and its Conceptual Development.* New York: Dover.
Breger, Herbert. 1986. "Leibniz, Weyl und das Kontinuum," 316-330, in *Beiträge zur Wirkungs- und Rezeptionsgeschichte von Gottfried Wilhelm Leibniz, Studia Leibnitiana Supplementa* **26**, ed. Albert Heinekamp. Stuttgart : Franz Steiner.
Breger, Herbert. 1992. "Le continu chez Leibniz," 76-84, in *Le labyrinthe du continu*, ed. Jean-Michel Salanskis et Hourya Sinaceur. Paris: Springer Verlag; reprinted in (Breger 2016), 127–135.
Breger, Herbert. 2016. *Kontinuum, Analysis, Informales—Beiträge zur Mathematik und Philosophie von Leibniz*, ed. Wenchao Li. Berlin-Heidelberg, Springer Spektrum.
Cardano, Girolamo. 1545. *Ars Magna, Sive de Regulis Algebraicis Liber Unus*. Nüremberg: Petreius.
Caramuel y Lobkowitz, Juan. 1670. *Mathesis biceps*. Leiden: Anisson.
Cavalieri, Bonaventura. 1635. *Geometria indivisibilibus continuorum nova quadam ratione promota.* Bologna: Clemens Ferronius.
Cavalieri, Bonaventura. 1647. *Exercitationes geometricae sex.* Bologna: Jacobus Montius.
Child, J. M. 1920. *The Early Mathematical Manuscripts of Leibniz.* Chicago: Open Court.
Cortese, João, and Rabouin, David. 2019. "Sur les indivisibles chez Pascal," 425–440, in *Passions géométriques. Mélanges en l'honneur de D. Descotes*, ed. Agnès Cousson. Paris: Champion.
Costabel, Pierre. 1990. "Idée et fiction : sondages dans la mathématique de l'art analytique," 137–153, in *Idea*, VI Colloquio Internazionale , ed. par M. Fattori et M.L. Bianchi. Rome: Edizioni dell'Ateneo.
Crippa, Davide. 2018a. *The Impossibility of Squaring the Circle in the Seventeenth Century.* Basel: Birkhaüser.
Crippa, Davide. 2018b. "On Leibniz's Theorem about the Impossibility of Squaring The Circle and its Relation with James Gregory's *Vera Circuli Quadratura*". *Quaderns d'Història de l'Enginyeria* **16**: 209–232.
D'Alembert. 1765. *L'Encyclopédie ou Dictionnaire raisonné des sciences, des arts et des métiers.* Paris: Briasson, David l'aîné, Le Breton, Durand. Art. « Négatif ».
Descartes, René. 1954. *The Geometry of René Descartes.* Engl. Transl. by Smith and Latham. New-York: Dover.
Descartes, René. 1964-76. *Oeuvres de Descartes.* ed. Charles Adam & Paul Tannery. Paris: J. Vrin. Cited as **AT**.
Diophantus. 1575. *Diophanti alexandrini rerum arithmeticarum libri sex, et de numeris multangulis.* Ed. Wilhelm Xylander. Basel: Eusebius Episcopius and the heirs of Nicolaus Episcopius.
Esquisabel, Oscar and Federico Raffo Quintana. 2017. "Leibniz in Paris: A Discussion Concerning the Infinite Number of All Units," *Revista Portuguesa de Filosofia* **73** (3-4): 1319-1342.
Esquisabel, Oscar and Federico Raffo Quintana. 2021. "Fiction, possibility and impossibility: Three kinds of mathematical fictions in Leibniz," *Archive for History of Exact Sciences* **75**: 613–647.
Fabry, Honoré. 1669. *Synopsis geometrica, cui accessere tria opuscula, nimirum, de linea sinuum et cycloide, de maximis et minimis centuria, et synopsis trigonometriae planae.* Leiden: Anthony Molin.

Fatio de Duillier, Nicolas 1699. *Lineæ Brevissimi Descensus Investigatio Geometrica Duplex: cui addita est Investigatio Geometrica Solidi Rotundi in quod Minima Fiat Resistentia*. London: R. Everingham, for J. Taylor.
Fermat, Pierre de. 1679. *Methodus ad disquirendam maximam et minimam et de tangentibus linearum curvarum*; published in *Varia opera mathematica*. Toulouse: apud Johannem Pech.
Gerhardt, Carl Immanuel. (ed.). 1846. *Historia et Origo a G.G. Leibnitio*. Hanover: Hahn.
Gerhardt, Carl Immanuel. (ed.). 1848. *Die Entdeckung der Differentialrechnung durch Leibniz*. Halle: Schmidt.
Gerhardt, Carl Immanuel. 1876. "Zum zweihundertjährigen Jubiläum der Entdeckung des Algorithmus der höheren Analysis durch Leibniz," *Mon.-ber. Kön. Preuss. Akad. Wiss.* Berlin (28 Oct 1875): 588–608.
Goethe, Norma, Philip Beeley, and David Rabouin (eds.). 2015. *G.W. Leibniz: Interrelations between mathematics and philosophy*. Archimedes series, vol. 41. Dordrecht: Springer.
Goldenbaum, Ursula, and Douglas Jesseph (eds.). 2008. *Infinitesimal differences: Controversies between Leibniz and his contemporaries*. Berlin/New York: De Gruyter.
Gottignies, G. F. de. 1687. *Logistica universalis*. Naples, Novelli de Bonis.
Gouye, Thomas. 1701. "*Nouvelle méthode pour déterminer aisément les rayons de la developée dans toute sorte de courbe algebraique*," *Journal de Trévoux*, *422–*430 (in the 1702 edition; 5–15 in supplement to the 1701 edition).
Grosholz, Emily. 2007. *Representation and productive ambiguity in mathematics and the sciences*. Oxford: Clarendon Press.
Hess, Heinz-Jürgen. 1991. "Maturing in Retirement. The unknown period of the Leibnizian Calculus between Paris and publication (1676–1684)". *Giornate di Storia della Matematica, Commenda di Rende: Editoria Elettronica*, 1991, 247–288.
Hobbes, Thomas. 1655. *Elementorum philosophiae sectio prima De corpore*. London.
Hofmann, J. E. 1974. *Leibniz in Paris 1672-1676: His growth to mathematical maturity*. Revised translation of the German edition of 1949. Cambridge: Cambridge Univ. Press.
Huygens, Christiaan. 1673. *Horologium oscillatorium sive de motu pendulorum ad horologia aptato demonstrationes geometricæ*. Paris: F. Muguet.
Ishiguro, Hidé. 1990. *Leibniz's Philosophy of Logic and Language*. 2nd edition. Cambridge: Cambridge University Press.
Jesseph, Douglas M. 1998. "Leibniz and the Foundations of the Calculus: The Question of the Reality of Infinitesimal Magnitudes," *Perspectives on Science* **6**: 6–40.
Jesseph, Douglas M. 1999. "The Decline and Fall of Hobbesian Geometry," *Studies in History and Philosophy of Science* **30**: 425–453.
Jesseph, Douglas. 2008. "Truth in Fiction: Origins and Consequences of Leibniz's Doctrine of Infinitesimal Magnitudes," in (Goldenbaum and Jesseph 2008), 215–233.
Jesseph, Douglas. 2015. "Leibniz on the elimination of infinitesimals," in (Goethe, Beeley and Rabouin 2015), 189–205.
Katz, Mikhail G. and David Sherry. 2012. "Leibniz's laws of continuity and homogeneity", *Notices of the American Mathematical Society*, **59** (11): 1550-1558.
Katz, Mikhail G. and David Sherry. 2013. "Leibniz's infinitesimals: Their fictionality, their modern implementations, and their foes from Berkeley to Russell and beyond." *Erkenntnis* **78** (3): 571–625.
Katz, Mikhail G., Karl Kuhlemann, David Sherry, and Monica Ugaglia, 2021. "Two-track depictions of Leibniz's fictions," *The Mathematical Intelligencer*. Published online first 02 December 2021: 1–6; and 2022, **44** (3): 261-266.

Keill, John. 1708. "*Epistola ad Clarissimum Virum Edmundum Hallejum Geometriæ Professorum Savillianum: De Legibus Virum Centripetarum,*" *Philosophical Transactions* **26**: 174-88. (Issued in 1710).
Kline, Morris. 1972. *Mathematical Thought from Ancient to Modern Times. Volume 1.* Oxford: Oxford University Press.
Knobloch, Eberhard. 1976. *Ein Dialog zur Einführung in die Arithmetik und Algebra.* Stuttgart: Frommann–Holzboog.
Knobloch, Eberhard. 2001. "*Déterminants et élimination chez Leibniz,*" *Revue d'histoire des sciences* **54** (2): 143-164.
Knobloch, Eberhard. 2002. "Leibniz's rigorous foundation of infinitesimal geometry by means of Riemannian sums," *Synthese* **133**: 59–73.
Kracht, Manfred and Erwin Kreyszig. 1990. "E. W. von Tschirnhaus: His Role in Early Calculus and His Work and Impact on Algebra," *Historia Mathematica* **17**, 16-35.
Lagrange, Joseph-Louis. 1797. *Théorie des fonctions analytiques.* Paris: Imprimerie de la République.
Leibniz, G. W. 1682. "*De vera proportione circuli ad quadratum circumscriptum in numeris rationalibus expressa,* [On the True Proportion of the Circle to the Circumscribed Square Expressed in Rational Numbers]," *Acta eruditorum*, February 1682, 41–46.
Leibniz, G. W. 1684a. "*De dimensionibus figurarum inveniendis* [On finding the measures of figures]," *Acta eruditorum*, May 1684, 233–36.
Leibniz, G. W. 1684b. "*Nova methodus pro maximis et minimis, itemque tangentibus, quae nec fractas, nec irrationales quantitates moratur, et singulare pro illis calculi genus* [A New Method for Maxima et Minima, as well as tangents, unhindered by either fractions or irrational quantities, and a singular kind of calculus for finding them]," *Acta eruditorum*, October 1684, 467–473.
Leibniz, G. W. 1689. "*Tentamen de motuum coelestium causis* [An Essay on the Causes of the Celestial Motions]," *Acta eruditorum*, February 1689, 82–96.
Leibniz, Gottfried. 1705. "*Nouvelle Arithmetique Binaire* [A New binary arithmetic]," *Histoire de l'Academie des Sciences, avec les Mémoires, Année MDCCIII*, (published in the third person). Paris: Jean Boudot, 58–63.
Leibniz, G. W. 1768. *Opera Omnia. Nunc primum collecta....* Ed. Ludovicus Dutens. 6 vols. Geneva: De Tournes; repr. Hildesheim: Georg Olms, 1989.
Leibniz, G. W. 1846. "*Historia et Origo Calculi Differentialis,*" in C. I. Gerhardt (1846).
Leibniz, G. W. 1849-63. *Leibnizens Mathematische Schriften.* Ed. C. I. Gerhardt. Berlin and Halle: Asher and Schmidt, 1849-63; reprint ed. Hildesheim: Georg Olms, 1971. 7 vols. Cited as **GM**.
Leibniz, G. W. 1875-90. *Die Philosophische Schriften von Gottfried Wilhelm Leibniz.* Ed. C.-I. Gerhardt. Berlin: Weidmann; reprint ed. Hildesheim/New York: Georg Olms, 1978. 7 vols. Cited as **GP**.
Leibniz, G. W. 1895. *Die Leibniz-Handschriften der königlichen öffentlichen Bibliothek zu Hannover.* Ed. Eduard Bodemann. Hannover and Leipzig: Hann'sche Buchhandlung. Cited as **LH**.
Leibniz, G. W. 1923–. *Sämtliche Schriften und Briefe,* ed. Berlin-Brandenburgischen Akademie der Wissenschaften und der Akademie der Wissenschaften zu Göttingen, Reihe 1-8, Darmstadt, Leipzig, Berlin; cited as **A**.
Leibniz, G. W. 1956. *The Leibniz-Clarke Correspondence* Ed. H. G. Alexander. Manchester: Manchester University Press.
Leibniz, G. W. 1969. *Philosophical Papers and Letters,* 2nd. ed., 1976. Edited and translated by Leroy Loemker. Dordrecht: D. Reidel. Cited as **L**.
Leibniz, G. W. 1981. *New Essays on Human Understanding.* Ed. and tr. Peter Remnant and Jonathan Bennett. Cambridge: Cambridge University Press. Cited as **NE**.

Leibniz, G. W. 1992. *De Summa Rerum: Metaphysical Papers 1675-1676*. Translated with an introduction by G. H. R. Parkinson. New Haven: Yale University Press. Cited as **DSR**.

Leibniz, G. W. 1993. *De quadratura arithmetica circuli ellipseos et hyperbolae cujus corollarium est trigonometria sine tabulis*. Edition and commentary by Eberhard Knobloch. Göttingen: Vandenhoek & Ruprecht.

Leibniz, G. W. 1995. *La Naissance du Calcul Différentiel*. Introduction, traduction et notes par Marc Parmentier. Paris: Vrin.

Leibniz, G. W. 1996. "*Geschichte des Kontinuumproblems*," herausgegen und eingeleitet von M. L. Alcoba, *Studia Leibnitiana* 37 (2): 183–198.

Leibniz, G. W. 2001. *The Labyrinth of the Continuum: Writings of 1672 to 1686*. Selected, edited and translated, with an introductory essay, by R. T. W. Arthur. New Haven: Yale University Press. Cited as **LLC**.

Leibniz, G. W. 2007. *The Leibniz-Des Bosses Correspondence*. Selected, edited and translated, with an introductory essay, by Brandon Look and Don Rutherford. New Haven: Yale University Press. Cited as **LDB**.

Leibniz, G. W. 2018. *Mathesis Universalis*; écrits sur la mathématique universelle. Textes introduits, traduits et annotées sous la direction de David Rabouin. Paris: Vrin.

Leibniz, G. W. 2021. *Mathesis. Transkriptionen und Vorauseditionen mathematischer Schriften für die Leibniz-Akademie-Ausgabe*. Version 1. Bearbeitet von Sandra Bella, Mattia Brancato, Davide Crippa, Vincenzo De Risi, Siegmund Probst und Achim Trunk unter Verwendung von Vorarbeiten von Vincenzo De Risi, Javier Echeverría und den Editionsstellen in Hannover und Münster, hrsg. von der Leibniz-Forschungsstelle Hannover der Akademie der Wissenschaften zu Göttingen beim Leibniz-Archiv der Gottfried Wilhelm Leibniz Bibliothek-Niedersächsische Landesbibliothek. Hannover, 31. August 20. **Mathesis preprint.**https://www.gwlb.de/leibniz/digitale-ressourcen/repositorium-des-leibniz-archivs/laa-mathesis.

Leibniz, G. W. 2023. *Journal Articles in Natural Philosophy*. Ed. with an introduction by Richard T. W. Arthur. Oxford: Oxford University Press.

Levey, Samuel. 2008. "Archimedes, infinitesimals and the law of continuity: On Leibniz's Fictionalism," in (Goldenbaum and Jesseph 2008), 107–133.

L'Hôpital, Guillaume, Marquis de. 1696. *Analyse des Infiniment Petits pour l'Intelligence des Lignes Courbes*. Paris: Imprimerie Royale.

Mancosu, Paolo. 1989. "The metaphysics of the calculus: A foundational debate in the Paris Academy of Sciences, 1700-1706," *Historia Mathematica*, **18**: 224-248.

Mancosu, Paolo. 1996. *Philosophy of mathematics and mathematical practice in the seventeenth century*. Oxford/New York: Oxford University Press.

Mancosu, Paolo, and Ezio Vailati. 1990. "Detleff Clüver: An early opponent of the Leibnizian differential calculus," *Centaurus* 33: 325–344.

Mercator, Nicolaus 1668. *Logarithmo-technia, sive methodus construendi Logarithmos Nova, Accurata, & Facilis*. London: Moses Pitt.

Ottaviani, Osvaldo. 2021. "Leibniz's Imaginary Bridge: The Analogy between Pure Possibles and Imaginary Numbers in the Paris Writings," *Oxford Studies in Early Modern Philosophy* 20: 133-168.

Pappus. 1986. *Book 7 of the Collection*, transl. A. Jones. Berlin/Heidelberg: Springer.

Pascal, Blaise. 1659. *Lettres de A. Dettonville contenant Quelques-vnes de ses Inuentions de Geometrie*. Paris: Guillaume Desprez.

Pasini, Enrico. 1985-86. *La nozione di infinitesimo in Leibniz tra matematica e metafisica*, Dissertation, Torino.

Pasini, Enrico. 1988. "*Die private kontroverse des GW Leibniz mit sich selbst. Handschriften über die infinitesimalrechnung im Jahre 1702*," *Leibniz. Tradition und Aktualität*. Hannover: Leibniz-Gesellschaft, 695–709.

Pasini, Enrico. 1993. *Il reale e l'immaginario. La fondazione del calculo infinitesimale ner pensiero di Leibniz*. Milano/Torino: Edizioni Sonda.

Prestet, Jean. 1675. *Elemens des mathematiques ou principes generaux de toutes les sciences qui ont les grandeurs pour objet*. Paris: A. Pralard.

Probst, Siegmund. 2006. "*Zur Datierung von Leibniz' Entdeckung der Kreisreihe*". In *Einheit in der Vielheit*. Eds. Jürgen Herbst, Herbert Breger and Sven Erdner, (Hannover, Vorträge des VIII. Internationalen Leibniz-Kongresses, 2006), 813-817.

Probst, Siegmund. 2008. "Indivisibles and Infinitesimals in Early Mathematical Texts of Leibniz." In (Goldenbaum and Jesseph 2008), 95–106.

Probst, Siegmund. 2012. "*Leibniz und die Cartesische Geometrie (1673-1676)*". In *Zeitläufte der Mathematik Tagung zur Geschichte der Mathematik* (Freising 2011). Eds. Hans Fischer and Stefan Deschauer. Augsburg: Erwin Rauner Verlag, 149–158.

Proclus of Lycia. 1533. *Commentarium Procli editio prima quae Simonis Grynaei opera addita est Euclidis elementis graece editis*. Basle: J. Hervagius. Engl. Transl. by G. Morrow. 1970. Proclus: A Commentary on the First Book of Euclid's Elements. Princeton: Princeton Univeristy Press.

Quine, Willard V. O. 1953. "On What There Is," in *From a Logical Point of View*. Cambridge, Mass.: Harvard University Press.

Rabouin, David. 2011. "*Infini mathématique et infini métaphysique : d'un bon usage de Leibniz pour lire Cues (… et d'autres)*," *Revue de métaphysique et de morale* 70: 203–220.

Rabouin, David. 2015. "Leibniz's rigorous foundations of the method of indivisibles, or how to reason with impossible notions," 347–364, in *Seventeenth-century indivisibles revisited*, ed. Vincent Jullien, (Science Networks. Historical Studies, vol. 49), Cham, Switzerland: Birkhäuser.

Rabouin, David. 2022. "Can one be a fictionalist and a platonist at the same time? Lessons from Leibniz," *Noesis* 38: 161-194.

Rabouin, David. (2024). "Negatives as fictions in 16th and 17th Century mathematics". Historia mathematica **69**: 41–61.

Raffo Quintana, Federico. 2016. "*La infinitud actual de partes del continuo en la Theoria motus abstracti de Leibniz*," *Thémata. Revista de Filosofía* 53 : 289-310.

Raffo Quintana, Federico. 2018. "Leibniz on the requisites of an exact arithmetical quadrature," *Studies in History and Philosophy of Science* 67: 65–73.

Robinet, André. 1986. Architectonique disjonctive, automates systémiques et idéalité transcendentale dans l'oeuvre de Leibniz. Paris: Vrin.

Robinson, Abraham. 1966. *Nonstandard Analysis*, Amsterdam: North-Holland Publishing Company.

Schooten, Francis van. 1659. *Principia matheosos universalis seu introductio ad Cartesianæ Geometriæ Methodum, Conscripta ab Erasmio Bartholino*. Leiden: Elzevir.

Sereda, Kyle. 2015. "Leibniz's Relational Conception of Number," *Leibniz Review* 25: 31–54.

Stedall, Jacqueline A. 2000. "Catching Proteus: The Collaborations of Wallis and Brouncker. I. Squaring the Circle," *Notes and Records of the Royal Society of London*, Sept., **54**, 3, 293-316.

Stevin, Simon. 1585. *Arithmétique*. Leiden: Christophe Plantin.

Stifel, Michael. 1545. *Deutsche Arithmetica*. Nürnberg: Johan Petreius.

Struik, Dirk. 1986. *A Sourcebook in Mathematics, 1200-1800*. Princeton: Princeton University Press.

Tho, Tzuchien. 2012. "Equivocation in the foundations of Leibniz's infinitesimal fictions," *Society and Politics*, **8** (2): 63–87.

Tschirnhaus, Ehrenfried Walther von. 1683. "*Methodus datae figurae, rectis lineis & Curva Geometrica terminatae, aut Quadraturam, aut impossibilitatem ejusdem Quadraturae determinandi*," *Acta eruditorum*, October 1683, 433-437.

Tschirnhaus, Ehrenfried Walther von. 1687. *Medicina Mentis: sive Tentamen genuinæ logicæ in qua differetur Methodo detegendi incognitas veritates*. Amsterdam: apud Albertum Magnum et Joannum Rieuwerts Juniorem.

Uckelman, Sara L. 2015. "The logic of categorematic and syncategorematic infinity," *Synthese* 192: 2361–2377.

Ugaglia, Monica. 2022. "Possibility vs. Iterativity: Leibniz and Aristotle on the Infinite," 255-270, in F. Ademollo et al. (eds.), *Thinking and Calculating*, (Cham: Springer, 2022).

Wallis, John. 1656. *Arithmetica infinitorum*. Oxford: Leonard Lichfield for Thomas Robinson. Engl. Transl. by J. Stedall. 2004. *The Arithmetic of Infinitesimals*. New-York: Springer.

Wallis, John. 1657. *Mathesis universalis, Operum Mathematicarum Pars Prima*. Oxford: Thomas Robinson.

Wallis, John. 1685. *A Treatise of Algebra, both historical and practical: shewing the original, progress, and advancement thereof*. London: John Playford for Richard Davies.

Weil, André 1975. Review of *Leibniz in Paris 1672-1676, his growth to mathematical maturity*, by Joseph E. Hofmann, Cambridge University Press, 1974, *Bull. Amer. Math. Soc.* 81(4) (July 1975) : 676-688.

Whiteside, Derek Thomas. 1961. "Patterns of mathematical thought in the later seventeenth century." *Archive for history of exact sciences* 1: 179–388.

Author Index

A
Alexander, H.G., 48
Alsted, J.H., 53
Andersen, K., 81
Antognazza, M.R., 137
Archimedes of Syracuse
Arthur, R.T.W., 5, 16, 18, 23, 24, 29, 47, 49, 59, 80, 81, 127, 129, 133, 136, 140, 141, 153, 155, 158, 218, 229, 271

B
Bachet de Méziriac, Claude Gaspar
Basnage de Bauval, H., 94
Bassler, O.B., 15, 123, 124, 126, 215
Bayle, P., 6, 237, 248
Beeley, P., 24
Bell, J.L., 6, 12
Bella, S., 17, 130, 133, 143
Berkeley, G. Bishop, 157, 160
Bernoulli, Jacob, 107, 116, 234
Bernoulli, Johann, 9, 15–18, 24, 41, 47, 57, 58, 60, 72, 75, 93, 96, 97, 107, 112, 116, 119, 122, 123, 125, 126, 129, 133, 136, 143, 160, 161, 186, 220–229, 237, 238, 259
Billy, J. de, 53
Bläsjø, V., 11
Blay, M., 133
Bodenhausen, R.C. von, 12, 78, 82, 93, 94, 96, 149, 151, 195
Bos, H.J.M., 16, 80, 104, 105, 110, 146, 153, 154, 271
Bosinelli, F., 19, 137
Boyer, C., 4

Breger, H., 19, 137
Brouncker, W., 38, 140

C
Caramuel y Lobkowitz, J., 18, 55
Cardano, G., 52, 53, 65, 68
Caroline, Princess, 48
Catelan, François, A. de, 17, 107
Cavalieri, B., 28, 29, 57, 60, 62–64, 79, 81, 119, 143, 190, 193, 259, 262
Child, J.M., 3, 103–105, 149, 261, 270
Clarke, S., 48, 65
Clavius, C., 130, 259
Clüver, D., 17, 143, 259
Costabel, P., 51
Costantini, F., 114
Crafft, J.D., 99
Craig, J., 12
Crippa, D., 37, 42

D
D'Alembert, J. Le R., 53, 157
Dangicourt, P., 7, 112
De Volder, B., 96, 223, 224
Des Bosses, B., 4, 19, 20, 134, 137, 140–142, 232–234
Desargues, G., 90, 190
Descartes, R., 13, 21, 30, 52, 56, 60, 69, 94, 105, 143, 193, 237, 260, 262
Diophantus, 53, 54
Dutens, Louis, 112

E
Elawani, J., 38
Esquisabel, O., 5, 6, 27, 28, 51, 72, 73, 125
Euclid of Alexandria

F
Fabry, H., 30
Fatio de Duillier, N., 3
Faulhaber, J., 56
Ferguson, J.J., 94
Fermat, P. de, 53, 56, 95, 143, 151, 199, 249
Fontenelle, B. Le B. de, 17, 90, 112, 238
Forcadel, P., 53
Frisius, G., 53
Froidmont, Libert, 127, 128, 132, 175, 218–220

G
Galilei, G., 77
Gallois (or Galloys), A., 17, 20, 77, 134, 136, 143, 237, 242–244
Gallois (or Galloys), J., 27
Gerhardt, C.I., 3, 12, 14, 18, 22, 80, 92, 93, 103, 104, 128, 131, 259, 268
Girard, A., 52
Gosselin, G., 53
Gottignies, G.F. de, 53, 56, 249
Gouye, T. Father, 17, 133, 135, 233, 237, 238, 242
Grandi, G., 118, 134, 147, 148, 163, 245–246
Gregory, J., 37, 79, 89
Grosholz, E., 86

H
Hardy, C., 143
Hermann, J., 17, 259, 260
Hess, H.-J., 7, 130
Heuraet, H. van, 30
Hippocrates of Chios, 177
Hobbes, T., 24, 25, 41, 43, 63, 79, 177, 259
Hofmann, J.E., 80
Hudde, J., 30
Hume, J., 56
Huygens, C., 14, 25, 27, 29, 30, 37, 69, 79, 89, 102, 107, 111, 172, 207, 211, 236, 262

I
Ikeda, S., 68
Ishiguro, H., 18, 147

J
Jenisch, P.J., 36, 58, 77, 78, 253–257
Jesseph, D.M., 5, 79, 92, 96, 130, 141, 146, 151, 152
Jungius, J., 76, 245, 247, 248

K
Katz, K., 6, 51, 117, 125, 128, 146, 147
Katz, M.G., 74, 147
Keill, J., 3
Kline, M., 4
Knobloch, E., 5, 60, 61, 80, 81, 83, 89, 93
Kracht, M., 102
Kreyszig, E., 102
Kuhlemann, K., 6

L
La Hire, P. de, 17, 143, 238, 262
Lagrange, J.-L., 4, 53
Laugwitz, D.
Levey, S., 16, 80
l'Hôpital, G.M. de, 7, 17, 90, 133, 147, 152, 234, 235

M
Mahnke, D., 7
Maignan, E., 56, 64
Malebranche, N., 13, 16, 17, 109, 133, 205, 237, 260
Mancosu, P., 6, 12, 17, 19, 89, 133, 143
Mariotte, E., 60, 74
Mencke, O., 99, 197–200
Mercator, N., 32, 54, 55, 60, 62, 63, 79, 151, 250
Mersenne, M., 143

N
Napier, J., 55
Newton, I., 3, 4, 8, 81, 94, 108, 120, 158
Nieuwentijt, B., 14, 17, 20, 103, 109, 147, 160, 203, 248–250, 259, 260
Nizolius, M., 37

O
Ockham, William of, 18, 137
Osiander, A., 55
Ottaviani, O., 18, 59, 68, 72, 127, 218, 229

P

Pappus, 54, 65
Pardies, I.-G. Father, 173
Parkinson, G.H.R., 5
Parmentier, M., 7, 99, 103, 105, 110
Pascal, B., 25, 30, 31, 47, 79, 81, 83, 85, 118, 158, 190
Pasini, E., 7, 13, 20, 21, 30, 123, 145, 215, 220, 238, 257
Peletier, J., 53, 131
Pinsson, F., 4, 17, 19, 21, 83, 103, 133, 143, 234–235, 251–253
Pourciau, B., 158
Prestet, J., 69, 71, 74, 95, 200, 246
Probst, S., 28, 30, 38, 63, 68, 70, 80
Proclus, 14, 54

Q

Quine, W.V.O., 6

R

Rabouin, D., 16, 53, 54, 61, 62, 66, 71, 78–80, 83, 85, 95, 99
Raffo Quintana, F., 6, 28, 36–38, 125
Remond, N., 139
Rieuwerts, J., 93, 94
Robinet, A., 123, 215
Robinson, A., 4, 6, 8, 146
Rolle, M., 17, 143, 145, 160, 237
Roth, P., 56

S

Saint-Vincent, G. de, 25–27, 167, 168, 172, 224
Sanchez, F., 259
Schooten, F. van, 30, 52
Schott, K., 53
Sereda, K., 115
Sherry, D., 6, 51, 146, 147, 153
Sophie, Electress of Hanover, 139
Soubry, J., 80

Spinoza, B., 41, 46, 58, 94, 137
Stedall, J.A., 38
Stevin, S., 52, 56
Stifel, M., 51, 53, 56
Struik, D., 12
Sturm, J.C., 12, 95, 103, 197–200

T

Tho, T., 16, 91
Torricelli, E., 81, 83, 89, 172
Trenchant, J., 53
Tschirnhaus, E.W. von, 41, 47, 71, 73, 102, 103, 107, 211

U

Uckelman, S.L., 136
Ugaglia, M., 6, 140, 141
Ursus (Nicolaus Reimers Baer), 55

V

Vailati, E., 89, 143
Varignon, P., 4, 9, 15, 17–21, 57, 64, 69, 78, 87, 97, 112, 118, 126, 134–136, 143, 157, 220, 235–238, 244, 245, 251
Vieta, François

W

Wallis, J., 14, 20, 24, 25, 31, 32, 34, 36–38, 52–54, 57, 60, 62, 63, 66, 71, 79, 81, 87, 95, 112, 140, 143, 150, 151, 153–155, 160, 172, 199, 244, 250–251
Weil, A., 100, 101, 105
White, T., 124, 126, 127, 161, 217–218
Whiteside, D.T., 83, 85, 86
Wolff, C., 56, 64, 76, 78, 248

Y

Yablo, S., 78

Subject Index

A

Angle of contact, 14, 26, 27, 41, 56, 76, 121, 130–132, 162, 168, 171, 175, 214, 243, 245
Anthyphairesis, 46
Assignable/unassignable, 5, 10, 11, 16, 19–21, 25, 36, 41, 46, 47, 59, 62, 84, 87, 90, 91, 105, 108, 109, 111, 125, 126, 128–130, 132, 135, 141, 142, 145, 152–154, 158–162, 169, 174, 179, 180, 186, 192, 206, 212, 213, 215, 219–222, 225, 227, 233, 234, 237, 239, 242, 243, 249–251, 253, 259, 261, 262, 264, 265, 270, 271
Asymptotes, 34, 35, 77, 88, 89, 172, 192, 219, 229, 245, 248
Atoms, 4, 6, 48, 118, 122, 237
Axiom
  of Archimedes, Archimedean Axiom, 14, 23, 74, 109, 131, 159, 161, 178
  part-whole, 22, 26, 28, 33, 113, 114, 122, 131, 148, 158, 162

C

Congenious (= homogonous), 131, 132, 162, 218, 219
Congruence/congruent, 15, 44, 49, 113, 114, 116, 118, 121, 178, 182, 207, 209, 211, 213–215
Continued fractions, 38
Continuity
  law (principle) of, 8, 9, 11–13, 21, 22, 56, 67, 76–78, 81, 86, 87, 90, 109, 145, 148–150, 152, 153, 159, 162, 220, 237, 240, 246, 248, 252, 253, 259, 260

Continuum
  Archimedean, 147
  composition of the, 23, 175, 262
  labyrinth of, 175
Curves
  cycloidal, 93, 95, 107, 200
  logarithmic, 107, 124

D

Demonstration
  Archimedean method of, 20, 22, 150, 162, 257
Differential
  algorithm, 6, 23, 60, 92–94, 112, 148–155, 204, 248–271
  calculus, 3, 5, 8, 11, 12, 16, 18, 20, 22, 23, 31, 41, 60, 62, 77, 92–94, 102, 103, 108, 111, 112, 117, 120, 129, 131, 133, 142–145, 150, 153, 155, 157, 162, 201, 204, 232, 234–235, 239, 250, 255, 263, 264
Diorism, 66, 73

E

Endeavour (*conatus*), 5, 9, 24, 113, 120, 128
Exhaustion, 11, 12, 81, 82, 86, 110, 146, 159, 211–213

F

False position, method of, 53

Fiction, fictional, fictitious, 5, 6, 8–11, 13, 15, 16, 18, 19, 21–23, 27, 32–36, 39, 40, 43, 46, 49, 51–78, 85–89, 91, 94, 96, 97, 104, 112, 116–120, 125, 128, 129, 134–136, 142, 145, 146, 149, 154, 157–159, 161–163, 172, 173, 177, 180, 186–188, 190, 192, 193, 212, 213, 223, 229, 233, 234, 236, 238, 245–247, 258
Fluxions, 3
Foundations, 4, 8, 11, 20, 22, 23, 59, 62, 76, 80, 83, 92, 95, 108, 110–119, 133, 143, 146, 150–152, 157–159, 190, 197, 207–214, 229, 236, 244, 246, 250, 258, 260

H

Homogeneity/homogeneous, 14, 30, 63, 81, 90, 101, 109, 113, 115–118, 120, 132, 147, 208, 209, 212–214, 239, 243, 249, 251, 258, 266
Hyperbola, 5, 31, 32, 34, 35, 38, 39, 44, 45, 54, 62, 70, 76, 77, 79, 88, 90, 96, 106, 107, 114, 158, 161, 171, 172, 174, 179, 186–187, 191, 199, 216, 223, 248, 261

I

Imaginary
  quantity, 52, 56, 60, 68, 69, 71, 72, 75, 76, 187, 217, 244
  root, 15, 16, 56, 57, 59, 60, 66, 68, 69, 71, 73–76, 78, 96, 97, 128, 134–136, 142, 146, 157, 158, 163, 187, 216, 220, 223, 229, 232, 233, 236, 238, 239, 246, 258, 261, 263
Incommensurable/commensurable, 9, 10, 42, 44–46, 63, 66, 70, 114, 142, 175, 178, 179, 187, 211, 212
Incomparable
  Lemmas on Incomparables, 13, 20, 22, 49, 103, 108, 112, 117, 129, 133, 147, 148, 152, 162, 201–207, 235, 239, 243, 250, 257
Indefinite, 18, 36, 37, 42, 45, 107, 110, 116, 137, 140, 187, 195, 196, 208, 230, 231, 239
Indistant parts, 25
Indivisibles
  method of, 11, 21, 39, 64, 79, 80, 82, 83, 85, 117, 159, 193, 213
Infinite
  actual, 15, 18, 19, 23, 36, 41, 45, 122, 137–141, 161, 162, 222, 224, 226, 232–234
  bounded/unbounded, 15, 21, 34, 39–42, 46, 48, 89–91, 96, 119, 120, 122–129, 141, 161, 162, 173, 175, 181, 182, 216, 217, 219, 220, 222, 223, 227, 228, 230, 233, 235, 262
  categorematic, 18, 136, 137, 232, 234

  hypercategorematic, 234
  number, 4, 5, 7, 9, 15, 18, 19, 21, 26–28, 30, 34–36, 38–40, 42, 44, 46, 56–59, 63, 66, 76, 78, 91, 105, 112, 114, 116, 117, 119, 120, 125–127, 134–142, 157, 158, 160–162, 167, 168, 173–175, 178–181, 186, 210, 212, 215, 217, 221, 222, 224, 225, 227–234, 236, 239–241, 245–249, 253
  potential, 19, 137–141, 233
  quantities, 5, 10, 14–16, 19, 20, 22, 31, 35, 39, 45, 46, 56–59, 62, 64, 66, 67, 75–78, 82, 83, 85, 89–91, 111–114, 116, 117, 119, 120, 125, 129, 133, 134, 136, 139, 148, 150, 151, 157, 158, 161–163, 167, 173, 174, 181, 186, 187, 190, 192, 202, 212, 215, 217, 221, 225, 226, 228, 231, 232, 235, 238–241, 244–247, 249, 255, 262
  syncategorematic, 18, 19, 91, 136–138, 141, 142, 232, 233, 236
Infinitely small, 4–6, 8–16, 18–20, 22, 23, 26, 28–49, 51, 54, 56, 58–60, 62–64, 66, 67, 69, 71, 73, 75–78, 81–83, 85–91, 96, 97, 101, 103–105, 108, 111, 113, 114, 116–128, 131–137, 139, 142, 143, 146–149, 151, 157–163, 170–183, 185–188, 190, 192, 193, 204–206, 212, 213, 215–217, 219, 220, 222–228, 230, 232–240, 245, 246, 248–251, 254–258, 261, 262, 264, 270
Infinite series
  power series expansion, 32, 33
  sum of, 29, 105
Infinitesimal
  quantities, 22, 78, 110–112, 116, 241, 244–245, 254, 259
Integration, 5, 32, 33, 80, 83, 159

M

Magnitude
  infinite, 17, 20, 21, 38, 39, 45, 88, 112, 114, 137, 141, 142, 161, 163, 172, 173, 205, 210, 213, 224, 227, 233, 235, 236, 249
  taken *lato sensu*, 116, 121
  taken *stricto sensu*, 53, 59, 121
Minima
  quasi-, 63, 111–119, 207–214
Mean proportional, 15, 39, 48, 62, 71, 77, 119, 120, 123–125, 129, 161, 173, 174, 185, 215, 240, 256

N

Nonstandard, 4, 8, 14, 123, 129
Number

Subject Index

(actually) infinite, 15, 38, 41, 43, 45, 48, 97, 137–139, 141, 161, 168, 218, 234
greatest, 44, 46, 138, 176, 178, 180, 181, 212, 225, 228, 229
infinitesimals, 5, 7, 14, 15, 17, 21, 26, 42, 64, 74, 112, 116, 119, 126, 159–161, 213, 249, 250, 253
irrationals, 14, 161, 209, 211
maximum, 75, 186, 231, 249
negatives, 53, 54, 57, 66, 68, 73, 244
number of all numbers, 27, 75, 176

P

Plagiarism, 3
Point
  as endpoints, 25, 196, 225
  as extensionless, 25
  metaphysical, 4, 41, 43, 44, 48, 62, 72, 122, 175, 235
  as possessing parts, 139
  taken *latu senso*, 122
Polygons, 5, 21, 35, 37, 43, 45, 48, 49, 63, 81, 82, 84, 85, 94, 103–105, 118, 161, 176–180, 193, 194, 198, 204, 212, 214, 215, 253, 262
Principle
  of activity, 138
  difference principle, 30, 35, 80, 109, 110, 158, 205, 219, 239, 253
  of sufficient reason, 118, 129
  of unassignable differences, 10, 19, 35, 81, 109, 117, 135, 158

Q

Quadratrix, 81–84, 88, 89, 95, 100, 107, 192, 200
Quadrature
  method of, 10, 11, 28, 76, 79, 84, 96, 102, 106, 107, 143, 159, 177, 195, 225, 259
Quantity
  intensive, 15

lato sensu, 14, 115, 120, 162
non-Archimedean, 14, 131, 163
stricto sensu, 15, 113, 120, 162

R

Rigour, rigorous, 4, 8, 11, 18, 20, 21, 43, 49, 58, 76, 78, 80, 81, 83, 91, 103, 135, 146, 158, 175, 181, 182, 229, 232, 234–237, 239, 243, 245, 246, 249, 253, 254, 256, 258, 259, 262, 263

S

Signs, Scholastic doctrine of, 25
Status transitus, 131
Sum of infinite series, 18

T

Transcendent
  curve, 99, 107, 144, 204, 252
  number, 115
Transformation
  continuous, 116
  quasi-, 118
Transmutation
  transmutation theorem, 31, 32, 44

V

Variables, 5, 12–14, 31–33, 35, 39, 40, 58, 59, 76, 77, 79, 83, 100, 101, 104, 106, 110, 115, 120, 125, 146, 155, 158, 159, 251, 257, 261, 270, 271

W

Whole
  collective, 19, 59, 138, 162, 229
  distributive, 19, 59, 138, 162, 229

The manufacturer's authorised representative in the EU is Springer Nature Customer Service Centre GmbH, Europaplatz 3, 69115 Heidelberg, Germany. If you have any concerns regarding our products, please contact ProductSafety@springernature.com

Printed and bound by CPI Group (UK) Ltd, Croydon, CR0 4YY

26/03/2026

02078970-0001